BALLOON FLYING HANDBOOK

Balloon Flying Handbook

FAA-H-8083-11A

Federal Aviation Administration

Skyhorse Publishing

Skyhorse Publishing books may be purchased in bulk at special discounts for sales promotion, corporate gifts, fund-raising, or educational purposes. Special editions can also be created to specifications. For details, contact the Special Sales Department, Skyhorse Publishing, 307 West 36th Street, 11th Floor, New York, NY 10018 or info@skyhorsepublishing.com.

Visit our website at www.skyhorsepublishing.com.

10 9 8 7 6 5 4 3 2 1

Library of Congress Cataloging-in-Publication-Data is available on file.
ISBN-13: 978-1-61608-715-9

AC 00-45E, Aviation Weather Services, is published jointly by the Federal Aviation Administration and the National Weather Service (NWS). This document supplements the companion manual AC 00-6A, Aviation Weather, that deals with weather theories and hazards.

This advisory circular, AC 00-45E, explains weather service in general and the details of interpreting and using coded weather reports, forecasts, and observed and prognostic weather charts. Many charts and tables apply directly to flight planning and inflight decisions. It can also be used as a source of study for pilot certification examinations.

The AC 00-45E was written primarily by Kathleen Schlachter with contributions from Jon Osterberg, Doug Streu, and Robert Prentice. A special thanks to Sue Roe for her help and patience in editing this manual.

Comments and suggestions for improving this publication are encouraged and should be direction to:

National Weather Service Coordinator, W/SR64
Federal Aviation Administration
Mike Monroney Aeronautical Center
P.O. Box 25082
Oklahoma City, OK 73125-0082
Advisory Circular AC 00-45E supersedes AC 00-45D, Aviation Weather Services, revised 1995.

Printed in China

Preface

This Balloon Flying Handbook introduces the basic pilot knowledge and skills that are essential for piloting balloons. It introduces pilots to the broad spectrum of knowledge that will be needed as they progress in their pilot training. This handbook is for student pilots, as well as those pursuing more advanced pilot certificates.

Student pilots learning to fly balloons, certificated pilots preparing for additional balloon ratings or who desire to improve their flying proficiency and aeronautical knowledge, and commercial balloon pilots teaching balloon students how to fly should find this handbook helpful. This book introduces the prospective pilot to the realm of balloon flight and provides information and guidance to all balloon pilots in the performance of various balloon maneuvers and procedures.

This handbook conforms to pilot training and certification concepts established by the Federal Aviation Administration (FAA). There are different ways of teaching, as well as performing flight procedures and maneuvers, and many variations in the explanations of aerodynamic theories and principles. This handbook adopts a selective method and concept to flying balloons. The discussions and explanations reflect the most commonly used practices and principles. Occasionally, the word "must" or similar language is used where the desired action is deemed critical. The use of such language is not intended to add to, interpret, or relieve a duty imposed by Title 14 of the Code of Federal Regulations (14 CFR).

It is essential for persons using this handbook also to become familiar with and apply the pertinent parts of 14 CFR and the Aeronautical Information Manual (AIM). Performance standards for demonstrating competence required for pilot certification are prescribed in the appropriate balloon practical test standard.

This handbook supersedes FAA-H-8083-11, Balloon Flying Handbook, dated 2001.

This handbook is available for download, in PDF format, from the FAA website, www.faa.gov.

This handbook is published by the United States Department of Transportation, Federal Aviation Administration, Airman Testing Standards Branch, AFS-630, P.O. Box 25082, Oklahoma City, OK 73125.

Comments regarding this publication should be sent, in email form, to the following address:

AFS630comments@faa.gov

Acknowledgements

This handbook was produced as a combined Federal Aviation Administration (FAA), industry, and individual contributor effort. The FAA wishes to acknowledge the following individual and corporate contributors:

Aerostar International, Inc. for the performance planning chart in Chapter 3 and the burner ratings sidebar on page 7-2

FireFly Balloons for imagery provided in Chapter 2

Lindstrand Balloons Ltd for the imagery in Chapter 2

J. Neils Enterprises, Inc. for imagery provided in Chapter 2

Lindan Hot Air Service Center for imagery provided in Chapter 2

Department of Atmospheric Sciences, University of Illinois Urbana-Champaign, for many images used in Chapter 4

Art Rangno, Sky Guide, for many of the cloud illustrations used in Chapter 4

Gordon Schwontkowski, author of Hot Air Balloon Crewing Essentials, for editorial content and contribution of material relating to the role of the ground crew in ballooning

Mike Bauwens, of Balloon the Rockies, for many of the maneuver descriptions and original graphics used in Chapters 6 and 7

Chapter 11, The Gas Balloon, was a collaborative effort of Peter Cuneo and Barbara Fricke, Andy Cayton, and Kevin Knapp.

Individual content and editorial contributions by: Raymond Bair, Philip Bryant, Stephen Turner, Jim Barnett, Kay West, Marian Deeney, Laura Hoeve, Tom Hamilton (for the False Lift discussion, originally printed in Balloon Life Magazine, and reprinted here in Chapter 6), Pat Cannon (for the pibal plotting procedure originally printed in Ballooning Magazine, and reprinted here in Chapter 3), Andrew and Barbara Ziolo, Chris Krowchuck, and Stuart Enloe

Individual photographic contributions have been made by Jane English (multiple lenticular cloud in Figure 4-23), Candy Ecker (Figure 3-15), and C. Tina Orvin (Figure 8-7).

The FAA acknowledges with appreciation the contributions of the Balloon Federation of America for technical support and input.

Table of Contents

Chapter 3
Preflight Planning..3-1

Chapter 4
Weather Theory and Reports............................4-1

Chapter 1

Introduction to Balloon Flight Training

Purpose of Balloon Flight Training

As outlined in this handbook, the purpose of balloon training is to learn, develop, and refine basic balloon flight skills. These skills include:

- Knowledge of the principles of flight;
- The ability to launch, operate, and land a balloon with competence and precision; and
- The use of good judgment that leads to optimal operational safety and efficiency.

Learning to fly a balloon requires a specific set of motor skills:

- Coordination—the ability to take physical action in the proper sequence to produce the desired results while launching, flying, and landing the balloon.
- Timing—the application of muscle coordination at the proper time to make the flight, and all maneuvers incident to it, a constant smooth process.
- Control touch—the ability to interpret, evaluate, and predict the actions and reactions of the balloon with regard to attitude and speed variations, by interpreting and evaluating varying visual cues and instrument readings.
- Situational awareness—the ability to sense instantly any reasonable variation of altitude, airspeed, and directional change, as well as a constant perception of relative position to ground-based structures and planned flight track.

A skilled pilot becomes one with the balloon and learns to assess a situation quickly and accurately. He or she also develops the ability to select the proper procedure to follow in a situation, to predict the probable results of the selected procedure, and to exercise safe practices. In addition, a skilled pilot learns to gauge the performance of the balloon being flown and to recognize not only personal limitations, but also the limitations of the balloon. This knowledge helps the pilot to avoid reaching personal or machine critical points.

Developing the skills needed to fly a balloon requires time and dedication on the part of the student pilot, as well as the flight instructor. Each balloon has its own particular flight characteristics, and it is not the purpose of balloon flight training to learn how to fly a particular model balloon. The purpose of balloon flight training is to develop skills and safety habits that can be transferred to any balloon. The pilot who acquires the necessary flight skills during training, and demonstrates these skills by flying with precision and safe flying habits, easily transitions to different model balloons. Student pilots should also remember that the goal of flight training is to develop a safe and competent pilot. To that end, it is important for the flight instructor to insure the student pilot forms the proper flying habits by introducing him or her to good operating practices from the first training flight.

Role of the Federal Aviation Administration (FAA)

The United States Congress has empowered the Federal Aviation Administration (FAA) to promote aviation safety by establishing safety standards for civil aviation. The FAA accomplishes this goal through the Code of Federal Regulations (CFR). Title 14 of the Code of Federal Regulations (14 CFR) part 61 pertains to the certification of pilots, flight instructors, and ground instructors. 14 CFR part 61 defines the eligibility, aeronautical knowledge, flight proficiency, as well as training and testing requirements for each type of pilot certificate issued.

14 CFR part 91 contains general operating and flight rules. The section is broad in scope and provides general guidance in the areas of general flight rules, visual flight rules (VFR), instrument flight rules (IFR), aircraft maintenance, and preventive maintenance and alterations.

Within the FAA, the Flight Standards Service promotes safe air transportation by setting the standards for certification and oversight of airmen, air operators, air agencies, and designees. It also promotes safety of flight of civil aircraft and air commerce by:

- Accomplishing certification, inspection, surveillance, investigation, and enforcement;
- Setting regulations and standards; and
- Managing the system for registration of civil aircraft and all airmen records.

The focus of interaction between the FAA Flight Standards Service and the aviation community and general public is the Flight Standards District Office (FSDO). *[Figure 1-1]* The FAA has approximately 130 FSDOs. These offices provide information and services for the aviation community. FSDO phone numbers are listed in the blue pages of the telephone directory under United States Government Offices, Department of Transportation, Federal Aviation Administration. Another convenient method of finding a local office is to use the FSDO locator available on the Regulatory Support Division's website: http://www.faa.gov/about/office_org/headquarters_offices/avs/offices/afs/afs600.

In addition to accident investigation and the enforcement of aviation regulations, the FSDO is also responsible for the

Figure 1-1. *Atlanta Flight Standards District Office (FSDO).*

certification and surveillance of air carriers, air operators, flight schools and/or training centers, and airmen including pilots and flight instructors. Each FSDO is staffed by aviation safety inspectors (ASIs) who play a key role in making the United States aviation system safe. They administer and enforce safety regulations and standards for the production, operation, maintenance, and/or modification of aircraft used in civil aviation. They also specialize in conducting inspections of various aspects of the aviation system, such as aircraft and parts manufacturing, aircraft operation, aircraft airworthiness, and cabin safety. ASIs must complete a training program at the FAA Academy in Oklahoma City which includes airman evaluation and pilot testing techniques and procedures. Inspectors also receive extensive on-the-job training and they receive recurrent training on a regular basis. The FAA has approximately 3,700 inspectors located in its FSDO offices. All questions concerning pilot certification (and/or requests for other aviation information or services) should be directed to the local FSDO.

Role of the Designated Pilot Examiner (DPE)

Among other duties, ASIs are responsible for administering FAA practical tests for pilot and flight instructor certificates and associated ratings. The administration of these tests is normally carried out at the FSDO level, but the agency's highest priority is making air travel safer by inspecting aircraft that fly in the United States. To satisfy the need for pilot testing and certification services, the FAA delegates certain responsibilities to private individuals who are not FAA employees, but designated pilot examiners (DPEs).

A DPE is an individual, appointed in accordance with 14 CFR part 183, section 183.23, who meets the qualification requirements of FAA Order 8710.3, Pilot Examiner's Handbook, and who:

- Is technically qualified;
- Holds all pertinent category, class, and type ratings for each aircraft related to their designation;
- Meets the requirements of 14 CFR part 61, sections 61.56, 61.57, and 61.58, as appropriate;
- Is current and qualified to act as pilot-in-command (PIC) of each aircraft for which they are authorized;
- Maintains at least a third-class medical certificate, if required; and
- Maintains a current flight instructor certificate, if required.

Designated as a representative of the FAA Administrator to perform specific pilot certification tasks on behalf of the FAA, a DPE may charge a reasonable fee. Generally, a DPE's authority is limited to accepting applications and conducting practical tests leading to the issuance of specific pilot certificates and/or ratings. The majority of FAA practical tests at the private and commercial pilot level are administered by FAA DPEs, following FAA-provided practical test standards (PTSs).

Only highly qualified individuals are accepted as DPEs. DPE candidates must have good industry reputations for professionalism, integrity, a demonstrated willingness to serve the public, and adhere to FAA policies and procedures in certification matters. The FAA expects the DPE to administer practical tests with the same degree of professionalism, using the same methods, procedures, and standards as an FAA ASI.

Since there are few DPEs for balloon pilot certification, it is important to determine early in flight training the availability of a DPE in a particular area. It may be necessary to make arrangements through the local FSDO for an appropriately rated FAA ASI to administer the test for a pilot certificate.

Role of the Flight Instructor

Unlike the rest of the aviation community, ballooning has no certificated flight instructor. This role is filled by commercially rated balloon pilots who choose to instruct and meet the provisions of 14 CFR part 61, Commercial Pilot Privileges and Limitations for a Balloon. In this discussion, the term "flight instructor" is understood to mean a commercial balloon pilot who provides instruction.

The flight instructor is the cornerstone of aviation safety and the FAA places full responsibility for student training on the authorized flight instructor. It is the job of the instructor to train the student pilot in all the knowledge areas and teach the skills necessary for the student pilot to operate safely and competently as a certificated pilot in the National Airspace System (NAS). The training includes airmanship skills, pilot judgment and decision-making, and good operating practices.

A pilot training program depends on the quality of the ground and flight instruction the student pilot receives. The flight instructor must possess a thorough understanding of the learning process, knowledge of the fundamentals of teaching, and the ability to communicate effectively with the student pilot. He or she uses a syllabus and teaching style that embodies the "building block" method of instruction. In this method, the student progresses from the known to the unknown via a course of instruction laid out in such a way that each new maneuver embodies the principles involved in the performance of maneuvers previously learned. Thus, with the introduction of each new subject, the student not only learns a new principle or technique, but also broadens his or her application of those principles or techniques previously learned. Insistence on correct techniques and procedures from the beginning of training by the flight instructor ensures that the student pilot develops proper flying habit patterns. Any deficiencies in the maneuvers or techniques must immediately be emphasized and corrected.

A flight instructor serves as a role model for the student pilot who observes the flying habits of his or her flight instructor during flight instruction, as well as when the instructor conducts other pilot operations. Thus, the flight instructor becomes a model of flying proficiency for the student who, consciously or subconsciously, attempts to imitate the instructor. For this reason, a flight instructor should observe recognized safety practices, as well as regulations during all flight operations.

The student pilot who enrolls in a pilot training program commits considerable time, effort, and expense to achieve a pilot certificate. Many times a student judges the effectiveness of the flight instructor and the success of the pilot training program based on his or her ability to pass the requisite FAA practical test. A competent flight instructor stresses to the student that practical tests are a sampling of pilot ability compressed into a short period of time. The goal of a flight instructor is to train the "total" pilot.

Sources of Flight Training

Flight training in the United States is conducted by FAA-approved pilot schools and training centers, non-certificated (14 CFR part 61) flying schools, and independent flight instructors. There are a limited number of part 141 balloon training programs in the United States with most balloon flight training being conducted by certificated commercial balloon pilots authorized to instruct under the authority of 14 CFR section 61.133 (a)(2)(ii)(a).

FAA-approved schools are flight schools certificated by the FAA as pilot schools under 14 CFR part 141. *[Figure 1-2]* Application for certification is voluntary and the school must meet stringent requirements for personnel, equipment, maintenance, and facilities; and teach an established curriculum which includes a training course outline (TCO) approved by the FAA. A list of FAA certificated pilot schools and their training courses can be found in Advisory Circular (AC) 140-2, FAA Certificated Pilot School Directory.

As noted above, the major source of balloon flight training in the United States is conducted by certificated commercial balloon pilots. Many of these individuals offer excellent training and meet or exceed the standards required of FAA-approved pilot schools, but not all flight instructors are equal. It is important for a student pilot to choose a flight instructor wisely, because a balloon flight training program is dependent upon the quality of the ground and flight instruction the student pilot receives.

Practical Test Standards (PTS)

The FAA has developed Practical Test Standards *[Figure 1-3]* for FAA pilot certificates and associated ratings. These practical tests are administered by FAA ASIs and DPEs. 14 CFR part 61 specifies the areas of operation in which knowledge and skill must be demonstrated by the applicant. Since the FAA requires that all practical tests be conducted in accordance with the appropriate PTSs and the policies set forth in the introduction section of the PTS book, the pilot applicant should become familiar with this book during training.

The PTS book is a testing document and not intended to be a training syllabus. An appropriately rated flight instructor is responsible for training the pilot applicant to acceptable standards in all subject matter areas, procedures, and maneuvers. Descriptions of tasks and information on how to perform maneuvers and procedures are contained in reference and teaching documents, such as this handbook. A list of reference documents is contained in the introduction section of each PTS book.

UNITED STATES OF AMERICA
DEPARTMENT OF TRANSPORTATION
FEDERAL AVIATION ADMINISTRATION

Air Agency Certificate

Number TG00R68L

This certificate is issued to
SAFETY RESEARCH CORPORATION OF AMERICA
whose business address is
807 DONNELL BOULEVARD, SUITE O
DALEVILLE, AL 36322
upon finding that its organization complies in all respects with the requirements of the Federal Aviation Regulations relating to the establishment of an Air Agency, and is empowered to operate an approved Pilot School.

with the following ratings:
Private Pilot Certification Course (Examining Authority-Flight Test Only)
Commercial Pilot Certification Course (Examining Authority-Flight Test Only)
Instrument Rating Course (Examining Authority-Flight Test Only)
Airline Transport Pilot Certification Course
Additional Aircraft Rating Course
Flight Instructor Certification Course

This certificate, unless canceled, suspended, or revoked, shall continue in effect UNITL MARCH 31, 2007

Date issued:
FEBRUARY 28, 2001
REISSUED MARCH 22, 2005

JOHN Q. SMITH
MANAGER, ALABAMA NORTHWEST FLORIDA FSDO

This Certificate is not Transferable, AND ANY MAJOR CHANGE IN THE BASIC FACILITIES, OR IN THE LOCATION THEREOF, SHALL BE IMMEDIATELY REPORTED TO THE APPROPRIATE REGIONAL OFFICE OF THE FEDERAL AVIATION ADMINISTRATION

Any alteration of this certificate is punishable by a fine of not exceeding $1,000, or imprisonment not exceeding 3 years, or both

FAA Form 8000-4 (1-67) SUPERSEDES FAA FORM 390.

Figure 1-2. *FAA-approved pilot school certificate.*

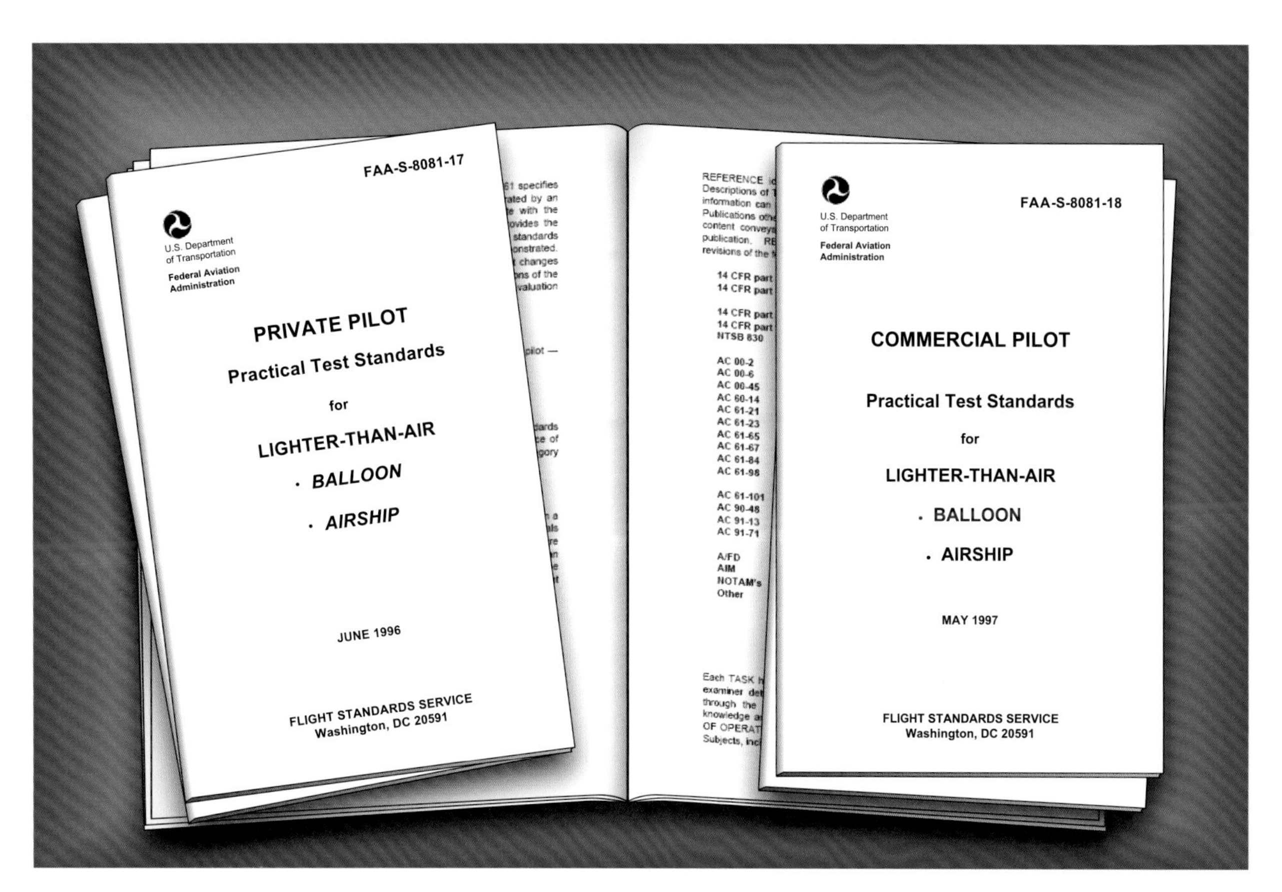

Figure 1-3. *Practical Test Standards.*

The PTS for lighter-than-air aircraft are the FAA-S-8081-17 (Private Pilot) and FAA-S-8081-18 (Commercial Pilot). Copies may be obtained by:

- Downloading from the Regulatory Support Division, AFS-600, website at http://afs600.faa.gov.
- Purchasing printed copies from the Superintendent of Documents, U.S. Government Printing Office, Washington, DC 20402.

The official online bookstore web site for the United States Government Printing Office is http://bookstore.gpo.gov.

Flight Safety Practices

In the interest of safety and the development of good flight habits, the flight instructor and student pilot should follow certain basic flight safety practices and procedures in every flight operation. These include, but are not limited to, collision avoidance procedures including proper scanning techniques, use of checklists, runway incursion avoidance and other airspace operations, positive transfer of controls, and workload management.

Collision Avoidance

All pilots must be alert to the potential for midair collision and near midair collisions. The general operating and flight rules in 14 CFR part 91 set forth the concept of "See and Avoid." This concept requires that vigilance shall be maintained at all times, by each person operating an aircraft. Pilots should also keep in mind their responsibility for continuously maintaining a vigilant lookout regardless of the type of balloon being flown and the purpose of the flight. Most midair collision accidents and reported near midair collision incidents occur in good VFR weather conditions and during the hours of daylight.

With regards to balloon operations, the argument can be made that any discussion of collision avoidance applies when dealing with operations close to the ground. When contour flying, or during an approach to a landing site, the potential of collision with trees, power lines, and other obstacles is increased. *[Figure 1-4]* The techniques used in collision avoidance can be extremely valuable, particularly in the evolution of a balloon flight, as the pilot is perhaps exposed more to the dangers of collision than any other aircraft.

The "See and Avoid" concept relies on knowledge of the limitations of the human eye, and the use of proper visual scanning techniques to help compensate for these limitations. The importance of, and the proper techniques for, visual scanning should be taught to a student pilot at the very beginning of flight training. The competent flight instructor should be familiar with the visual scanning and collision

Figure 1-4. *When flying or landing, always be aware of the potential for collision with trees, powerlines, and other obstacles.*

avoidance information contained in AC 90-48, Pilot's Role in Collision Avoidance, and the Aeronautical Information Manual (AIM).

Runway Incursion Avoidance

A runway incursion is any occurrence at an airport involving an aircraft, vehicle, person, or object on the ground that creates a collision hazard with an aircraft taking off, landing, or intending to land.

Most balloon flight operations are conducted away from an airport or at airports without an operating control tower. There may be circumstances that require the use of airport property, either for launch or landing and recovery of the balloon. These activities can be safely conducted at an airport, if the balloon pilot remains aware of the movement and location of other aircraft and ground vehicles, and also complies with standard operating procedures and practices. The absence of an operating airport control tower creates a need for increased vigilance on the part of any pilot operating at those airports.

Planning, clear communications, and enhanced situational awareness during airport surface operations reduces the potential for surface incidents. Safe balloon operations can be accomplished and incidents eliminated if the pilot is properly trained early on and, throughout his or her flying career, complies with standard operating procedures and practices when operating on airport property. This requires the development of the formalized teaching of safe operating practices during ground operations. The flight instructor is the key to this teaching. The flight instructor should instill in the student an awareness of the potential for runway incursion.

Use of Checklists

Checklists are the foundation of pilot standardization and safety. Checklists aid the memory and help ensure that critical items necessary for the safe operation of the balloon are not overlooked or forgotten. Checklists have no value if they are not used. Pilots who fail to use checklists at the appropriate times are relying instead on memory, become complacent, and increase the odds of making a mistake.

The consistent use of checklists in primary flight training establishes habit patterns that will serve the pilot well throughout his or her flying career. It is important that the flight instructor promote a positive attitude toward the use of checklists so the student pilot recognizes their importance. At a minimum, prepared checklists should be used for the following phases of flight:

- Crew Briefing and Preparation
- Layout and Assembly
- Preflight Inspection
- Inflation
- Passenger Briefing
- Prelaunch Check
- Emergency Procedures
- Postlanding
- Recovery, Deflation, and Packing

Checklists are covered in greater detail in Chapter 6, Layout to Launch.

Positive Transfer of Controls

It is imperative that a clear understanding exists between the student and flight instructor of who has control of the balloon during flight training. The flight instructor should conduct a briefing prior to any dual training flight that includes the procedure for the exchange of flight controls. The following three-step process for the exchange of flight controls is highly recommended.

1. When a flight instructor wishes the student to take control of the balloon, he or she should say "You have the flight controls."

2. The student acknowledges immediately by saying, "I have the flight controls."
3. The flight instructor confirms transfer of controls by saying, "You have the flight controls."

Both the flight instructor and student pilot should make a visual check to ensure the designated person actually has the flight controls. When the student pilot wishes to return the controls to the flight instructor, he or she follows the same procedure and stays on the controls until the flight instructor says, "I have the flight controls." There should never be any doubt as to who is in control of the balloon. The establishment of positive transfer of control during initial training ensures the formation of a good flying habit.

Aeronautical Decision-Making

Aeronautical decision-making (ADM) is a systematic approach to the mental process used by pilots to consistently determine the best course of action in response to a given set of circumstances. Learning effective ADM skills can help a pilot offset the one unchanging factor that remains despite all the changes in improved flight safety—the human factor. It is estimated that 90 percent of balloon accidents are human factors related.

ADM builds on the foundation of conventional decision-making, but enhances the process to decrease the probability of pilot error. ADM provides a structure to analyze changes that occur during a flight and determine how these changes might affect a flight's safe outcome. This process includes identifying personal attitudes hazardous to safe flight, learning to recognize and cope with stress, developing risk assessment skills, and evaluating the effectiveness of one's ADM skills.

Hazardous Attitudes and Antidotes

A hazardous attitude, which contribute to poor pilot judgment, can be effectively counteracted by redirecting that hazardous attitude so that correct action can be taken. Recognition of a hazardous thought is the first step toward neutralizing it. After recognizing a thought as hazardous, the pilot should label it as hazardous, then state the corresponding antidote. The antidotes for each hazardous attitude should be memorized so it automatically comes to mind when needed. Each hazardous attitude with its appropriate antidote or learning modification is shown in *Figure 1-5.*

Learning How To Recognize and Cope With Stress

An important component of the ADM system is the ability to recognize stress. Stress is a term used to describe the body's nonspecific response to demands placed on it. Stress can be emotional, physical, or behavioral, and it is important for a pilot to become knowledgeable about stress and how to cope with it.

Developing Risk Assessment Skills

An examination of the National Transportation Safety Board (NTSB) reports and other accident research can help a pilot to assess risk more effectively. For example, studies indicate the types of flight activities that are most likely to result in the most serious accidents. For balloons, landing accidents consistently account for over 90 percent of the total number of accidents in any given year.

These accidents consistently account for the majority of injury to pilots and damage to balloons. Accidents are more likely during landing because the tolerance for error is greatly diminished and opportunities for pilots to overcome errors

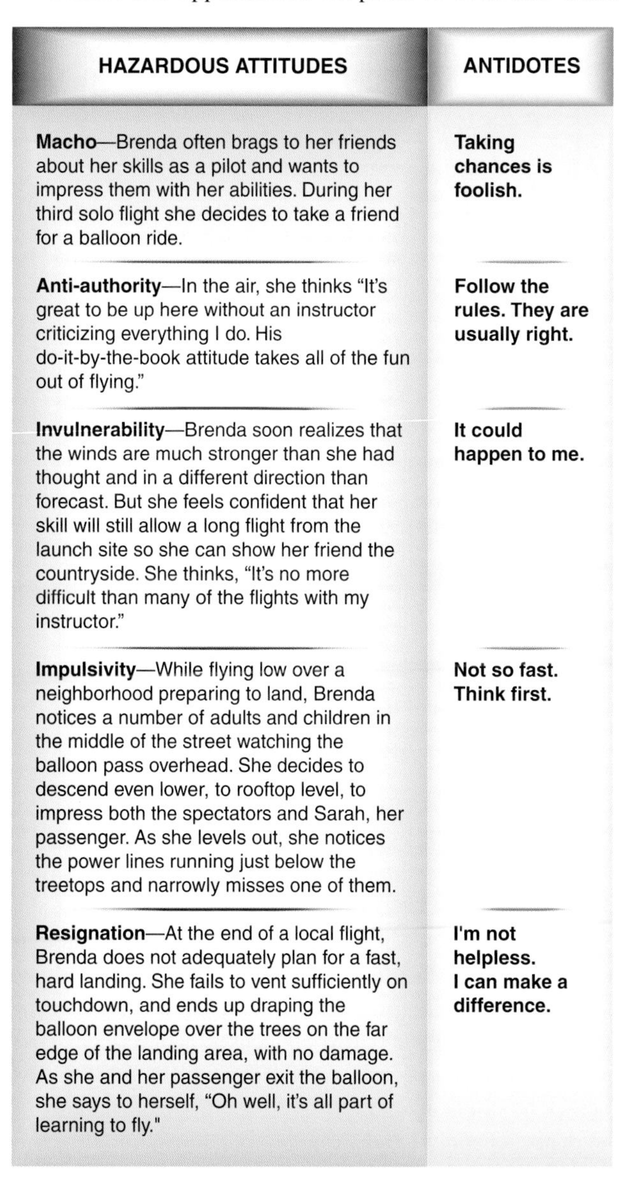

HAZARDOUS ATTITUDES	ANTIDOTES
Macho—Brenda often brags to her friends about her skills as a pilot and wants to impress them with her abilities. During her third solo flight she decides to take a friend for a balloon ride.	**Taking chances is foolish.**
Anti-authority—In the air, she thinks "It's great to be up here without an instructor criticizing everything I do. His do-it-by-the-book attitude takes all of the fun out of flying."	**Follow the rules. They are usually right.**
Invulnerability—Brenda soon realizes that the winds are much stronger than she had thought and in a different direction than forecast. But she feels confident that her skill will still allow a long flight from the launch site so she can show her friend the countryside. She thinks, "It's no more difficult than many of the flights with my instructor."	**It could happen to me.**
Impulsivity—While flying low over a neighborhood preparing to land, Brenda notices a number of adults and children in the middle of the street watching the balloon pass overhead. She decides to descend even lower, to rooftop level, to impress both the spectators and Sarah, her passenger. As she levels out, she notices the power lines running just below the treetops and narrowly misses one of them.	**Not so fast. Think first.**
Resignation—At the end of a local flight, Brenda does not adequately plan for a fast, hard landing. She fails to vent sufficiently on touchdown, and ends up draping the balloon envelope over the trees on the far edge of the landing area, with no damage. As she and her passenger exit the balloon, she says to herself, "Oh well, it's all part of learning to fly."	**I'm not helpless. I can make a difference.**

Figure 1-5. *A pilot must be able to identify hazardous attitudes and apply the appropriate antidote when needed.*

in judgment and decision-making become increasingly limited, particularly in high wind conditions. The most common causal factors for landing accidents include collision with obstructions in the intended landing area. Prior to a flight, a pilot should assess personal fitness. The "I'm Safe Checklist" helps a pilot determine his or her ability to fly. *[Figure 1-6]*

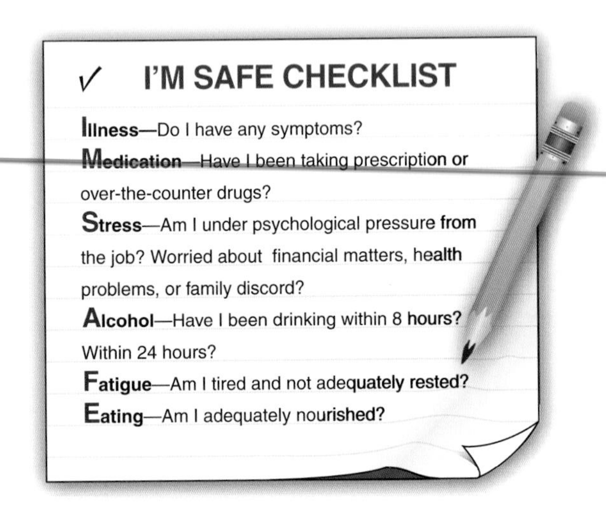

Figure 1-6. *Prior to flight, a pilot should assess personal fitness, just as he or she evaluates the balloon's airworthiness.*

Evaluating ADM Skills

The "What If" discussions an instructor pilot has with a student pilot are designed to accelerate development of decision-making skills by posing situations for the trainee to ponder. Research has shown that these types of discussions help build judgment and offset low experience. Once a student pilot has obtained his or her certification, it is important that he or she continue to evaluate flight decisions. To self-evaluate:

- Pose an open-ended question about the situation encountered during flight,
- Examine the decision made,
- Explore other ways to solve the problem, and
- Evaluate whether or not the best solution was used.

Crew Resource Management

ADM originated with the airline industry in an attempt to reduce human factors in aircraft accidents. The airlines developed a training program for flight crews called Crew Resource Management (CRM). It focuses on the effective use of all available resources to prevent accidents. While CRM focuses on pilots operating in crew environments, many of the concepts apply to single-pilot operations, but are not a "best fit" for balloon operations.

Single-Pilot Resource Management

A variant of the CRM model that may be of more practical application to the balloon pilot is Single-Pilot Resource Management (SRM), which may be defined as "the art and science of managing all resources (both from on-board and external sources) available to the single-pilot (prior to and during flight) to ensure the successful outcome of the flight." Virtually all ballooning is done as a single-pilot operation; there is no "crew resource" available from the perspective of having a co-pilot to assist in workload management.

For any single pilot, the primary emphasis of SRM is to integrate the underlying thinking skills needed by the pilot to consistently determine the best course of action to take in response to a given set of circumstances. SRM integrates the following concepts:

- Human Resources
- Risk Management
- Situational Awareness
- Training
- Decision-Making Process

Human Resources

Balloons differ from general aviation aircraft in the balloon pilot's reliance on diverse human resources for flight. Human resources include all groups working with pilots to ensure flight safety. A safe balloon flight includes, but is not limited to, a crew chief and ground crew, weather briefers, volunteers, spectators, "locals" with current and often unpublished information on roads and landing sites, landowners, and others who contribute assistance or information. Balloons differ from airplanes in their reliance on unlicensed, non-FAA-certified/recognized, and even first time volunteers to assemble and support ground handling of the balloon. Crew action—or inaction—at any stage of flight can contribute as much or more to flight safety than pilot input. Balloon flight safety often relies on many people beyond those onboard.

For example, a routine inflation on most balloons requires several sets of hands; moderate winds can quickly mean more help is needed. Having someone to handle a drop line offers a pilot landing site options inaccessible through onboard maneuvering. Added weight or "hands on" allows a pilot to choose a smaller landing site than when landing unassisted, or it can mean avoiding trees, power lines, or other obstacles.

Crew members make important information contributions to flight safety because crew can access real time flight related information before a pilot. For example, precipitation is often visible on the chase vehicles long before it compromises a balloon's inflight performance or gains a pilot's attention. The crew can also warn a pilot who is contour flying into the sun of power lines downwind or of livestock behind trees or buildings. A crew report on the current state of variable surface conditions can alert a pilot who is descending or landing into winds different from those of launch or flight. Crew action can easily mean the difference between a safe flight and an accident.

The essential and decisive roles crew and other human resources play in ballooning also create an ironic dilemma/dynamic between legal and operational realities. 14 CFR part 91 requires a pilot to act as the sole and final authority regarding operation of the balloon, yet every pilot must rely on crew who are not trained, certified, or even recognized by any governing body for a flight to occur. Each pilot thus requires and leads this integral, yet legally invisible team on each flight. Overlooking, minimizing, or dismissing the crew's role opens the door to mishaps. Safety often lies in recognizing how the crew's skill, knowledge, and experience complement and enhance the pilot's own. While all final decisions and the responsibility for safety still rest with the pilot, this broader than usual SRM model recognizes the human resources upon which every pilot relies for safe flight planning and decision-making.

Risk Management

Flying involves risk. To stay safe, a pilot needs to know how to judge the level of risk, how to minimize it, and when to accept it. The risk management decision path is best seen through the Perceive-Process-Perform model *[Figure 1-7]* which offers a structured way to manage risk.

Perceive hazards by looking at:

- Pilot experience, currency, condition;
- Aircraft performance, fuel;
- Environment (weather, terrain); and
- External factors.

Process risk level by considering:

- Consequences posed by each hazard,
- Alternatives that eliminate hazards,
- Reality (avoid wishful thinking), and
- External factors (get-home-itus).

Perform risk management:

- Transfer—can someone be consulted?
- Eliminate—can hazards be removed?

Figure 1-7. *The Perceive-Process-Perform model.*

- Accept—do benefits outweigh risk?
- Mitigate—can the risk be reduced?

During each flight, decisions must be made regarding events that involve interactions between the four risk elements—the pilot in command, the aircraft, the environment, and the operation. *[Figure 1-8]* One of the most important decisions a pilot in command makes is the go/no-go decision. Evaluating each of these risk elements can help a pilot decide whether a flight should be conducted or continued. Below is a review of the four risk elements and how they affect decision-making.

- Pilot—a pilot must continually make decisions about his or her competency, condition of health, mental and emotional state, level of fatigue, etc. For example, a pilot may plan for an early morning flight after an all night drive, which means little sleep. Tired, achy, congested from the beginnings of a cold, is that pilot safe to fly?
- Balloon—a pilot frequently bases decisions on the evaluations of the balloon, such as performance, equipment, or airworthiness. A pilot is on an afternoon flight in a rural area. Landing areas are becoming sparse because the terrain is mostly swampland. The wind is decreasing and sunset is only half and hour away. Should he or she continue to fly over this terrain?
- Environment—this encompasses many elements not pilot or balloon related. It includes, but is not limited to, such factors as weather, terrain, launch and landing areas, and surrounding obstacles. Weather is one element that can change drastically over time and

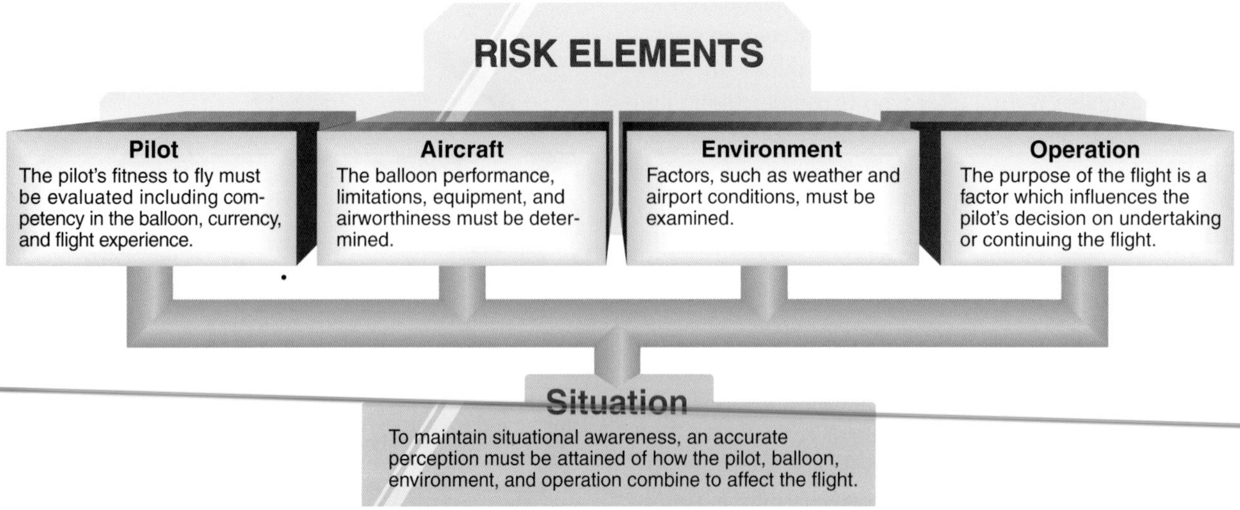

RISK ELEMENTS

Pilot
The pilot's fitness to fly must be evaluated including competency in the balloon, currency, and flight experience.

Aircraft
The balloon performance, limitations, equipment, and airworthiness must be determined.

Environment
Factors, such as weather and airport conditions, must be examined.

Operation
The purpose of the flight is a factor which influences the pilot's decision on undertaking or continuing the flight.

Situation
To maintain situational awareness, an accurate perception must be attained of how the pilot, balloon, environment, and operation combine to affect the flight.

Figure 1-8. *When situationally aware, a pilot has an overview of the total operation and is not fixated on one perceived significant factor.*

distance. During an afternoon flight with an indefinite ceiling, slight precipitation and the rumble of thunder is encountered. Should the pilot stay aloft, trusting the weather briefing's assertion that "there is no precipitation in the area," or land at the first available site as soon as possible?

- Operation—the interaction between the pilot, the balloon, and the environment is greatly influenced by the purpose of each flight operation. The pilot must evaluate the three previous elements to decide on the desirability of undertaking or continuing the flight as planned. It is worth asking why the flight is being made, how critical it is to maintain the original intent, and if the continuation of the flight is worth the risks?

Situational Awareness

Situational awareness is the accurate perception and understanding of all the factors and conditions within the four fundamental risk elements that affect safety before, during, and after the flight. To maintain situational awareness, a pilot needs to understand the relative significance of these factors and their future impact on the flight. When a pilot is situationally aware, he or she has an overview of the total operation.

Some obstacles to maintaining situational awareness include (but are not limited to) fatigue, stress, and work overload; complacency; and classic behavioral traps such as the drive to meet or exceed flight goals. Situational awareness depends on the ability to switch rapidly between a number of different, and possibly competing, information sources and tasks while maintaining a collective view of the environment. Experienced pilots are better able to interpret a situation because of their base of experience, but newer pilots can compensate for lack of experience with the appropriate fundamental core competencies acquired during initial and recurrent flight training. SRM training helps the pilot maintain situational awareness, which enables the pilot to assess and manage risk and make accurate and timely decisions. To maintain situational awareness, all of the skills involved in ADM are used.

The Decision-Making Process

Understanding the decision-making process provides a foundation for developing the necessary ADM skills. Some situations, such as an extinguished pilot light, require an immediate response using established procedures. While pilots are well trained to react to emergencies, they are not as prepared to make decisions that require a more reflective response. The ability to examine any changes that occur during a flight, gather information, and assess risk before reaching a decision constitutes the steps of the decision-making process.

Defining the Problem

Problem definition is the first step in the decision-making process. Defining the problem begins with recognizing a change has occurred or an expected change did not occur. A problem is perceived first by the senses and then is distinguished through insight and experience. This "gut" reaction, coupled with an objective analysis of all available information, determines the exact nature and severity of the problem.

Choosing a Course of Action

After the problem has been identified, the pilot must evaluate the need to react to it and determine the actions that need to be taken to resolve the situation in the time available. The expected outcome of each possible action should be considered and the risks assessed before deciding on a response to the situation.

Implementing Decisions and Evaluating the Outcomes

Although a decision may be reached and a course of action implemented, the decision-making process is not complete. It is important to think ahead and determine how the decision could affect other phases of the flight. As the flight progresses, a pilot must continue to evaluate the outcome of the decision to ensure that it is producing the desired result.

The DECIDE Model

A common approach to decision-making for the last decade has been the rational choice model. This concept holds that good decisions result when a pilot gathers all the information related to a particular scenario, reviews it, analyzes the options available, and decides on the best course of action to follow.

The DECIDE Model, a six-step process intended to provide the pilot with a logical way of approaching decision-making, is an example of this concept. The six elements of the DECIDE Model represent a continuous loop process to assist a pilot in decision-making. If a pilot uses the DECIDE Model in all decision-making, it becomes natural and results in better decisions being made under all types of situations. *[Figure 1-9]*

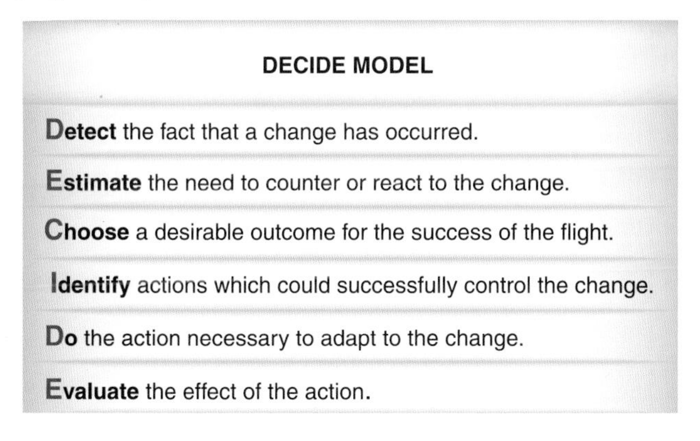

DECIDE MODEL
Detect the fact that a change has occurred.
Estimate the need to counter or react to the change.
Choose a desirable outcome for the success of the flight.
Identify actions which could successfully control the change.
Do the action necessary to adapt to the change.
Evaluate the effect of the action.

Figure 1-9. *The DECIDE Model can provide a framework for effective decision-making.*

The OODA Loop

Colonel John Boyd, USAF (Retired), coined the term and developed the concept of the "OODA Loop" (Observation, Orientation, Decision, Action). *[Figure 1-11]* The ideas, words, and phrases contained in Boyd's briefings have penetrated not only the United States military services, but the business community and academia around the world. The OODA Loop is now used as a standard description of decision-making cycles.

The OODA Loop is an interlaced decision model which provides immediate feedback throughout the decision-making process. For SRM purposes, an abbreviated version of the concept *[Figure 1-10]* provides an easily understood tool for the balloon pilot.

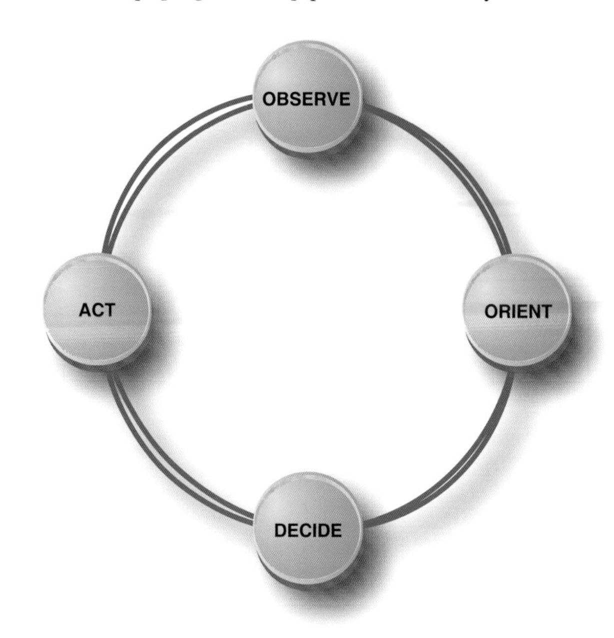

Figure 1-10. *The four nodes of OODA Loop decision-making.*

The first node of the OODA Loop, **Observe,** reflects the need for situational awareness. A balloon pilot must be aware of those things around him or her that may impact the flight. Continuous monitoring of wind, weather, ground track, balloon responses, and so forth provide a constant reference point by which the pilot knows his or her starting point on the loop. This permits the ability to immediately move to the next step.

Orient, the second node of the OODA Loop, focuses the pilot's attention on one or more discrepancies in the flight. For example, while contour flying over trees, the balloon passes the edge of the tree line and begins a gradual descent. The pilot is aware of this deviation and considers available options in view of potential hazards to continued flight.

The pilot then moves to the third node, **Decide,** in which he or she determines the action to create a specific desired effect. That decision is made based on experience and knowledge of potential results of a particular action. The pilot then **Acts** on that decision, making a physical input to cause the balloon to react in the desired fashion.

Once the OODA Loop has been completed, the pilot is once again in the Observe position. Assessment of the resulting action is added to the previously perceived aspects of the

Dave has received his weather information for an afternoon flight, and finds the winds are forecast to stay above 7 mph until just before sunset. He goes ahead with planning the flight, believing the winds will decrease, and will soon be going out to the launch site to start preparations for launch.

In this example, Dave is in the ***Observe*** part of the decision cycle. He will continue to ***Observe*** until he detects a condition that would cause him to cancel the flight. Dave has completed a full cycle of the OODA Loop, but has returned to the ***Observe*** node to continue to monitor the weather.

Twenty minutes into the flight, Dave notices frost forming on the outside of the fuel fitting leading to the burner. Since he cannot tighten the fitting (it is a fixed fitting), he decides to shut off the fuel on that side, vent the line, and land as soon as practicable in accordance with the established "emergency" procedure.

Here, Dave has completed an entire ***Decision*** cycle in a very short time due to the circumstances of the situation. He will now move to the ***Act*** node, and will take action to resolve the issue before it causes danger to him or his passenger.

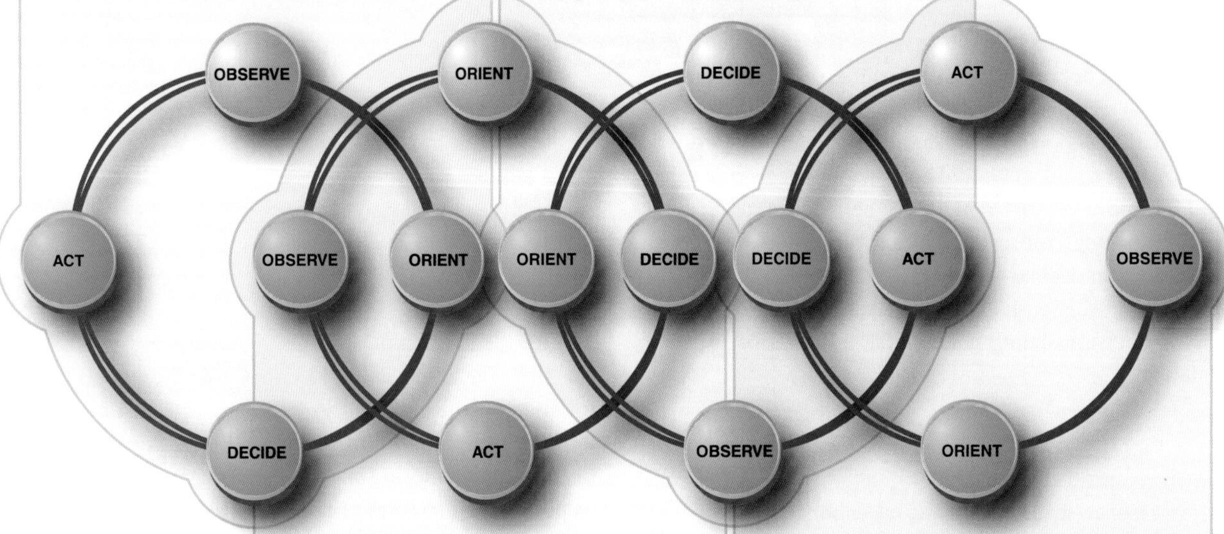

Upon reaching the launch site, Dave checks the winds using a pibal and finds that winds are decreasing. He believes a flight can be safely conducted, and goes ahead with his launch plans. After launch, Dave finds that the winds are in a direction that is not quite what was forecast, and are taking him into an area with few good landing areas. He continues the flight, and watches the direction closely to detect any further change of direction.

In this example, Dave has completed an OODA decision-making cycle, and has returned to the ***Orient*** node. He will continue to monitor the situation, being aware of changes, and is prepared to make a decision when the situation warrants.

While making an approach to land, Dave's passenger, Pat, alerts him to powerlines running across the near side of the small field. Dave evaluates his approach and ***Acts*** to land accordingly, knowing that winds may shift somewhat as he descends. He touches down well past the powerlines, never having put him or his passenger at risk, and has an uneventful deflation.

Dave again completed an OODA Loop decision-making cycle. Had there been an indication that the approach could not be safely completed, Dave could have made a decision to abort the landing ***(Decide),*** and then performed an ***Act*** to effect a change.

Figure 1-11. *Using the OODA Loop as a model, it is possible to have multiple decision-making cycles in progress, in different stages of completion. While these examples show a sequence, this is not always the case; the OODA Loop cycles may overlap in any stage of execution.*

flight to further define the flight's progress. The advantage of the OODA Loop Model is that it may be cumulative. Also, multiple cycles in different stages of completion may be occurring at any given point in the flight.

Balloon Certificate Eligibility Requirements

To be eligible to fly a balloon solo, an applicant must be at least 14 years of age and demonstrate satisfactory aeronautical knowledge on a test developed by a flight instructor. Flight training must be received and logged for the maneuvers and procedures in 14 CFR part 61 that are appropriate to the make and model of aircraft to be flown. Only after all of these requirements are met can a flight instructor endorse a student's certificate and logbook for solo flight.

To be eligible for a sport pilot certificate with a balloon endorsement, an applicant must be at least 16 years of age, complete the specific training and flight time requirements described in 14 CFR part 61, pass a knowledge test, and successfully complete a practical test. A sport pilot certificate with balloon endorsement authorizes piloting a balloon with a maximum takeoff gross weight of 1,320 pounds or less. Satisfactory proficiency and safety must also be demonstrated.

To be eligible for a private pilot certificate with a balloon rating, an applicant must be at least 16 years of age, complete the specific training and flight time requirements described in 14 CFR part 61, pass a knowledge test, and successfully complete a practical test. A private pilot certificate allows a pilot to carry passengers, but he or she may not receive compensation. The FAA has specified the aeronautical knowledge and flight proficiency that must be demonstrated to earn a private certificate as listed below.

Aeronautical Knowledge, Private Pilot, 14 CFR Part 61, Section 61.105.

(a) General. A person who is applying for a private pilot certificate must receive and log ground training from an authorized instructor or complete a home-study course on the aeronautical knowledge areas of paragraph (b) of this section that apply to the aircraft category and class rating sought.

(b) Aeronautical knowledge areas.

1. *Applicable Federal Aviation Regulations of this chapter that relate to private pilot privileges, limitations, and flight operations;*
2. *Accident reporting requirements of the National Transportation Safety Board;*
3. *Use of the applicable portions of the Aeronautical Information Manual and FAA Advisory Circulars;*
4. *Use of aeronautical charts for VFR navigation using pilotage, dead reckoning, and navigation systems;*
5. *Radio communication procedures;*
6. *Recognition of critical weather situations from the ground and in flight, windshear avoidance, and the procurement and use of aeronautical weather reports and forecasts;*
7. *Safe and efficient operation of aircraft, including collision avoidance, and recognition and avoidance of wake turbulence;*
8. *Effects of density altitude on takeoff and climb performance;*
9. *Weight and balance computations;*
10. *Principles of aerodynamics, powerplants, and aircraft systems;*
11. *Stall awareness, spin entry, spins, and spin recovery techniques for the airplane and glider category ratings;*
12. *Aeronautical decision-making and judgment; and*
13. *Preflight action that includes—*
    - *i. How to obtain information on runway lengths at airports of intended use, data on takeoff and landing distances, weather reports and forecasts, and fuel requirements; and*
    - *ii. How to plan for alternatives if the planned flight cannot be completed or delays are encountered.*

Electronic Code of Federal Regulations. (1997, July 30). 14 CFR part 61, section 61.105. Retrieved on 19 January 2007 from http://www.faa.gov/regulations_policies.

Flight Proficiency, Private Pilot, 14 CFR Part 61, Section 61.107

(a) General. A person who applies for a private pilot certificate must receive and log ground and flight training from an authorized instructor on the areas of operation of this section that apply to the aircraft category and class rating sought.

(b) Areas of operation.

NOTE: Steps 1–7 have been omitted from this reference. To view these steps, reference the manual listed in the title of the section text.

8. *For a lighter-than-air category rating with a balloon class rating:*
   - i. *Preflight preparation;*
   - ii. *Preflight procedures;*
   - iii. *Airport operations;*
   - iv. *Launches and landings;*
   - v. *Performance maneuvers;*
   - vi. *Navigation;*
   - vii. *Emergency operations; and*
   - viii. *Postflight procedures.*

Electronic Code of Federal Regulations. (1997, July 30). 14 CFR part 61, section 61.107. Retrieved on 19 January 2007 from http://www.faa.gov/regulations_policies.

To be eligible for a commercial pilot certificate with a balloon rating, an applicant must be 18 years of age, complete the specific training requirements described in 14 CFR part 61, pass the required knowledge tests, and pass another practical test.

The FAA has specified the aeronautical knowledge and flight proficiency that must be demonstrated to earn a commercial certificate as listed below.

Aeronautical Knowledge, Commercial Pilot, 14 CFR Part 61, Section 61.125.

(a) *General. A person who applies for a commercial pilot certificate must receive and log ground training from an authorized instructor, or complete a home-study course, on the aeronautical knowledge areas of paragraph (b) of this section that apply to the aircraft category and class rating sought.*

(b) *Aeronautical knowledge areas.*

1. *Applicable Federal Aviation Regulations of this chapter that relate to commercial pilot privileges, limitations, and flight operations;*
2. *Accident reporting requirements of the National Transportation Safety Board;*
3. *Basic aerodynamics and the principles of flight;*
4. *Meteorology to include recognition of critical weather situations, windshear recognition and avoidance, and the use of aeronautical weather reports and forecasts;*
5. *Safe and efficient operation of aircraft;*
6. *Weight and balance computations;*
7. *Use of performance charts;*
8. *Significance and effects of exceeding aircraft performance limitations;*
9. *Use of aeronautical charts and a magnetic compass for pilotage and dead reckoning;*
10. *Use of air navigation facilities;*
11. *Aeronautical decision-making and judgment;*
12. *Principles and functions of aircraft systems;*
13. *Maneuvers, procedures, and emergency operations appropriate to the aircraft;*
14. *Night and high-altitude operations;*
15. *Procedures for operating within the National Airspace System; and*
16. *Procedures for flight and ground training for lighter-than-air ratings.*

Electronic Code of Federal Regulations. (1997, July 30). 14 CFR part 61, section 61.125. Retrieved on 19 January 2007 from http://www.faa.gov/regulations_policies.

Flight Proficiency, Commercial Pilot, 14 CFR Part 61, Section 61.127.

(a) *General. A person who applies for a commercial pilot certificate must receive and log ground and flight training from an authorized instructor on the areas of operation of this section that apply to the aircraft category and class rating sought.*

(b) *Areas of operation.*

8. *For a lighter-than-air category rating with a balloon class rating:*
   - i. *Fundamentals of instructing;*
   - ii. *Technical subjects;*
   - iii. *Preflight preparation;*
   - iv. *Preflight lesson on a maneuver to be performed in flight;*
   - v. *Preflight procedures;*
   - vi. *Airport operations;*
   - vii. *Launches and landings;*
   - viii. *Performance maneuvers;*
   - ix. *Navigation;*
   - x. *Emergency operations; and*
   - xi. *Postflight procedures.*

Electronic Code of Federal Regulations. (1997, July 30). 14 CFR part 61, section 61.127. Retrieved on 19 January 2007 from http://www.faa.gov/regulations_policies.

NOTE: A commercial pilot with a balloon category and class rating may instruct, and is authorized to do so under the provisions of his or her certificate.

If a pilot currently holds a pilot certificate for a powered aircraft and wants to add a balloon category rating on that certificate, he or she is exempt from the knowledge test, but must satisfactorily complete the flight training and practical test. Certificated balloon pilots are not required to hold an airman medical certificate to operate a balloon within the United States. Flights in foreign countries fall under the International Civil Aviation Organization (ICAO) regulations, and a medical certificate is generally required.

For private pilots working toward the commercial pilot certificate, the regulatory requirement is ten additional hours of flight training, per the provisions of 14 CFR part 61. Flight time logged as instructional time towards the initial certification may not, under most circumstances, be counted towards that time which must be accumulated toward the commercial rating.

Introduction to Integrated Airman Certification and/or Rating Application (IACRA)

The Integrated Airman Certification and/or Rating Application (IACRA) is an online application system that allows for the issuance of student, private, and commercial certificates without generating paperwork; all certificate application and approval is done through the use of electronic signatures. IACRA interfaces with multiple databases to validate data and verify specific fields of information, and upon completion of the approval process, forwards the data automatically to the Airman Certification Branch.

In order to utilize the system, an applicant must first register at http://acra.faa.gov/iacra/. This registration process results in the issuance of an FAA Tracking Number (FTN). This number becomes the applicant's registration number for all future utilization of the system, so it is important to keep it for future reference and action.

Upon completion of all training requirements, the student pilot (which refers to both the student applying for a private certificate, as well as the private pilot upgrading to the commercial certificate) makes application through the IACRA web site. The FAA Form 8710-1 is completed online in several steps. When complete, the form is submitted to the central server at the FAA Airman Certification Branch in Oklahoma City. If the form is accepted at the server, the applicant is issued an application ID number for future reference.

The instructor, who has previously enrolled in the IACRA system as a recommending instructor, uses the applicant's FTN to retrieve the application for review. It is also the responsibility of the instructor to attach the results of the student's written test to the IACRA file, using the Exam ID, which is usually found in the upper right portion of the printed results. Once the recommending instructor has reviewed the application, and the results are attached, the application is again sent to the central server at the Airman Certification Branch.

At the appropriate time, the examiner retrieves the applicant's application from the central server, verifies the individual's identification, and administers the practical test. IACRA has provisions for the issuance of a temporary airman certificate, notice of discontinuance, or notice of disapproval. When the evaluation is completed, and the examiner has completed the appropriate processing, the file is again returned to the Airman Certification Branch. Upon successful completion, the permanent certificate is prepared and forwarded to the pilot.

Chapter Summary

This chapter provides an overview of the FAA, the flight instructor, and DPE's roles in the qualification and licensing of lighter-than-air pilots. Additionally, the regulatory requirements for both ground and aeronautical training have been outlined.

The reader has also been given some basic information and techniques concerning flight safety practices and aeronautical decision-making. This information is not exhaustive; the student pilot would be well advised to avail himself or herself of other materials published by the FAA and other organizations.

Chapter 2

Hot Air Balloon Design, Systems, and Theory

Introduction

This chapter presents an introduction to the history of flying balloons, the physics of balloon flight, balloon components, balloon terminology, support equipment, and how to choose a balloon.

History

Hot air balloons are the oldest successful human flight technology. The first recorded manned balloon flight was made on November 21, 1783 in a hot air balloon developed by the Montgolfier brothers of France. *[Figure 2-1]* Flown by Pilatre de Rozier and the Marquis d'Arlandes, the flight lasted 23 minutes and covered 5.5 miles. Although the Montgolfiers are given credit for the first documented flight, there are some earlier claims. The Chinese are credited with using manned kites, and perhaps hot air balloons, some 2,000 years ago, and the Nazcas of Peru may have used smoke-filled balloons.

FAI Category	Cubic Meters	Number of People	Cubic Feet
AX5	900–1,200	1	31,779–42,372
AX6	1,200–1,600	1–2	42,372–56,372
AX7	1,600–2,200	2–4	56,496–77,682
AX8	2,200–3,000	4–7	77,682–105,930
AX9	3,000–4,000	6–10	105,930–141,240

Figure 2-1. *Model of the Montgolfier brothers balloon.*

Ten days after the successful flight of the Montgolfier balloon, a young physicist, Professor Jacques Charles flew the first gas balloon made of a varnished silk envelope filled with hydrogen. His flight lasted two hours and covered 27 miles, reaching an altitude of 9,000 feet. Begun as an attempt to duplicate and validate the achievements of the Montgolfier brothers, Charles based his experiment on misinformation. He mistakenly believed the Montgolfier brothers used hydrogen to inflate their balloon, so he used hydrogen. Thus, the two kinds of balloons flown today—hot air and gas—were developed in the same year.

Gas ballooning became a sport for the affluent and flourished on a small scale in Europe and the United States. Since ballooning drew crowds, one way to offset the cost of a flight was to charge admission. Ballooning was a perilous undertaking that drew male and female daredevils eager to court danger. The parachute, invented by balloonist Andre Garnerin in 1797 as the means of performing a daring stunt, is probably ballooning's most significant contribution to flight.

At the turn of the century, the smoke balloon (a canvas envelope heated by fire on the ground) was a common county fair opening event. As the smoke balloon ascended, a man or woman rider balanced on a trapeze attached to the balloon. After the initial climb (about 3,000 feet per minute (fpm)) the hot air cooled and the rider separated from the balloon, deploying a parachute to return to earth.

Balloons also found a home with the military. Napoleon used anchored observation balloons in some of his battles and considered using balloons to ferry troops in his proposed invasion of England. During the American Civil War, both the North and South used tethered observation balloons. In Europe, balloons were used during the 1870 siege of Paris (Franco-Prussian War) to carry messages and important people out of Paris. World War I saw balloons used by both sides for artillery spotting. By World War II, airplanes had replaced balloons for observation and reconnaissance purposes although barrage balloons (several large balloons tethered close together) were often used to discourage low level bombers or dive bombers. The United States Navy contracted with the General Mills Company in the 1950s to develop a small hot air balloon for military purposes. The Navy never used the balloon, but the project created the basis for the modern hot air balloon. With the use of modern materials and technology, hot air ballooning has become an increasingly popular sport.

Physics

In concept, the balloon is the simplest of all flying machines. It consists of a fabric envelope filled with a gas that is lighter than the surrounding atmosphere. Since air in the envelope is less dense than its surroundings, it rises, taking the basket filled with passengers or payload with it. A balloon is distinct from other aircraft in that it travels by moving with the wind and cannot be propelled through the air in a controlled manner.

There are two main types of balloons, hot air and gas, but other specialty type balloons are also flown. The Rozier balloon is an example of a less common balloon. A hybrid balloon that utilizes both heated and unheated lifting gases for long distance record flights, a Rozier was flown by Steve Fossett in his record-setting first solo circumnavigation in 2002. A recent addition to the hot air balloon field is the solar balloon, which uses heat radiation from the sun to provide lift. This handbook primarily covers hot air balloons.

Why Do Balloons Fly?

The physics of balloon flight is based on the principles of fluid dynamics and associated theorems. Therefore, it is helpful to think of the air, the medium of balloon flight, as a fluid when discussing the concept of "buoyancy" as applied to balloon flight. In physics, buoyancy is the upward force of an object produced by the surrounding fluid (i.e., liquid or gas) in which it is fully or partially immersed, due to the pressure difference of the fluid between the top and bottom

Figure 2-2. *The air inside the balloon envelope is heated to create buoyancy.*

of the object. The net upward buoyancy force is equal to the magnitude of the weight of fluid displaced by the object. This force enables the object to float or at least to appear lighter. An object must make room for its own volume by pushing aside, or displacing, an equal volume of liquid. For example, an aircraft carrier exerts downward force on the water and the water exerts upward force on the aircraft carrier. A solid object floats when it has displaced just enough water, or air in the case of a balloon, to equal its own original weight.

To create the necessary buoyancy for flight, the air inside the balloon envelope is heated which causes the air to expand, making it less dense. *[Figure 2-2]* Once the interior air weighs less than the non-heated ambient air (air that surrounds an object), the balloon becomes lighter in weight and rises in an effort to find a level where the interior air density matches that of the exterior air density. The envelope is carried along "for the ride," as it does little more than contain the heated air mass. The balloon rises to a point where the lift created by the action of heating the air is equal or greater than that of the balloon itself. The balloon rises because it has reached a state of "positive buoyancy" and the amount of lift is greater than the weight of the balloon.

The greater the heat differential between the air inside the envelope and the ambient air, the faster the balloon rises. Hot air is constantly being lost from the top of the envelope by leaking through the fabric, seams, and deflation port. Heat is also lost by radiation. Only the best and newest fabrics are nearly airtight. Some fabrics become increasingly porous with age and some colors radiate heat faster than others do. Under certain conditions, some dark colored envelopes may gain heat from the sun. To compensate for heat loss, prolonged flight is possible only if fuel is carried on board to make heat.

The internal temperature of the air in the envelope is raised or lowered to change altitude. To climb, the temperature in the envelope is raised by heating the air which creates more lift. To descend, the air in the envelope is allowed to cool. Cooling of the envelope is also possible by allowing hot air to escape through a vent. This temporary opening closes and seals automatically, due to the upward pressure, when it is not in use.

A balloon's weight when in flight is not only the figure as stated in the flight manual, but also includes the weight of the air within the inflated envelope, the balloon components and equipment, as well as the pilot and passengers. The average 77,000 cubic foot hot air balloon contains an air mass that weighs over 3,000 pounds. By adding all these factors together, it is easy to understand how weight influences the balloon's response to pilot actions during flight maneuvers. The weight and sheer momentum of a balloon in flight make it difficult for a pilot to effect rapid changes.

Balloon Components

A hot air balloon consists of three main components: envelope, heater system, and basket. In addition, flight instruments, fuel tanks, and other support equipment are needed for a safe balloon flight. *[Figure 2-3]* The most common ballooning terms are used in the following text, in

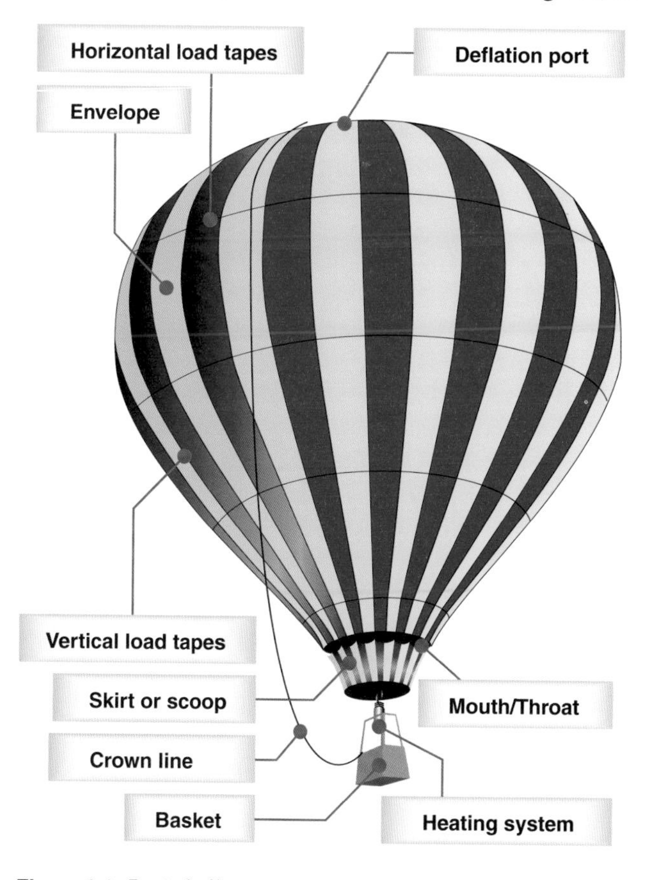

Figure 2-3. *Basic balloon terms.*

the generic illustrations, and are also listed in the glossary which contains balloon and aeronautical terminology. Some terms and names used by manufacturers are also included.

Envelope

The envelope is usually made of light-weight and strong synthetic fabrics such as ripstop nylon or Dacron®. The material is cut into panels which are sewn together in vertical rows that are called gores due to their triangular shape. The traditional envelope shape is a teardrop. The gores are reinforced with sewn-in webbing called horizontal and vertical structural load tapes which are continuous to the top center of the balloon where they are sewn into a load ring. Galvanized, stainless steel, or Kevlar® cables transfer basket loads to load tapes which in turn support the load. The nylon "skirt" at the base of the envelope is coated with special fire resistant material to keep the flame from igniting the balloon.

The deflation port is located at the top of the envelope and allows for the controlled release of hot air. It is covered by the deflation panel sometimes called a top cap, parachute top, or spring top. *[Figure 2-4]* In a balloon with a parachute top, partial opening of the parachute valve is the normal way to cool the balloon. Balloons with other types of deflation panels may have a cooling vent in the side or the top. Many balloons are also equipped with turning vents, which allow for the pilot to turn the balloon on its vertical axis while in flight. Turning vents help a pilot align the basket for landing, or in the case of commercial balloons, align the balloon's logo toward the crowd.

Figure 2-4. *Deflation system.*

Special Shape Balloons

Balloons that do not have a traditional "teardrop" shape are called special shape balloons. *[Figure 2-5]* They may be completely engineered systems which have been designed to resemble cans, sports balls, cartoon characters, cars, etc.

Figure 2-5. *Special shape balloons.*

Some balloons have appendages added to the envelope. Appendages are pieces added to a balloon envelope in order to create a particular shape or rendition, not necessarily keeping with a standard shape balloon. To be designated an appendaged envelope, less than 10 percent of the total capacity of the balloon is contained within the appendage. While the appendaged envelope has the same general flight characteristics as a standard balloon shape, there are some differences. For example, the added weight of the appendage may cause the overall envelope to weigh significantly more than teardrop balloons of equal size. Appendage balloons also have the tendency to rotate during aggressive climbs and descents.

A special shape envelope requires a substantial amount of engineering to ensure the envelope is properly stressed, and the balloon has no undesirable flight characteristics due to the shape. Special shape balloons built in the United States or the United Kingdom are normally issued Standard Airworthiness Certificates, but special shape balloons imported from other manufacturers in other countries may be issued an Experimental Airworthiness Certificate. A balloon with an Experimental Airworthiness Certificate usually may not be flown for compensation or hire, which negates the marketability of such a balloon. Additionally, an experimental balloon may not be flown over congested areas, per Title 14 of the Code of Federal Regulations (14 CFR) part 91, section 91.319. Experimental balloons also require specific documentation when flown outside of an area of 50 miles from its home port. Pilots of special shape balloons with an Experimental Airworthiness Certificate should coordinate their activities with their local Flight Service District Office to avoid problems.

Thermal Airships

A thermal airship combines the characteristics of a hot air balloon, with respect to lifting force, and an airship, with respect to its capacity of being steered while in flight. To develop proficiency in this aircraft, knowledge is required of not only hot air balloon operations and physics, but also of airship operations. It is estimated that there are currently less than 10 of these aircraft in the United States, and there is no specific pilot certificate for thermal airships. These aircraft are extremely expensive to purchase, and have some significant operating limitations with reference to winds. Any further discussion is outside the scope of this handbook.

Heater System

The heater system consists of propane burners (one or more), fuel tanks that store liquid propane, and fuel lines that carry the propane from the tanks to the burners. The burners convert ambient air into hot air, which in turn provides the lift required for flight. *[Figures 2-6 and 2-7]*

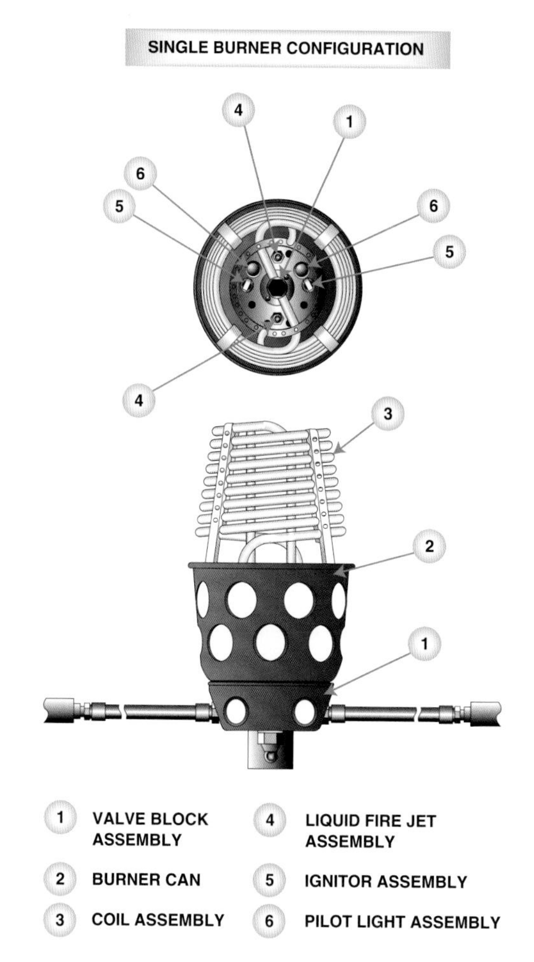

Figure 2-6. *Typical single heater (burner).*

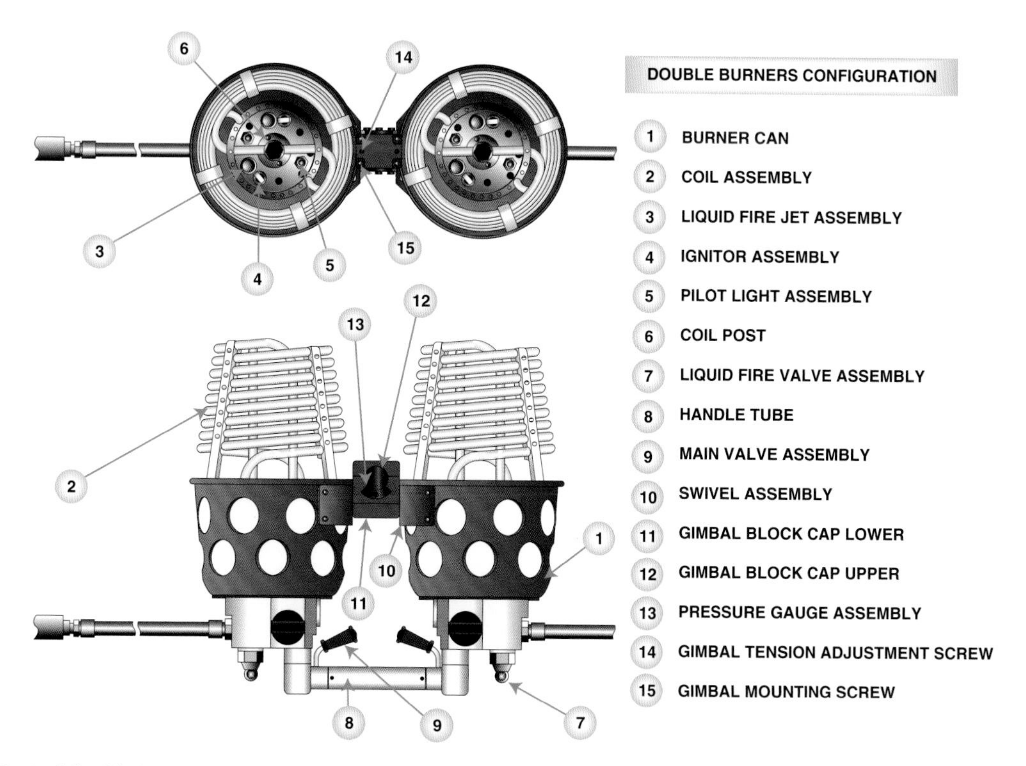

Figure 2-7. *Typical double heater (burner).*

Propane fuel is used to heat the air which generates buoyancy for flight. The propane is stored in one or more fuel tanks located in the basket. A withdrawal tube attached to the liquid tank valve permits liquid propane to be drawn from the bottom of the fuel tanks. The liquid propane is supplied to the burner assembly through the fuel hoses that connect the fuel tanks to the heater assembly (commonly referred to as the burner). The fuel system also provides propane to the pilot light.

There are two types of pilot light systems: liquid and vapor. In a liquid pilot light system, liquid propane is diverted from the main supply line at the heater via a pilot shut-off valve. The fuel goes through a vapor converter and regulator, and is distributed through a pilot head. A piezo-electric system ignites the vapor at inflation for most heaters, but many balloonists choose to use a striker.

In a vapor pilot light, a second fuel hose is used to supply vapor to the heater assembly pilot light from the pilot light tank valve located on top of the fuel tank. A regulator is used to decrease the pressure of the propane vapor for proper pilot operation. A pilot light valve located on the heater controls the flow of propane vapor to the pilot light.

The main liquid tank valve controls the flow of liquid propane to the burner, while the blast valve controls fuel flow at the heater. With the liquid tank valve open, opening the burner blast permits liquid propane to enter the heat exchange coil where it is either completely or partially vaporized. After exiting the heat exchange coil through the orifices in the lower portion of the coil, the propane is ignited by the pilot light. *[Figure 2-8]*

To meet redundancy requirements, heaters have a secondary system which allows for operation at a reduced efficiency should a problem develop with the main blast valve. These backup, or bypass systems generally have proprietary names unique to the individual manufacturers. Pilots should consult individual flight manuals for an explanation of their use.

The heaters typically have an output of approximately 20 million British thermal units (BTU) in use. There is a power loss associated with altitude, generally considered to be four percent per thousand feet of altitude. This is particularly important when dealing with higher density altitudes. Pilots accustomed to flying at lower altitudes are frequently surprised by the decreased performance of their balloon upon their first flight at a higher altitude.

Basket

Wicker is the preferred material for the passenger compartment basket of a hot air balloon because it is sturdy, flexible, and relatively lightweight. *[Figure 2-9]* The flexibility of wicker

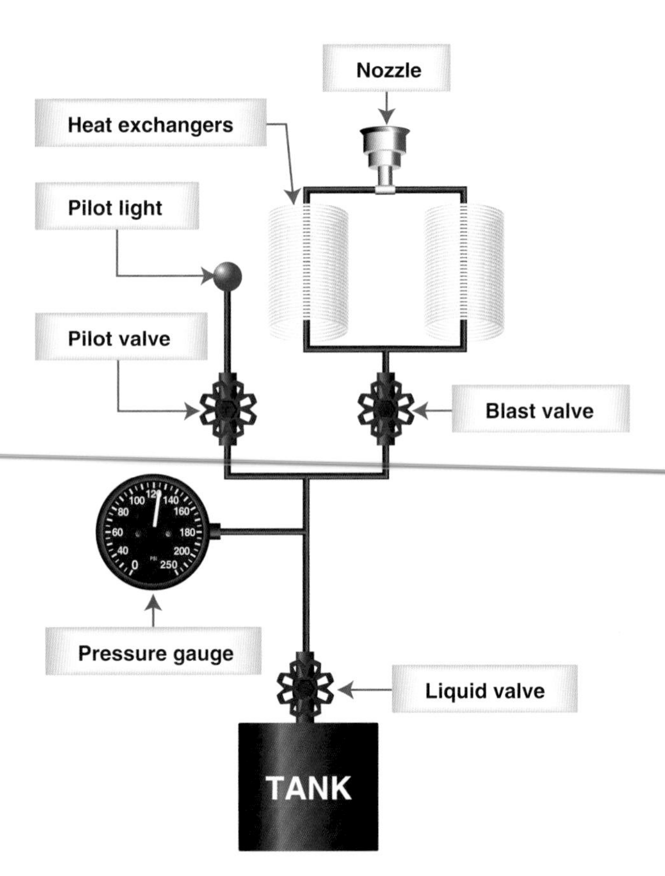

Figure 2-8. *Fuel system schematic.*

Figure 2-9. *Representative sport basket configuration.*

helps with balloon landings and cushions some of the impact force at landing. The basket contains the fuel tanks, instruments, pilot, and passengers.

Aluminum, stainless steel, or flexible nylon poles (in conjunction with stainless steel cables) located on the upper portion of the structure transfer the basket load to the envelope attachment points and support the burner assembly. Quick pins or aircraft bolts connect the support tubes, with nylon rods usually inserted into the sockets, and the cables attached around them. Lower frame tubes support the floor, permitting the floor load to be transferred to the lower frames. Oak skids, usually affixed to the floor, add rigidity and provide a point of abrasive resistance to the floor. Rattan sidewalls of varying thickness, design, and color surround and protect the passengers, equipment, and fuel tanks. Larger baskets, usually found on large ride balloons, may have padded basket dividers to form passenger compartments.

Instruments

As required by 14 CFR part 31, balloons are equipped with an altimeter, a rate of climb indicator, fuel quantity gauges, and an envelope temperature gauge. Many newer balloons use some type of electronic instrument system, but older balloons may still be equipped with traditional, pressure driven analog instruments.

The most common arrangement of instruments is a small pod or package which includes the altimeter, rate of climb indicator (variometer or vertical speed indicator), and the envelope temperature gauge (pyrometer). A cable is plugged into the instrument package during the preflight or layout process which connects to a sensor located in the top of the balloon and operates the envelope temperature gauge. Fuel quantity gauges, located on the top portion of each fuel tank, provide a reference for the quantity of fuel remaining in the tank. As these gauges are mechanical, they are sometimes inaccurate, and in most configurations do not read from 0 to 100 percent. This needs to be taken into consideration during the flight planning process.

In recent years, many manufacturers have added a wireless system that transmits the temperature signal to a receiver in the basket via a radio or infrared signal. This eliminates the necessity of a wire being located in the balloon envelope. These instrument systems are popular, but have reliability issues because radio interference or the thermal "plume" from the burner can degrade the signal transmission.

Fuel Tanks

Balloons generate heat through the use of propane. The propane is contained in aluminum or stainless steel tanks mounted inside the basket. These tanks may be either vertical or horizontal, and contain 10, 15, or 18 gallons of propane. Larger tanks are available for larger ride balloons. *[Figure 2-10]*

The tanks, sometimes referred to as pressure vessels, are commonly equipped with a service valve (or main liquid valve), a fixed maximum liquid level gauge (or "spit valve"), a float gauge, and a pressure relief valve. The service valve regulates the flow of the liquid propane to the heater. The fixed maximum liquid level gauge provides an indication that the tank is filled to its maximum design quantity, or 80 percent of the total capacity of the tank. The float gauge provides a reading of the remaining capacity of the tank in a percentage. The pressure relief valve allows for the release of propane from the tank should the tank exceed the maximum design pressure. A pressure relief valve normally releases at 375 pounds per square inch (psi). *Figure 2-11* provides additional information on propane management.

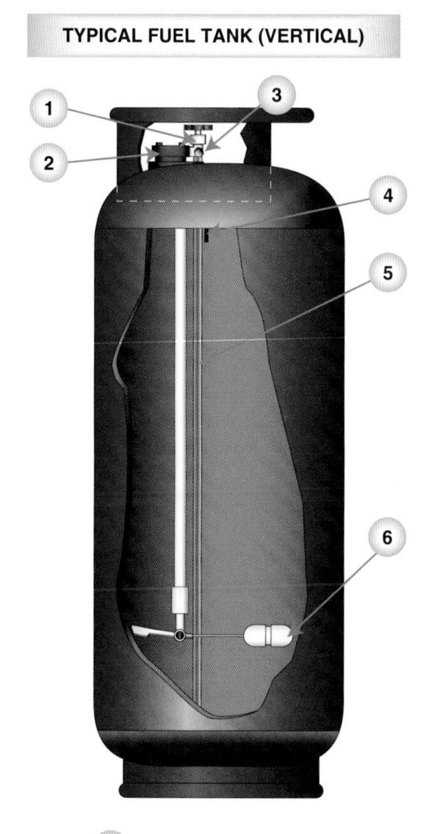

Figure 2-10. *Typical fuel tank.*

Propane Primer

Propane is a liquefied petroleum gas. Propane is preferred over butane and other hydrocarbons in balloon design because propane has a lower boiling point (propane -44 °F, butane 32 °F), and, therefore, a consistently higher vapor pressure for a given temperature. Under ideal circumstances, a gallon of propane produces 91,600 BTU of heat.

In its gaseous form, propane is odorless. However, an odorant (a strong smelling chemical compound) is added to propane to indicate the leakage of even small quantities of gas. The odorant normally added to propane, ethyl mercaptan, loses its odor when burned. Tanks should never be stored in an enclosed area, or near a heater and/or a device with a pilot light. Propane vapor is heavier than air, and will collect in low areas. Concentrated propane vapor constitutes an explosive hazard.

There is a popular misconception that propane is always at -44 °F when stored. This is incorrect—the propane, in the tank, is the same temperature as the ambient air. Propane turns cold when vaporizing, due to heat exchange.

The combustion of propane yields carbon dioxide and water.

The balloonist should be aware of the large volume of combustible mixture that will result from the escape of a small amount of liquid propane. As a rough approximation, a given volume of liquid propane produces a combustible mixture 6,800 times the original liquid propane volume.

The propane cylinder raises the boiling point of propane by trapping the pressure built up by the vaporized propane. Any given temperature will produce a specific pressure within the propane cylinder. This balance between temperature and pressure is referred to as the point of equilibrium. As long as the propane cylinder is neither completely full nor completely empty of the liquefied fuel, and no foreign substance such as air is present, the pressure within the cylinder is dependent upon the prevailing temperature of the liquid, and not upon the amount of liquid propane within the cylinder. Therefore, maintaining a proper propane temperature is necessary to supply sufficient fuel pressure to the burner. The temperature-pressure point of equilibrium is only applicable to containers containing vapor over the liquid. While propane vapor is easily compressible, propane liquid is practically incompressible.

Butane-propane mixtures can be a problem. At certain times of the year in certain areas, butane will be mixed with propane to increase its boiling point. A butane-propane mixture will produce a more yellow, sooty flame, may tend to go out when the blast valve is pulled, and the pressure available for a given temperature will be less.

Burners commonly used in hot air balloons are vaporizer burners. A vaporizer burner combines a vaporizer with a burner into a single unit where burner heat is used to vaporize the liquid propane being fed into the burner. The vaporizer normally consists of a coil in direct contact with the burner flame. If there is no fuel flow through one of the coils during burner operation, the coil may be damaged as a result of overheating.

The heat output of a burner is one of the parameters that indicate how a given balloon will perform in flight. A long sustained burn will produce a lower rating that the same time increment broken up into short burns. The lower output for a sustained burn is due to the inertia of the fuel, the friction hindering fuel flow, and pressure decay due to the withdrawal of liquid propane from the cylinder.

Periodic inspections of the airborne heating system should include a visual inspection of the hoses and fittings, and a high pressure leak check with a test gauge adaptor.

Source: Saum, Nick. "Propane and Fuel Management."
Joint publication of the Safety and Education Committees of the Balloon Federation of America, 1991.

Figure 2-11. *Propane primer. (Propane pressures at different temperatures are listed in Appendix A.)*

Support Equipment

Standard support equipment for ballooning includes an inflation fan, transport/chase vehicle, and small miscellaneous items, such as igniters, drop lines, gloves, spare parts, and helmets.

Inflation Fans

The inflation fan is one of the most dangerous pieces of equipment in ballooning. *[Figure 2-12]* Keep this fact in mind when purchasing and operating any inflation fan. Fan blades have been known to shatter or break, throw rocks at high velocity, and inadequate cages or guards fail to protect fingers and hands. Any fan considered for purchase by a prospective pilot should be evaluated for potential safety hazards. Also, remember that the blade spinning at high revolutions per minute (rpm) generates a significant gyroscopic effect. Fans should not be moved while running. If the fan must be moved, it should be shut off, repositioned, and restarted.

Fans come in different styles and sizes. Personal finances, style of inflation, and size of the balloon determines the best fan. Points to consider in selecting a fan are:

- Weight—someone has to lift the fan into and out of the transport vehicle. Wheels help one person move the fan, but they add to the weight and are not helpful on soft ground. One person can carry a small fan, but a larger fan may require two people.
- Safety—fan blades today can be wood, aluminum, fiberglass, or composite, with wood being the most popular. Wood or aluminum blades designed specifically for balloon fan use are best. The fan should have a cowling of fiberglass or metal because a cage or grill alone is not sufficient to stop rocks or pieces of blade from being thrown.
- Transport—available space in a pickup truck, the back of a van, or on a trailer may determine the size of the fan.
- Cubic feet per minute (CFM)—fan blade design, duct design, and engine speed determine the amount of air moved in a given time. Do not confuse engine size with CFM. Larger engines do not necessarily push more air. The volume of air moved is primarily a function of blade design and performance. Moving a high volume of air is not necessarily the ultimate goal in fan performance. Some people prefer a slower cold inflation to accommodate a thorough preflight inspection.
- Fuel—gasoline degrades in storage. Do not store gasoline in the fan due to fire hazard and the formation of varnish, which can clog fuel passages.
- Fan maintenance—a good fan requires little maintenance and should be easy to maintain. Check the oil periodically and change it once a year. Check hub bolts and grill screws for tightness on a regular basis.

Figure 2-12. *Balloon inflation using a typical inflation fan.*

Transport/Chase Vehicle

Balloon ground transportation varies. *[Figure 2-13]* The most common vehicles are a van with the balloon carried inside, a pickup truck with the balloon carried in the bed, or a van or pickup truck with a small trailer (flatbed or covered). Some considerations in selecting a transport/chase vehicle are:

- Finances—if costs are an issue, a trailer hitch on the family sedan and a small flatbed trailer may work just fine.
- Convenience—for ease of handling the balloon, a small flatbed trailer low to the ground makes the least lifting demands on the pilot and crew. One consideration is that volunteer crew members may have little or no experience in backing a small trailer.
- Number of crew members—if the number of crew members is small, handling the balloon should be made as easy as possible. If the number of crew members is large, the size of the chase vehicle and other factors may be more important.
- Storage—some balloonists, who do not have room for inside storage and want security on the road, choose an enclosed trailer. If an enclosed trailer is used for storage of the balloon, the trailer should be a light color to help reduce the heat inside. Keeping the trailer cool keeps the tank pressure within reasonable limits (so as not to aggravate a potential fuel leak), and reduces the vaporization of gasoline in the fan tank (the fumes

Figure 2-13. *This is an example of a transport vehicle which carries a small balloon, three adults, a 20-inch inflation fan, and all other necessary equipment.*

can attack the composition of the balloon's fabric and render it unairworthy).

- Vehicle suitability—terrain, vehicle road clearance, and number of chase crew members are factors that determine the suitability of a transport/chase vehicle.

Quick/Safety Release

Safety restraints, referred to as "quick releases" or "safety tie downs," are used in balloon inflations. They are designed to restrain the balloon from movement in breezy or windy conditions.

There are several different types of safety restraints available, but none are part of the aircraft certification process. This lack of aircraft certification has led to controversy over the use of safety restraints among ballooning enthusiasts. Since event participation often requires their use for safety reasons, the use of safety restraints is now recommended for balloon launches. Each type of restraint has its own advantages and disadvantages which a pilot can learn via observation and discussion with an instructor and/or other balloon pilots. When a pilot decides to utilize a safety restraint, it is important to follow the balloon manufacturer's recommendations on how to attach it to the balloon superstructure. Many balloons have been seriously damaged by using an improperly attached restraint in excessive winds.

It is also important to insure that all personnel involved with the inflation, whether pilot, crew or spectator, be aware of the dangers of a safety restraint. The quick release rates with the inflation fans as one of the most hazardous pieces of equipment on the launch field. Early release under load, or breaking of the safety restraint may cause serious injury. All personnel involved should be briefed and made aware of the potential hazards.

Miscellaneous Items

- Radios—most pilots use some kind of two-way radio for air to ground communication. There are many choices available, ranging from Family Radio Service (FRS) and General Mobile Radio Service (GMRS) radios, which are relatively low cost, to the more sophisticated FM business band systems, which can be expensive.

  The GMRS and FM radios require licensing by the Federal Communications Commission (FCC). FRS radios do not. Using cell phones for air-ground communications is a violation of FCC rules.

- Igniters—most manufacturers provide at least two sources of ignition on board. The best igniter is the simple welding striker. Nearly all balloons have built-in piezo ignition systems.
- Fueling adapter—adapters are required to connect the balloon fuel tanks to the propane source. Pilots should carry their own adapters to ensure the adapters are clean and not worn. Dirty and worn adapters may damage a fuel system.
- Compass—compasses are used to track pibals, check map orientations, and navigate the balloon. While almost any good quality compass will do, the best kind to use is probably the sighting compass.
- Fire extinguisher—most balloons now come equipped with small fire extinguishers affixed to the basket. If one is present, it will be inspected during the annual inspection. These fire extinguishers are often too small to extinguish grass fires or serious basket fires caused by a propane leak. In the case of a propane-leak fire, turning off a valve usually extinguishes the fire. This is a better use of pilot time than fumbling for a fire extinguisher that might not extinguish the fire.
- First aid kit—the location and contents of first aid kits vary. Some pilots keep a small first aid kit in their balloon; some keep one in the chase vehicle. A frequent topic at Safety Seminars, the contents of the kit often depend on the area of the country in which the balloon is flown.

- Drop line—drop lines allow ground crew to assist the pilot in landing in a confined area, or to move the balloon to an area better suited for deflation and retrieving. A good drop line has a quick release provision; is easy to deploy, recover and store; and is easy for a person on the ground to handle. Webbing is a popular drop line material because it is strong. Webbing is hard to roll up, but easy to store. Half-inch nylon braid is strong and is easily rolled into a ball and put in a bag.
- Gloves—pilots and crew members should develop the habit of wearing gloves anytime they handle the balloon and associated equipment. A well fitting pair of gloves can reduce the injuries that occur while handling balloon equipment, such as rope, cables, bag handles, etc. In the case of a small fuel leak at a burner fitting, gloves can minimize a potentially disastrous situation.

  Gloves should be made of light colored smooth leather to reflect/deflect propane, and gauntlet style to cover the wrist. Avoid synthetic material which melts in heat and ventilated gloves which let in flame or gas. A second pair of gloves, of appropriate rubberized material and looser fit, can be used to conduct refueling operations.
- Helmets—balloon manufacturers usually mandate protective headgear be worn, especially in high wind conditions to protect heads from impact injury. Store helmets in a bag that can be carried inside or outside the basket, depending on number of passengers and available room.
- Spares—the following are recommended spares to carry in the chase vehicle or to have on hand:
  * Local and aeronautical maps
  * Helium tank and pibals (pilot balloons)
  * Quick pins and carabiners
  * Gloves and helmets
  * Envelope fabric and/or patches
  * Refueling adapters
  * Spare tire for the trailer
  * Extra fuel for the fan
  * Extra strikers/igniters

Aircraft Documents

Airworthiness Certificate

An Airworthiness Certificate is issued by a representative of the FAA after the balloon has been inspected, is found to meet the requirements of 14 CFR part 31, and is in condition for safe operation. The Airworthiness Certificate must be displayed in the aircraft so it is legible to the passengers and crew whenever it is operated. *[Figure 2-14]* The Airworthiness Certificate is transferred with the aircraft except when it is sold to a foreign purchaser.

A Standard Airworthiness Certificate is issued for aircraft type certificated in the normal category for manned free balloons. A Standard Airworthiness Certificate remains in effect as long as the aircraft receives the required maintenance and is properly registered in the United States. Flight safety relies,

UNITED STATES OF AMERICA
DEPARTMENT OF TRANSPORTATION—FEDERAL AVIATION ADMINISTRATION
STANDARD AIRWORTHINESS CERTIFICATE

1. NATIONALITY AND REGISTRATION MARKS	2. MANUFACTURER AND MODEL	3. AIRCRAFT SERIAL NUMBER	4. CATEGORY
N63308	Aerostar RX-8	RX-8-3094	Balloon

5. AUTHORITY AND BASIS FOR ISSUANCE
This airworthiness certificate is issued pursuant to the Federal Aviation Act of 1958 and certifies that, as of the date of issuance, the aircraft to which issued has been inspected and found to conform to the type certificate therefor, to be in condition for safe operation, and has been shown to meet the requirements of the applicable comprehensive and detailed airworthiness code as provided by Annex 8 to the Convention on International Civil Aviation, except as noted herein.
Exceptions:
NONE

6. TERMS AND CONDITIONS
Unless sooner surrendered, suspended, revoked, or a termination date is otherwise established by the Administrator, this airworthiness certificate is effective as long as the maintenance, preventative maintenance, and alterations are performed in accordance with Parts 21, 43, and 91 of the Federal Aviation Regulations, as appropriate, and the aircraft is registered in the United States.

DATE OF ISSUANCE	FAA REPRESENTATIVE	DESIGNATION NUMBER
11-29-91	John C. Curtice	DAR 32-AC-CE

Any alteration, reproduction, or misuse of this certificate may be punishable by a fine not exceeding $1,000, or imprisonment not exceeding 3 years, or both. THIS CERTIFICATE MUST BE DISPLAYED IN THE AIRCRAFT IN ACCORDANCE WITH APPLICABLE FEDERAL AVIATION REGULATIONS.

FAA Form 8100-2 (7-67) FORMERLY FAA FORM 1362 ☆U.S. Government Printing Office — 1976-675-526

Figure 2-14. *Standard Airworthiness Certificate.*

in part, on the condition of the aircraft, which is determined by inspections performed by mechanics, approved repair stations, or manufacturers who meet specific requirements. A Special Airworthiness Certificate is issued for all aircraft certificated in other than the Standard classifications, such as Experimental or Restricted. When purchasing an aircraft classified as other than Standard, it is recommended that the local FAA Flight Standards District Office (FSDO) be contacted for an explanation of the pertinent airworthiness requirements and the limitations of such a certificate.

Certificate of Aircraft Registration

Before an aircraft can be flown legally, it must be registered with the FAA Civil Aviation Registry. The Certificate of Aircraft Registration, which is issued to the owner as evidence of the registration, must be carried in the aircraft at all times. *[Figure 2-15]* The Certificate of Aircraft Registration cannot be used for operations when:

- The aircraft is registered under the laws of a foreign country.
- The aircraft's registration is canceled at the written request of the holder of the certificate.
- The aircraft is totally destroyed or scrapped.
- The ownership of the aircraft is transferred.
- The holder of the certificate loses United States citizenship.

When one of the events listed in 14 CFR part 47, section 47.41 occurs, the previous owner must notify the FAA by filling in the back of the Certificate of Aircraft Registration, and mailing it to:

Federal Aviation Administration
Civil Aviation Registry, AFS-750
P.O. Box 25504
Oklahoma City, OK 73125

After compliance with 14 CFR part 47, section 47.41, the pink copy of the application for a Certificate of Aircraft Registration is authorization to operate an unregistered aircraft for a period not to exceed 90 days. Since the aircraft is unregistered, it cannot be operated outside the United States until a permanent Certificate of Aircraft Registration is received and placed in the aircraft.

NOTE: For additional information concerning the Aircraft Registration Application or the Aircraft Bill of Sale, contact the nearest FSDO.

Aircraft Owner/Operator Responsibilities

The registered owner/operator of an aircraft is responsible for certain items, such as:

- Having a current Airworthiness Certificate and a Certificate of Aircraft Registration in the aircraft.

REGISTRATION NOT TRANSFERABLE

UNITED STATES OF AMERICA
DEPARTMENT OF TRANSPORTATION - FEDERAL AVIATION ADMINISTRATION
CERTIFICATE OF AIRCRAFT REGISTRATION

This certificate must be in aircraft when operated.

NATIONALITY AND REGISTRATION MARKS N63308

AIRCRAFT SERIAL NO. RX-8-3094

MANUFACTURER AND MANUFACTURER'S DESIGNATION OF AIRCRAFT
Aerostar RX-8

ICAO Aircraft Address Code:

ISSUED TO
John A. Doe
900 ELM St.
Anytown, AL
12345

This certificate is issued for registration purposes only and is not a certificate of title. The Federal Aviation Administration does not determine rights of ownership as between private persons.

It is certified that the above described aircraft has been entered on the register of the Federal Aviation Administration, United States of America, in accordance with the Convention on International Civil Aviation dated December 7, 1944, and with the Federal Aviation Act of 1958, and regulations issued thereunder.

U.S. Department of Transportation
Federal Aviation Administration

DATE OF ISSUE
February 1, 1994

Richard Edward ADMINISTRATOR

AC Form 8050-3 (11/93) Suspersedes previous editions

Figure 2-15. *Certificate of Aircraft Registration.*

- Maintaining the aircraft in an airworthy condition, including compliance with all applicable Airworthiness Directives.
- Assuring that maintenance is properly recorded.
- Keeping abreast of current regulations concerning the operation and maintenance of the aircraft.
- Notifying the FAA Civil Aviation Registry immediately of any change of permanent mailing address, or the sale or export of the aircraft, or the loss of the eligibility to register an aircraft. (Refer to 14 CFR part 47, section 47.41.)

Aircraft Maintenance

Maintenance is defined as the preservation, inspection, overhaul, and repair of an aircraft, including the replacement of parts. A properly maintained aircraft is a safe aircraft. In addition, regular and proper maintenance ensures that an aircraft meets an acceptable standard of airworthiness throughout its operational life. Although maintenance requirements vary for different types of aircraft, experience shows that aircraft need some type of preventive maintenance every 25 hours of flying time or less, and minor maintenance at least every 100 hours. This is influenced by the kind of operation, climatic conditions, storage facilities, age, and construction of the aircraft. Manufacturers provide maintenance manuals, parts catalogs, and other service information that should be used in maintaining the aircraft.

Balloon Inspections

14 CFR part 91 places primary responsibility on the owner or operator for maintaining a balloon in an airworthy condition. Certain inspections must be performed on the balloon, and the owner must maintain the airworthiness of the balloon during the time between required inspections by having any defects corrected. This typically means that fabric damage outside the maximum allowable damage limits specified by the manufacturer must be repaired before the envelope can be deemed airworthy and returned to service.

Annual Inspection

Any balloon flown for business or pleasure and not flown for compensation or hire is required to be inspected at least annually. The inspection must be performed by a certificated and appropriately rated repair station, by the manufacturer, or by a certificated airframe and powerplant (A&P) mechanic who holds an Inspection Authorization (IA). The aircraft may not be operated unless the annual inspection has been performed within the preceding 12 calendar months. A period of 12 calendar months extends from any day of a month to the last day of the same month the following year.

100-Hour Inspection

All balloons used to carry passengers for hire must have received a 100-hour inspection within the preceding 100 hours of time in service and have been approved for return to service. Additionally, an aircraft used for flight instruction for hire, when provided by the person giving the flight instruction, must also have received a 100-hour inspection. This inspection must be performed by an appropriately rated FAA certificated repair station, the aircraft manufacturer, or by an FAA certificated A&P mechanic. An annual inspection, or an inspection for the issuance of an Airworthiness Certificate, may be substituted for a required 100-hour inspection.

Preflight Inspections

The preflight inspection is a thorough and systematic means by which a pilot determines if the aircraft is airworthy and in condition for safe operation. The balloon's Flight Manual contains a section devoted to a systematic method of performing a preflight inspection. For balloons, this inspection is usually a part of the layout and inflation process, and is greatly aided by the use of a checklist. Again, the pilot must also be aware of maximum damage limitations as published by the manufacturer.

Preventative Maintenance

Preventive maintenance is considered to be simple or minor preservation operations and the replacement of small standard parts, not involving complex assembly operations. Certificated pilots may perform preventive maintenance on any balloon that is owned or operated by them. According to 14 CFR part 43, appendix A, preventive maintenance may be performed by the owner/operator of an aircraft who holds at least an FAA Private Pilot Certificate with a balloon rating.

The following is a partial list of preventive maintenance that may be performed by the owner/operator of a balloon:

- Replacing defective safety wiring or cotter keys.
- Lubrication not requiring disassembly.
- The making of small fabric repairs to envelopes (as defined in, and in accordance with, the balloon manufacturers' instructions) not requiring load tape repair or replacement.
- Refinishing decorative coating of the basket when removal or disassembly of any primary structure or operating system is not required.
- Applying preservative or protective material to components where no disassembly of any primary structure or operating system is involved and where such coating is not prohibited or is not contrary to good practices.

- Repairing upholstery and decorative furnishings of the balloon basket interior when the repairing does not require disassembly of any primary structure or operating system or interfere with an operating system or affect primary structure of the aircraft.
- Replacing seats or seat parts with replacement parts approved for the aircraft, not involving disassembly of any primary structure or operating system.
- Replacing prefabricated fuel lines.
- Replacing and servicing batteries.
- Cleaning of balloon burner pilots and main nozzles in accordance with balloon manufacturers' instructions.
- Replacement or adjustment of nonstructural standard fasteners incidental to operations.
- The interchange of balloon baskets and burners on envelopes when the basket or burner is designated as interchangeable in the balloon Type Certificate Data Sheet (TCDS), and the baskets and burners are specifically designed for quick removal and installation.

Repairs and Alterations

Repairs and alterations are classified as either major or minor. 14 CFR part 43, appendix A, describes the alterations and repairs considered major. Major repairs or alterations shall be approved for return to service on FAA Form 337, Major Repair and Alteration, by an appropriately rated certificated repair station, an FAA certificated A&P mechanic holding an Inspection Authorization, or a representative of the Administrator. Minor repairs and minor alterations may be approved for return to service with a proper entry in the maintenance records by an appropriately certificated repair station or FAA certificated A&P mechanic.

For modifications of experimental aircraft, refer to the operating limitations issued to that aircraft. Modifications in accordance with FAA Order 8130.2, Airworthiness Certification of Aircraft and Related Products, may require the notification of the issuing authority.

Airworthiness Directives (ADs)

A primary safety function of the FAA is to require correction of unsafe conditions found in an aircraft, aircraft engine, propeller, or appliance when such conditions exist and are likely to exist or develop in other products of the same design. The unsafe condition may exist because of a design defect, maintenance, or other causes. 14 CFR part 39, Airworthiness Directives (ADs), define the authority and responsibility of the Administrator for requiring the necessary corrective action. ADs are the means used to notify aircraft owners and other interested persons of unsafe conditions and to specify the conditions under which the product may continue to be operated. ADs may be divided into two categories:

1. Those of an emergency nature requiring immediate compliance prior to further flight.
2. Those of a less urgent nature requiring compliance within a specified period of time.

ADs are regulatory and shall be complied with unless a specific exemption is granted. It is the aircraft owner or operator's responsibility to ensure compliance with all pertinent ADs. 14 CFR part 91, section 91.417 requires a record to be maintained that shows the current status of applicable ADs, including the method of compliance; the AD number and revision date, if recurring; the time and date when due again; the signature; kind of certificate; and certificate number of the repair station or mechanic who performed the work. For ready reference, many aircraft owners have a chronological listing of the pertinent ADs in the back of their aircraft maintenance records.

Choosing a Balloon

Many companies manufacture balloons that are type-certificated by the FAA. A type-certificated balloon has passed many tests, has been approved by the FAA, and conforms to the manufacturer's TCDS. Balloon size is rated by envelope volume with categories defined in metric units. *Figure 2-16* illustrates the most popular size ranges in use today (volumes are provided in cubic meters, as well as cubic feet).

FAI Category	Cubic Meters	Number of People	Cubic Feet
AX5	900–1,200	1	31,779–42,372
AX6	1,200–1,600	1–2	42,372–56,372
AX7	1,600–2,200	2–4	56,496–77,682
AX8	2,200–3,000	4–7	77,682–105,930
AX9	3,000–4,000	6–10	105,930–141,240

Figure 2-16. *Popular balloon size ranges.*

Advantages of Balloon Sizes

Different balloon sizes offer different advantages. The size of the balloon purchased should be determined according to planned use(s). Most pilots think smaller balloons are easier to handle, fly, and pack up. Bigger balloons use less fuel, operate cooler, and last longer. Higher elevations or hotter climates or passengers indicate a larger balloon. Balloon competitions and sport flying require a smaller balloon. *[Figure 2-17]*

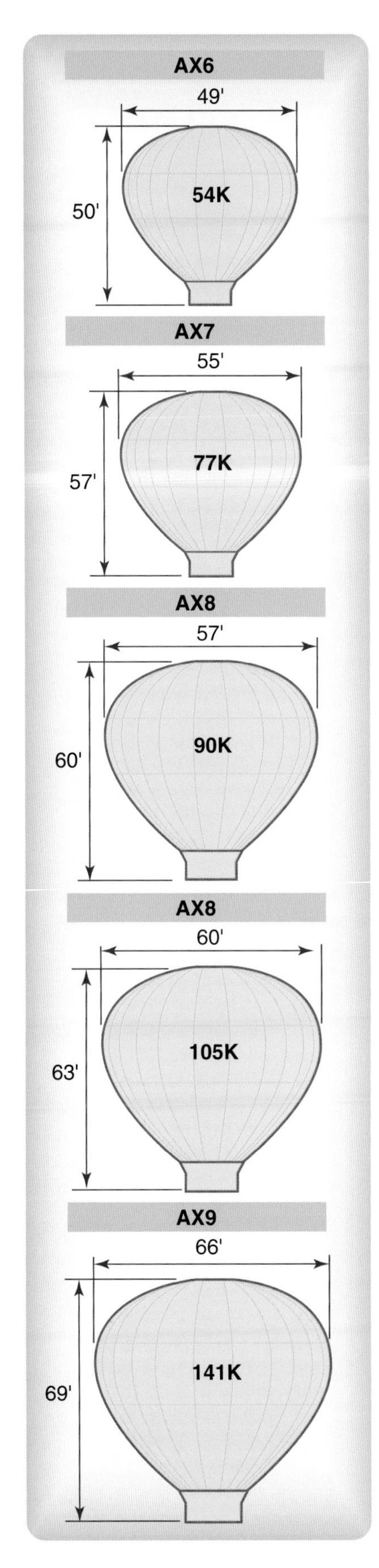

Figure 2-17. *Comparison of balloon sizes.*

Selecting a New or Used Balloon

The cost is the most obvious difference between new and used balloons. Some new pilots buy a used balloon to gain proficiency, and then purchase a new balloon when they have a better idea of what they want or need.

Prior to purchasing any used balloon, it is important to ensure that the balloon is airworthy, to avoid purchasing an aircraft which may be nearing, or perhaps past, its useful life. Most balloon envelopes are constructed of fabrics that last well into the 300–400 hour range, with some newer fabrics exceeding that life span. It would be prudent to have the balloon inspected by a reputable repair station or qualified inspector prior to purchase.

Balloon Brands

The level of after sales service available—locally and from the manufacturer—is an important criterion in deciding which brand of balloon to purchase. Talk to local pilots and ask questions. How does the local balloon repair station feel about different brands? Do they stock parts for only one brand? Does the manufacturer ship parts and fabric for balloons already in the field, or do they reserve these parts and fabric for new production? Do they ground older model balloons for lack of materials while new balloons are being built?

There are other criteria that could be considered, such as altitude at which the balloon will be flying, climate, and interchangeability of components, to give some examples. Before making the final decision, talk to people with different kinds of balloons who do different kinds of flying. Crewing for different balloons is an excellent way to learn about balloons and can help in the decision on what first balloon to purchase.

Chapter Summary

This chapter gives the reader common terminology for use in the ballooning community. Many times, confusion exists between the student pilot and the instructor, due to differences in terminology used, and it is hoped that the discussions here resolve those issues. The reader also should have an understanding of the physics of hot-air ballooning, as well as a good understanding of the support equipment involved with ballooning activities.

Propane information has also been included in this chapter, and it is recommended that all pilots review this on a recurring basis, perhaps as part of a yearly safety seminar. Additionally, proper documentation, and inspection requirements have been covered. Each pilot should become knowledgeable in these areas.

Chapter 3

Preflight Planning

Introduction

Flight planning starts long in advance of the few hours before the launch. Title 14 of the Code of Federal Regulations (14 CFR) part 91, section 91.103 states: "Each pilot in command (PIC) shall, before beginning a flight, become familiar with all available information concerning that flight…(to include) weather reports and forecasts, fuel requirements…(and) other reliable information appropriate to the aircraft, relating to aircraft performance under expected values of airport elevation,…aircraft gross weight, and wind and temperature."

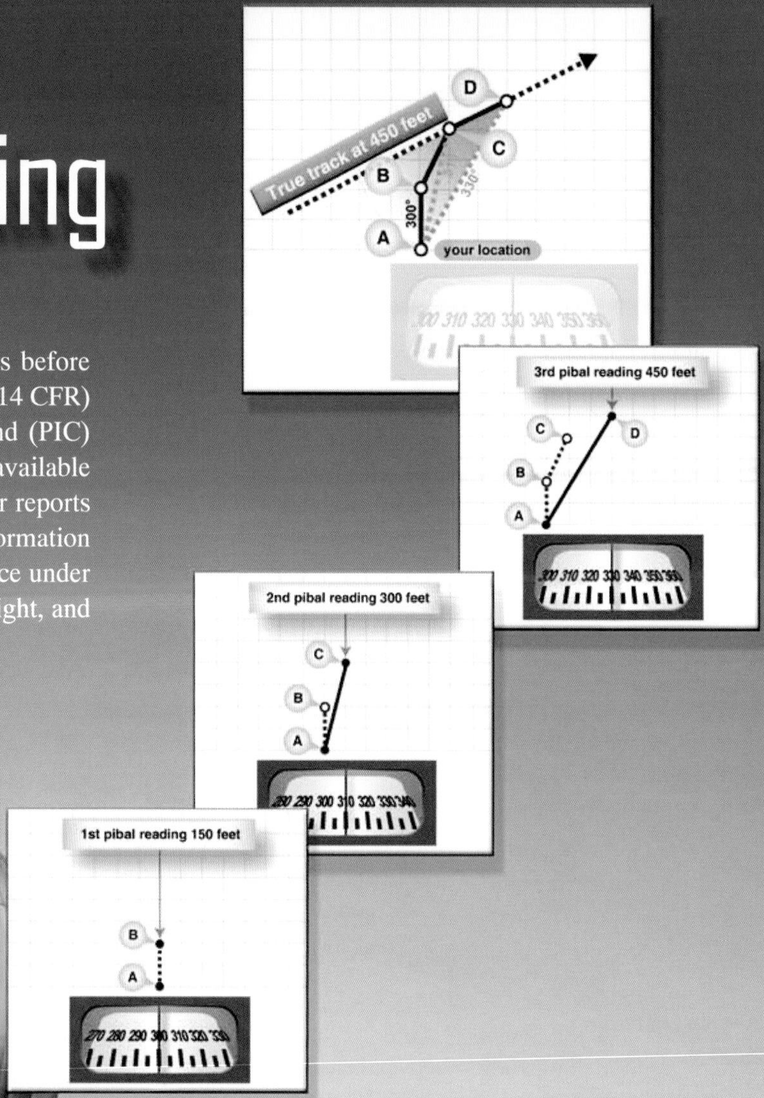

The practical test standards (PTS), for both Private and Commercial certificates, indicate a number of items that must be considered, evaluated, and planned in the execution of a safe flight. Some of these items are the use and interpretation of weather data to plan a flight, the use and interpretation of aeronautical charts and local area maps, and performance and limitation of the balloon.

Weather Theory and Reports will be covered in some detail in Chapter 4 and The National Airspace System (NAS) will be reviewed in Chapter 5. The following discussions assume familiarity with both subjects, and will introduce a number of other new subjects.

Purpose of Flight

Preflight planning will vary according to the flight's purpose. For example, if a training flight is planned, more detailed attention to map work and performance planning may be appropriate. If a passenger-carrying flight is being undertaken, a meeting point for the passengers and crew will need to be designated, and refreshments will need to be planned. If the flight is to participate in an organized rally, particular attention must be paid to weather trends and wind plotting, to ensure the pilot is able to reach the intended target and score points. These type of considerations are part of the initial balloon preflight planning process.

Weather

A good balloon pilot studies the weather several days before the day of the flight in order to understand the weather trends, cycles, and the correlation of weather report information with the actual weather in a particular flying area. Most, if not all, weather reporting information is computed for a large regional area, whereas balloon flying is generally conducted in an area about 15 square miles. When a balloon pilot makes the correlation between the weather outlooks and forecasts, and how that will impact winds and environment in the local flying area, he or she is well on the way to understanding the effects of weather on preflight planning, as well as the balloon flight.

Particular attention should be paid to the location and movement of pressure systems and the jet stream, frontal activity, temperatures in front of and behind frontal zones, and winds. As the proposed flight date draws closer, a reasonable prediction of possible weather can be forecast, but a pilot must remember that a weather forecast more than 72 hours prior to a flight is not an absolute. It is also worthwhile to watch local and nationally televised weather broadcasts to gain insights on the weather systems that may be affecting the desired flying area at the time of the flight.

Unofficial sources of weather information can also prove helpful for obtaining weather information about a particular area. It is beneficial to contact balloon pilots who fly in the area of intended flight to learn of nuances in the weather patterns, especially during initial training or when flying in a new area. Another source of information for weather is pilots who fly other types of aircraft in the proposed flight area. They can be located through the local airport's fixed base operator (FBO). People who make their living outside, particularly farmers, have a unique perspective on local weather. They often offer weather information on local weather that is unavailable through a commercial source.

When possible, it is valuable for a balloon pilot to visit the local National Weather Service (NWS) office. *[Figure 3-1]* NWS provides information and sources for a number of weather products, which must be considered in the weather planning process. A visit to the NWS office also gives a balloonist the opportunity to talk with the individuals who provide the weather information used in the briefings. NWS can provide the balloonist with a clear explanation of what products and information are required to make an intelligent flight decision.

Figure 3-1. *National Weather Service Office, Falcon Field, Peachtree City, Georgia.*

The night before a flight is anticipated (or in the morning, in the event of an afternoon flight), a call should be made to the Flight Service Station (FSS) for an outlook briefing. These are generally available 6 hours or more before a specific flight period. (There are three different types of briefings available: standard, abbreviated, and outlook. They will be discussed in more detail in Chapter 4, Weather Theory and Reports). This briefing information is used to make tentative decisions regarding the flight, such as go/no-go, and potential directions of travel. Additionally, a pilot should pay particular attention to local and regional forecasts in the media, as they may provide information specific to the area of flight.

Prior to flight, a standard briefing should be obtained from the FSS. This briefing will contain the most recent weather information and data, and will serve either to verify information obtained through other sources, or validate the possibility of a go/no-go decision. It is also helpful to check one or more automated weather reporting sites, such as the Automatic Terminal Information Service (ATIS) or Automatic Weather Observing System (AWOS) that are close to the intended flying area. ATIS and AWOS provide the advantage of a real-time, immediate information source. They may be contacted by telephone, or often monitored by aviation radio. Phone numbers for the ATIS and AWOS systems may be found in the Airport/Facility Directory (A/FD). Radio frequencies for the ATIS and AWOS are shown on aviation sectional charts.

Gathering weather information en route to the launch site can be done by searching for indications of current winds. For example, observe how the leaves on a tree move, track the smoke from a factory smokestack, or notice the direction a flag blows. All of these signs give good indications of the current winds, both on the ground and at low altitude. Once at the launch site, or possible launch site, most experienced pilots inflate and release a pibal (pilot balloon) to assess on site wind speeds and direction. *[Figure 3-2]*

Figure 3-2. *Preparing to release a pibal.*

Many pilots develop historical data on weather conditions in their home flying areas. When shared with the beginning pilot, this weather data provides a wealth of information on trends and cycles. The comparison of individual predictions with actual weather experienced offers understanding and insight into micro-area weather conditions. Comparison of weather reports from nearby weather reporting stations with the actual weather experienced is also be an excellent learning tool. This exercise provides insight into the weather patterns common in a particular flying area. See Appendix A for a sample weather briefing checklist that may be used as a guide to develop personal forms for recording weather briefings.

There are numerous sources of weather information available on the Internet. These include but are not limited to web sites operated by the NWS (www.nws.noaa.gov), Intellicast (www.intellicast.com), and Unisys (www.weather.unisys.com). Web sites devoted to weather and ballooning include but are not limited to Blastvalve.com (www.blastvalve.com/weather), US Airnet.com (www.usairnet.com), and the Balloon Federation of America (BFA) at www.bfa.net. Ballooning enthusiast Ryan Carlton has developed a wind forecasting site that is located at ryancarlton.com/wind.php. All of these sites provide resources and reference information on weather.

It should be remembered that none of these web sites provide an official weather briefing. It is necessary to call the FSS, or use an on-line briefing service such as Direct User Access Terminal System (DUATS) to receive an official briefing. Failure to receive a proper briefing may create a liability issue for a pilot in the event of an incident or accident.

Some weather related tips are:

- Forecasts are a good place to start, but are not the end of weather planning. Unforecast events happen continuously. Proficiency in understanding small area weather is necessary, and can only be developed with practice and experience.
- Balloons generally fly early in the morning, within the first two hours after sunrise, to avoid unstable conditions, which may prove to be hazardous to balloon flights and operations. It may be possible to fly in the late afternoon, within an hour or two of sunset, when thermal effects are calming down and winds are usually decreasing.
- Almost all balloon flying is done in relatively benign weather conditions and mild winds. Most pilots prefer to launch and fly in winds less than 7 knots. While balloon flying is performed in higher winds, pilots accept that the faster the winds, the more they are exposed to risk and injury. Balloon flight manuals list the maximum launch winds for a particular balloon; this information, as well as personal limitations, are considerations for any pilot.
- Balloons do not fly in significant (or unstable) weather. A balloon should not be launched in the face of a squall line, or during a tornado warning or watch.
- Flying in precipitation is a bad practice. Rainwater (or any frozen precipitation) on the balloon causes it

to get wet and become heavier, often to the point of being unable to maintain altitude without exceeding temperature limitations of the envelope. A wet envelope heated to flight temperatures can be seriously damaged because the heat often causes fabric coatings and treatments to degrade, decreasing the life of the fabric. If a balloon gets wet, it should never be dried out by the application of heat to the point of equilibrium, or neutral buoyancy.

- Precipitation also often causes the atmosphere to become increasingly unstable. Downdrafts, wind gusts, and the possibility of hail and lightning follow. The pilot may be the last one to know that it is raining because the balloon will shield him or her from the precipitation. Ground crew can detect the slightest trace of precipitation before a pilot does, and need to communicate this information immediately to the pilot. In the face of possible precipitation, cancellation of the flight is the best plan.

Navigation

Navigation of a balloon is unlike that of any other aircraft because it cannot be steered in the conventional sense. Directional control is achieved through the use of differing wind directions at different altitudes. With effort, study, and some practice, it is possible for a balloon pilot to determine a point on the ground at some distance, and fly to it with relative ease and accuracy.

The first step in learning balloon navigation is understanding the maps used in balloon flight. Two types of maps are used: sectional aeronautical charts and local topographical maps. Both have their uses and each has advantages and disadvantages. Another type of map may be available to the balloon pilot. This is a local area map developed by the local balloon club which shows prohibited zones and sensitive areas.

Sectional aeronautical charts (or sectionals) are published on a routine basis by the National Aeronautical Charting Office (NACO), a division of the Federal Aviation Administration (FAA). *[Figure 3-3]* These charts are at a scale of 1:500,000 (one inch representing 500,000 inches on the ground, or about 7.9 miles), are similar to an automobile road map, and provide useful information to a balloon pilot flying under visual flight rules (VFR). Charts are generally named for the most prominent city contained within the area of the sectional chart.

There are also sectional charts with a smaller scale, 1:250,000, to represent the areas immediately surrounding Class B airspace, which is the airspace surrounding major air traffic facilities in the United States. Airspace is discussed in detail in Chapter 5, The National Airspace System. These charts (commonly referred to as terminal area charts) show a significantly increased level of detail, and, if available, may be of more value than a standard sectional.

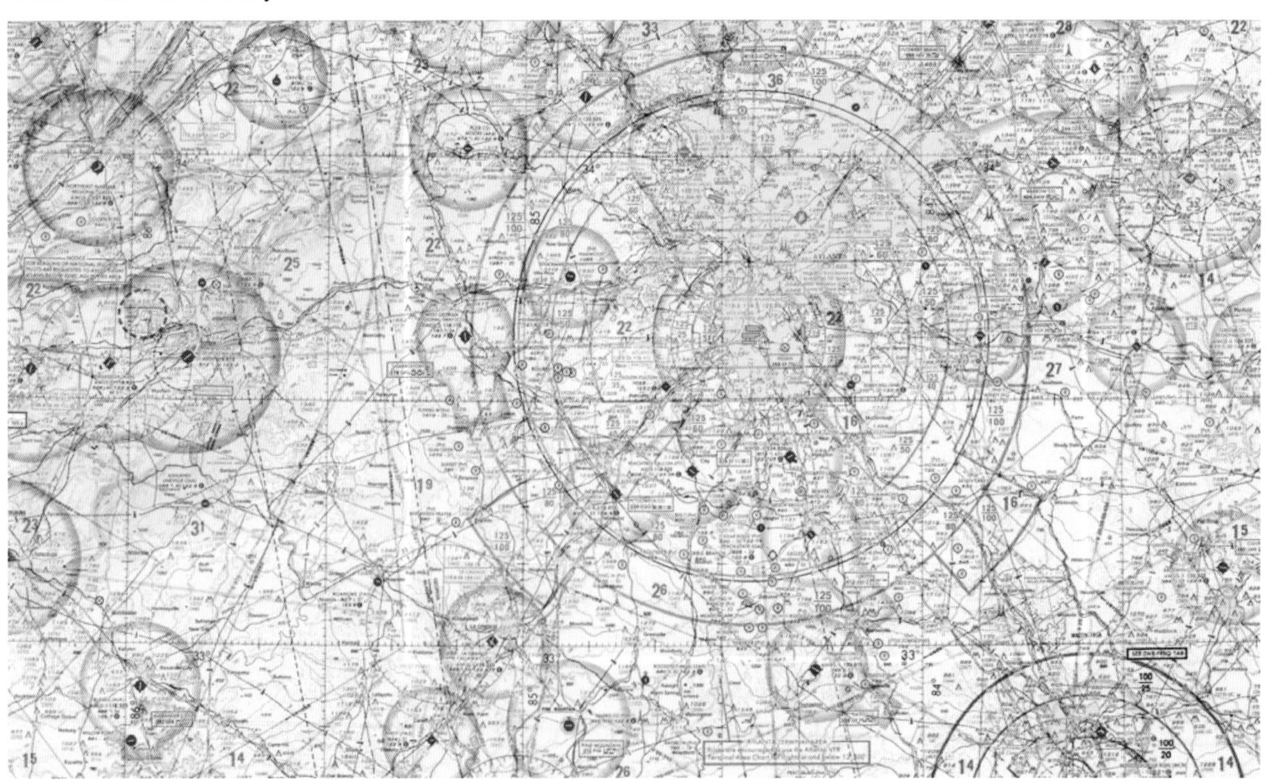

Figure 3-3. *Sectional chart depicting the Atlanta-Hartsfield-Jackson International Airport Class B airspace.*

Sectionals depict many different things, including controlled and uncontrolled airspace, airports, major roads and highways, cities and small towns, etc. They also indicate obstacles to flight, such as major transmission lines, radio, TV, and water towers, smokestacks, and other items. The legend of the sectional provides a means to identify these landmarks. A more detailed explanation of sectionals and the information they contain, is found in the FAA Aeronautical Chart User's Guide, a publication of the NACO. This publication may be found at many pilot supply stores where sectionals are sold, or may be purchased online, along with the maps themselves, at www.naco.faa.gov.

Pilots review the sectional chart and familiarize themselves with the airspace they may be using when flying in a new area or refreshing their memory of a frequently flown area. The sectional helps a pilot determine obstacles to flight (towers, powerlines, etc), as well as locating landmarks for use during the flight. While sectionals offer much valuable information on an area, their lack of resolution on a small scale means they do not provide enough information for a balloon flight. The length of the average balloon flight is 6 to 8 miles. On the sectional, this equates to the distance between the first joint and tip of one's thumb. This lack of significant detail is a disadvantage for navigation in a balloon, but sectionals are useful as a source of general information about a given area.

A good topographic chart, such as the commercially available United States Geological Survey (USGS) maps offer more value to the balloon pilot. These maps depict information on a relatively small scale and are more useful to the balloon pilot. They show individual terrain features such as roads and road networks, built up areas, schools and churches, and will indicate wooded areas, as well as open pastureland. *[Figure 3-4]*

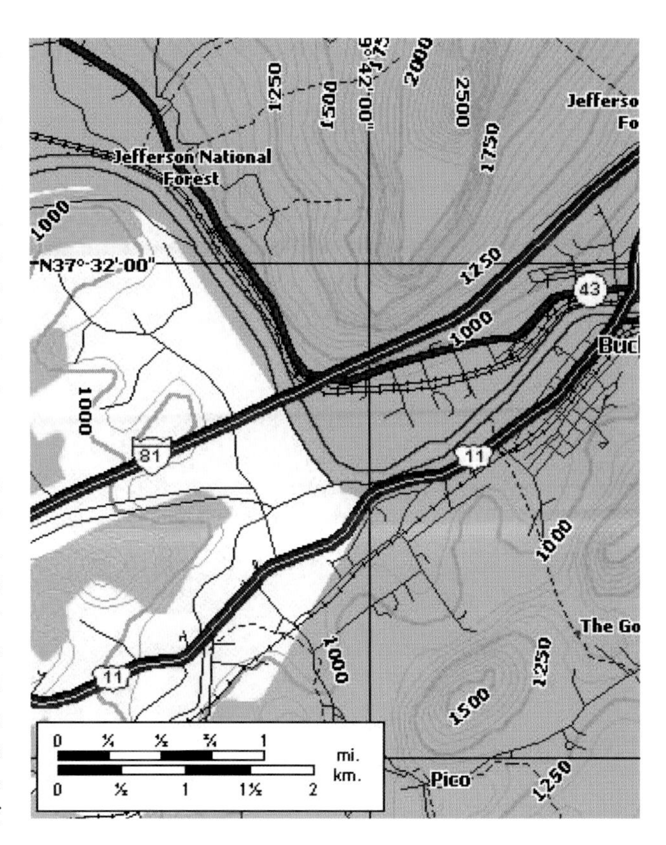

Figure 3-4. *Detailed topological map.*

With any map, it is important to insure the map is current and has an accurate depiction of north. To check the orientation of a map, select one particular road or feature with a specific directional orientation. Then orient the map to that feature, matching the direction of the road with the map. Place a compass (preferably a sighting compass) on the map to determine the azimuth. Use the same compass that will be used for computing the flight path. Ensure that nothing is affecting the compass reading. It is important to distinguish "true north," used by most cartographers, and "magnetic north," as indicated on the compass. Once the azimuth is established, sketch a compass rose, or place a "stick-on" type compass rose on the map.

Once the map is oriented and aligned to north, fill in other information as necessary as reminders. For example, airspace that may preclude balloon operations, local no-fly areas, or areas with potential landowner relations problems should be marked. If the pilot is flying competitively, he or she may elect to mark designated "targets" on the map for ease in identifying them at a later time.

Perhaps the most underutilized use of maps is predicting likely flight paths, landmarks, and potential landing sites. Using the simple technique outlined below, this field technique allows pilots accurate real time and on-site weather data for flight planning information. A pilot needs to know where he or she is going in order to plan how to get there. This is a necessary part of flight planning, and learning the basic skills and knowledge required to plot this information improves the flight experience.

Pat Cannon, a former BFA National Champion and competitive pilot, developed a technique derived from a NWS procedure (that was later modified) to plot the information obtained from a pibal reading. This procedure requires a pencil, large square graph paper, an aviation plotter, pibals, the compass used to calibrate the map, and a watch with a sweep second hand. Two assumptions are made with this procedure. First, most pibals rise at an average rate of 300 feet per minute (fpm). (A chart of pibal climb rates can be found in Appendix B.) Therefore, after 30 seconds, a pibal will be approximately 150 feet above ground level (AGL).

Second, for the purposes of this exercise, the winds do not have any significant speed changes.

Prior to starting the plot, a scale depicting the wind speed must be established. In this example illustrated, two squares on the graph paper will represent a wind speed of 5 miles per hour (mph). In the absence of a wind meter, or other accurate wind reading, a rough estimate of the wind speed may be made using the technique shown in *Figure 3-5*.

Figure 3-5. A *method for determining wind speed.*

To begin plotting the pibal recording information, release the pibal and track it with the compass. After 30 seconds, take a reading and make a mark on the graph paper to represent the start point. Make a second mark to represent the direction plotted. In *Figure 3-6,* a track of 300° at 5 mph is depicted. Label the first two points "A" and "B." *[Figure 3-6]*

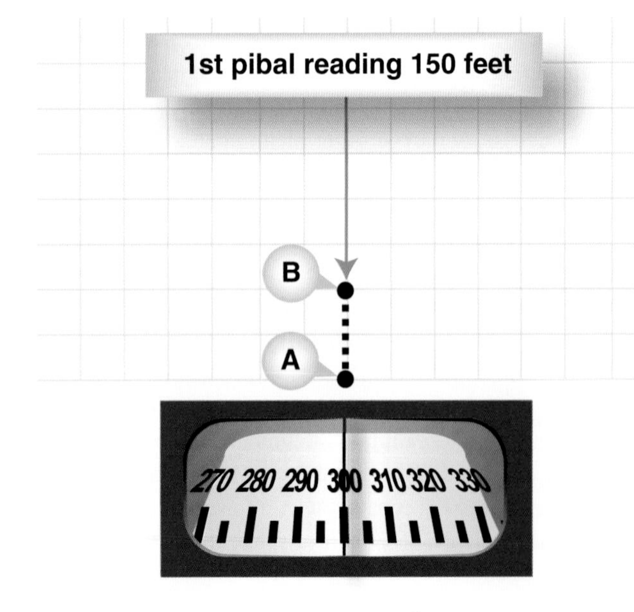

Figure 3-6. *First pibal plot showing 300° at 30 seconds.*

At 1 minute, take a second reading. The pibal will be at approximately 300 feet AGL. In this example, the reading taken is 310°. Using the plotter, draw a line 10° off the original azimuth (the A-B line), and make another mark approximately two squares away from the mark labeled "B." For clarity, this is be labeled "C". See the example in *Figure 3-7.* (NOTE: The angles in the successive graphics are exaggerated for clarity.)

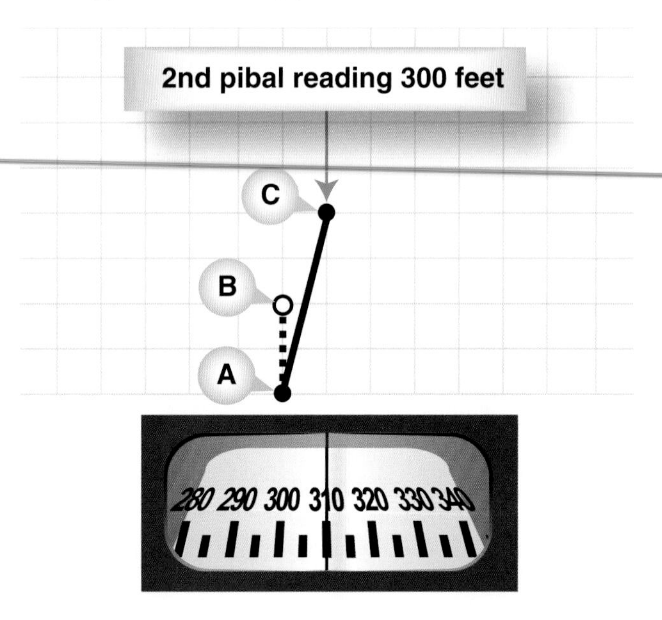

Figure 3-7. *Second pibal plot showing 310° at 1 minute.*

At 1:30 minutes, take another reading. The pibal will be at approximately 450 feet. Using the plotter, draw a line 30° off the original azimuth (the A-B line), and make another mark approximately two squares away from the mark labeled "C." This mark may be labeled "D" for clarity. *[Figure 3-8]*

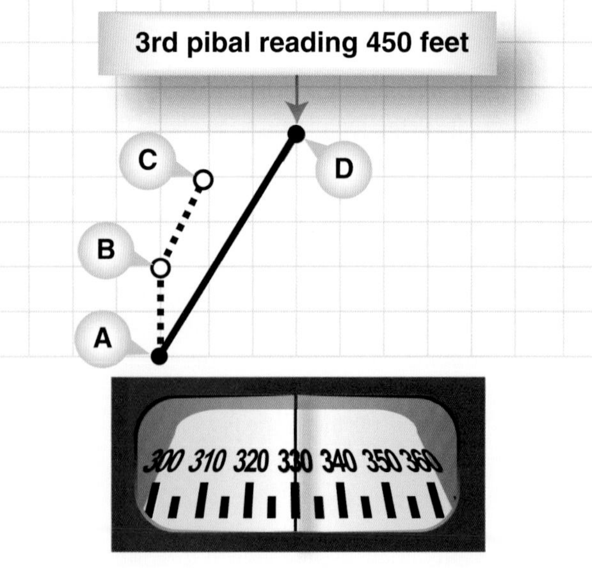

Figure 3-8. *Third pibal plot showing 330° at 1:30 minutes.*

Although plotting can be continued as long as the pibal remains in sight, only the three points marked will be used for this exercise. *Figure 3-9* illustrates the results of the above sequence.

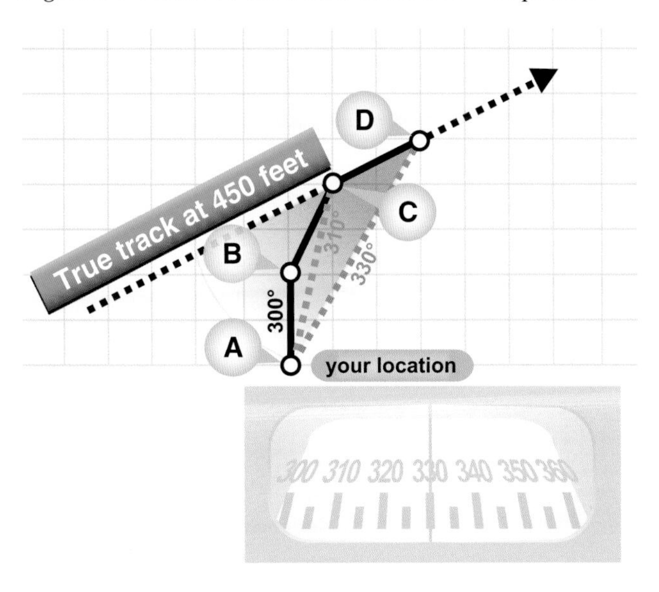

Figure 3-9. *A line drawn through the last two plots provides a basis to measure the angle and determine the wind at that altitude. In this case, it is 450 feet.*

To determine the wind directions at different altitudes, extend lines between the plotted points as shown in *Figure 3-9* back through the initial azimuth. Using the plotter, measure the angle between the lines (the angle between the A-B line and the C-D line). That angle, added to the original azimuth heading, gives a good approximation of the winds at that altitude. For the example shown in this sequence, the true track at 450 feet AGL is 005°. A grid appropriate for this computation is located in Appendix B.

This exercise demonstrates a practical method for determining approximate wind directions using items readily available to most pilots. It does not require expensive handheld calculators, laptop computers, or a theolodite that costs thousands of dollars. There is some error inherent in this process that can be lessened with experience and practice, but the readings obtained by this method can offer real time, on site weather data no forecast or briefer can provide. *[Figures 3-10* and *3-11]*

The information on basic surface winds and winds aloft readings gathered by this method can be used by a pilot to project a flight path and anticipated landing sites with a sectional or topographic map. This plot will form a "V," with the cone beginning at the launch site. The two legs will represent the extremes of the plotted measurements. The difference between these two extremes is called steerage. Flying higher will track the flight path closer to the winds aloft reading, while contour flying will put the balloon closer to the ground track leg. Varying altitude will allow the pilot to fly down the middle of the "v." Accuracy will depend on the consistency of the conditions, but flight paths and landing sites may be predicted, after practice, with a high degree of reliability.

Assume for exercises one and two that winds are at 5 miles per hour.

Exercise 1: The morning of the flight, the following pibal plots are taken:

30 seconds:	117°
1 minute:	122°
1:30 minutes:	135°
2 minutes:	140°
2:30 minutes:	144°
3 minutes:	150°

What is the wind direction at 600 feet above ground level (AGL)? At 900 feet AGL?

Exercise 2: On an afternoon flight, the following readings are taken:

30 seconds:	290°
1 minute:	273°
1:30 minutes:	277°
2 minutes:	279°
2:30 minutes:	282°
3 minutes:	284°

What is the wind direction at 600 feet?

Figure 3-10. *Practice pibal plots. These exercises are designed to assist the student pilot in devleoping proficiency in using the pibal plotting method. (answers on next page)*

The balloon pilot, more than pilots who fly other types of aircraft, must have the capability of visualizing the winds aloft in three dimensions. Continued spatial awareness (how the balloon is moving through the air), is important for maintaining control of the balloon and navigating to the desired point on the ground. Every other safety measure taken is compromised by inflating a balloon and taking off without proper planning and an understanding of the winds and terrain to be navigated. *[Figure 3-12]*

Exercise 1: The pibal is climbing at approximately 300 fpm. The winds between the third and fourth reading are for 600 feet AGL, and the winds between the fifth and sixth reading are for 900 feet AGL.

For the first computation, draw a line between the third and fourth plots, extending back through the original azimuth line of 117°. Measure the angle created between the original azimuth and the new line. That angle is 36°. Add 36° to the original azimuth of 117°. The resulting 153° should be the wind direction at approximately 600 feet AGL.

For the second computation, follow the same procedure. Draw a line from the fifth plot to the sixth plot, extending back through the original azimuth line. Measure the angle created between the original azimuth and the new line. Add the result of 68° to the original azimuth of 117°. The resulting 185° equals the wind direction at approximately 900 feet AGL.

Exercise 2: This exercise is a little more difficult, due to two factors: (1) the winds are starting out bearing to the left instead of to the right, which occurs during some weather and wind patterns; and (2) after the first reading, it is impossible to draw the lines back through the original azimuth line; therefore,another method must be used.

Draw a line through the first and second plots, the 300 feet AGL wind line. Measure the angle between that line and the original azimuth. That angle is 35°. Subtact 35° from the original azimuth of 290°, and the result is 255°. This line become a new baseline azimuth.

Draw a line between the fourth and fifth plots back through the new azimuth line. Measure the angle. Add the resulting 30° to the new azimuth line. The result of 285° is the apprximate wind direction at 600 feet.

Figure 3-11. *Additional practice pibal plots.*

Figure 3-12. *As the balloon ascends, the flightpath inclines to the right. Correlate this visualization to a map to determine the ground track of the balloon during flight.*

Performance Planning

Prior to a discussion of performance planning, a number of terms must be defined.

Maximum Allowable Gross Weight is that maximum amount of weight that the balloon may lift, under standard conditions. This figure is usually stipulated in design criteria, and addressed in the Type Certificate Data Sheet pertaining to that balloon. It can also be found on the weight and balance page of the flight manual for that particular balloon. An average of 1,000 cubic feet of air, when heated, will lift 20 pounds.

Useful lift (load) in aviation is the potential weight of the pilot, passengers, equipment, and fuel. It is the basic empty weight of the aircraft (found in the flight manual for each balloon) subtracted from the maximum allowable gross weight. This term is frequently confused with payload, which in aviation is defined as the weight of occupants, cargo, and baggage.

Density altitude is defined in the Pilot's Handbook of Aeronautical Knowledge (FAA-H-8083-25) as "pressure altitude corrected for nonstandard temperature." Density altitude is determined by first finding pressure altitude, and then correcting this altitude for nonstandard temperature variations. For example, when set at 29.92, the altimeter may indicate a pressure altitude of 5,000 feet. Under standard temperature conditions (59 °F), this may allow for a useful load of 1,050 pounds. However, if the temperature is 20° above standard, the expansion of the air raises the density altitude level (the air is less dense, thereby mimicking the density of the air at a higher altitude). Using temperature correction data from tables or graphs, it may be found that the density level is above 8,000 feet, and the useful load is then reduced to 755 pounds. This definition, however, has a tendency to confuse many new (and some not-so-new) pilots, so a more thorough explanation is justified.

The AIM explains density altitude as being nothing more than a way to comparatively measure aircraft performance. Paragraph 7-5-6 states, in part, "Density altitude is a measure of air density. It is not to be confused with pressure altitude, true altitude or absolute altitude. It is not to be used as a height reference, but as a determining criteria [sic] in the performance capability of an aircraft." With respect to ballooning, this is a more useful definition of the term.

How does density altitude affect balloon performance? Density altitude affects balloon performance in two ways. First and more important, as a balloon gains altitude, it loses capacity, insofar as its lifting capability is concerned. This means a balloon capable of lifting 1,400 pounds at sea level may only be able to lift 1,150 pounds or less at 4,000 feet. For a pilot who seldom leaves the local area, this rarely causes a problem. For the pilot who travels from the low area of the Southeast to fly in the mile-high altitudes of Albuquerque,

New Mexico, the changes in balloon capability and decrease in burner performance are important considerations while planning for the flight.

Second, heater performance is degraded at a rate of 4 percent per 1,000 feet of altitude. This means on a standard reference day, a particular heater will have lost 12 percent of its efficiency at 3,000 feet, or be performing at 88 percent of its capability. This is due to the loss of the partial pressure of oxygen, a necessary component of combustion.

Preflight planning requires consideration of balloon loading and performance with respect to altitude and expected temperatures. Balloon manufacturers have provided the information necessary to determine these factors in the form of a performance chart in the flight manual. Referred to as nomographs or nomograms, performance charts are simple to use and provide excellent planning information. *[Figure 3-13]*

If three of the above factors are known, a fourth may be determined. The performance charts may be used in many ways to determine performance of the balloon on a given day. This process does not have to be computed at the beginning of each flight. Many pilots develop a listing of possible weights, temperatures, and altitudes, depending on the average flying conditions for their home area. This is an acceptable practice

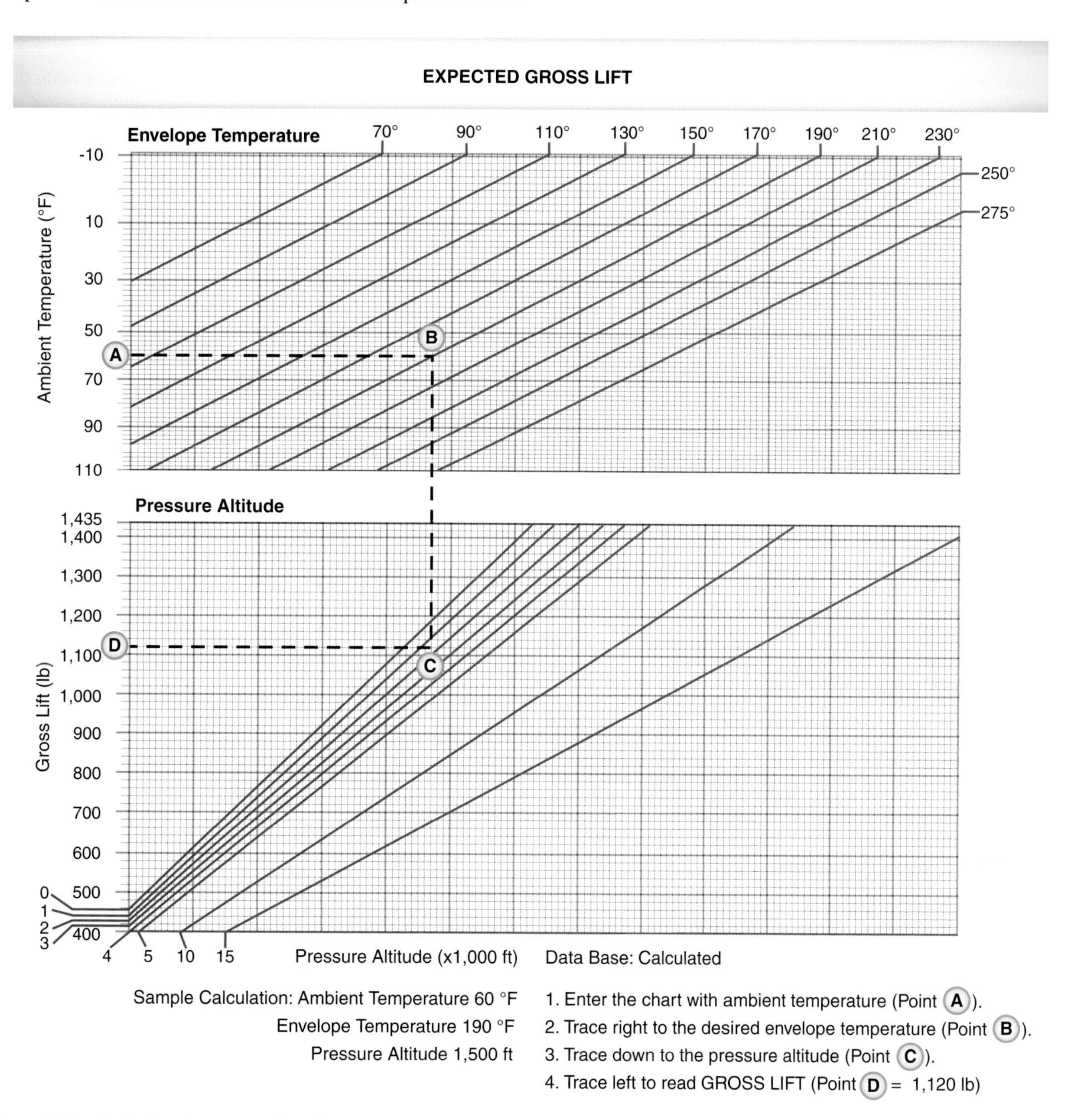

Figure 3-13. *Typical performance chart for a 77,000 cubic foot balloon.*

as long as the information is available and consulted when appropriate.

Using the chart in *Figure 3-13*, determine the maximum gross lift that may be expected on a 60 °F day, with decisions not to exceed 190 °F envelope temperature and 1,500 feet pressure altitude. In this example, the established parameters equate what many pilots consider when doing performance planning. They decide they do not want to exceed a given altitude or envelope temperature.) To determine the maximum gross lift available, the nomograph should be entered at point A, at the ambient temperature of 60 °F. Move right, to the line indicating an envelope temperature of 190 °F (point B). Then, move down vertically to a point equidistant between the lines denoting altitude of 1,000 and 2,000 feet (point C). Then, move horizontally to the left to the gross lift axis of the chart and read the result (point D). In the illustrated example, this computation results in a maximum gross lift of 1,120 pounds.

Using the chart again in *Figure 3-13*, determine the maximum altitude to which the balloon may climb, given the same maximum gross lift figure of 1,150 pounds. In this example, it is simply a matter of extending the lines appropriately. The A-B line would be extended to the diagonal line indicating a maximum temperature of 250 °F (which is the maximum continuous operating temperature for most balloons). Then, a perpendicular line would be drawn from that intersection point. After that line is drawn, extend the C-D line to the right, and the intersection of those two lines will indicate the maximum altitude. In this example, this computation results in a maximum altitude of 10,000 feet.

Special Conditions

Most balloon flying is done in non-hostile terrain and benign weather. There are, some instances in which the terrain may be more difficult, both for the pilot and the chase crew, and the weather may become a significant factor. With proper preflight planning, the problems inherent in mountain flying and cold weather flying can be resolved. While somewhat riskier than normal flying, this type of flying can be safely conducted.

Cold Weather Flying

Some pilots prefer flying in cold weather, which offers the advantages of more stable air and less fuel consumption to maintain flight. This means long, gentle flights for the pilot. There are two main disadvantages to cold weather flying: the need to maintain adequate pressure in the balloon's fuel system and the difficulty of keeping the pilot, crew, and passengers warm.

As propane gets colder, it has less vapor pressure. (See chart of propane and butane partial pressures, Appendix C). To ensure adequate pressure in cold weather, follow the manufacturer's recommended method, which will be described in the flight manual. Many manufacturers recommend the use of nitrogen, an inert gas that may be added to the fuel tanks by means of a regulator. This is perhaps the easiest way to pressurize tanks, as it may be done on site, and with little or no prior planning. It does require the use of a nitrogen tank and a two-stage regulator. These items must be available to the pilot before the flight. If the flight is cancelled after pressurization, and the anticipated rise in temperature is expected to be more than 30°, the pressure will need to be bled off by using the fixed liquid level gauge.

Balloon systems using a vapor feed pilot system may not be able to use nitrogen as it can result in an unreliable pilot light. Those systems commonly use heat tapes or heated tank covers in order to warm the propane. Heat tapes, similar to those used to prevent water pipes from freezing, are reliable. They do require frequent inspection, as normal wear and tear may cause an electrical short, with potential danger of damage to the fuel tanks. Among the aftermarket types of heat tapes, the ones with an internal thermostat will cycle once a particular temperature is reached, reducing the possibility of overheating the tanks.

Use tank heaters and heat tapes with extra caution. Tanks must not be heated in an area within 50 feet of an open flame, near an appliance with a pilot light, or in a closed area without natural ventilation.

Pilots, as well as crew, should be dressed appropriately for the environment. Layered clothing that entraps warm air is standard cold weather gear. Note that cold weather environments commonly promote static electricity. It is important that clothing of natural fibers be used, rather than synthetics. A hat is important, as significant body heat escapes from the head. Warm gloves and footwear are a must. Remember that certain types of hypoxia, or lack of oxygen to the brain (discussed in Chapter 9, Aeromedical Factors), may be aggravated by exposure to continued cold. Pilots and crew should be aware of the symptoms of hypothermia and frostbite, guard against them, and have a plan in place to deal with potential medical emergencies.

From an equipment standpoint, the balloon requires no special preparation, other than insuring proper pressure in the fuel system. Pilots should be aware that seals and O-rings may shrink somewhat or become brittle in cold weather. This may cause a propane leak, and special caution should be taken during the equipment preflight process to ensure that this will not be an issue. The pilot and chase crew should also be careful not to pack snow in the envelope, particularly if the balloon will be stored for a long period of time before the next flight. With respect to the chase vehicle, remember to have antifreeze in the cooling system. It is advisable to carry chains, a shovel, and a windshield wiper/scraper if there is a possibility of snow.

Mountain Flying

Flying in mountainous terrain can provide one of the most exhilarating flights imaginable, but there are numerous planning factors that must be considered. *[Figure 3-14]*

Figure 3-14. *Mountain flying.*

Weather, with its associated phenomena, is perhaps the most important to understand of the many factors involved in mountain flying. When inflating a balloon, drainage winds (a form of orographic wind) may cause the envelope to move from its planned position, and may even roll back over the basket, pilot, and crew. As most weather forecasts do not address this issue, consult with local pilots regarding these wind's formation, strength, and onset. In flight, winds flowing across mountain terrain set up features, such as rotors and standing waves (discussed in Chapter 4, Weather) which may cause a complete loss of control of the balloon. Other less violent winds may cause the balloon to proceed in unplanned directions, and require adjustments to landing and retrieval plans. It is important that any pilot contemplating flight in mountainous terrain be aware of these potential conditions, and plan to minimize their effects.

Communications in mountainous terrain can be a significant factor because most radios used by balloonists are line of sight, and will not work well, if at all, in particularly hilly or mountainous terrain. Cell phones may be used after landing, but again may be limited by the lack of cell towers and general reception problems. A good communications plan between pilot and ground crew includes a "lost balloon" contact with a common phone number that both parties call in order to find where the other is. This could possibly be a person at home, willing to relay the information as necessary, or perhaps an answering machine from which both the pilot and crew may retrieve messages.

Mountain flying that involves long distances requires appropriate clothing. Refer to the earlier paragraph on dressing for cold weather flying, which also applies to mountain flying. Good preflight planning will ensure that the pilot and passengers are prepared for a cold weather flight. It also prepares for the possibility of remaining out in the cold while the ground crew locates the balloon, since following a balloon can be difficult in mountainous terrain. Some pilots carry additional equipment in the balloon that they do not carry on flatland flights. Suggested provisions and equipment are water, additional warm clothing or a sleeping bag, a strobe, a radio, a compass, a lightweight shelter (a Mylar® sheet can be made into a simple tent, for example), and a good map or maps of the area.

The Ground Crew

As ground crew have no legal status or authority within 14 CFR, it is easy to overlook or downplay their role in the preflight process, as well as flight safety. Ground crew knowledge and skill bring both the brains and brawn necessary at every stage of a flight—from equipment set-up and flight path plotting to taking inflight wind readings and assisting challenging landings. Without sufficient crew, a pilot becomes rushed, distracted, uninformed, or in the midst of hazardous conditions. Ground crews serve not only as physical help and assistance, but also serve as a form of redundancy for a pilot's eyes, muscles, and mind. *[Figure 3-15]*

While crew requirements may vary from flight to flight, consider the following during the preflight process:

- Two crew members to assist is a good starting point. During the chase phase, one may drive while the other navigates; two may handle emergencies better than one. Words of caution: more is not always better. In some cases, passengers on board the balloon may be able and willing to help pack and unpack equipment. Another factor may be the capacity of the chase vehicle.

Figure 3-15. *Equipment set-up with a ground crew.*

- At least one crew member should be familiar with the balloon, the pilot's flight planning and routines, emergency procedures, vehicle operations (liftgate operations, trailer backing, etc.) and the local area. If out of town, a local person who knows the area and roads can prove to be invaluable.
- A crew chief should be designated early in the preflight planning process to avoid conflicts or communications breakdowns later. This person oversees all ground operations, under the direction of the pilot, directs other crew members, and speaks on behalf of the pilot when meeting landowners.
- The crew should be free of major disabilities (pre-existing back conditions, severe allergies/respiratory conditions), capable of lifting/moving heavy equipment, and in good physical shape.
- A meeting place should be designated well in advance of the launch time. It is prudent to err on the early side, as being late causes all other facets of the flight process to be rushed, increasing the potential for risks. If the meeting site is to serve also as the launch site, there should be a secondary location available in case of relocation.
- Crew members who are healthy, rested, and focused perform best. Minimize conditions or distractions that compromise these items.

While most crew members participate in ballooning for the fun and friendship it offers, it is essential they are committed to their ongoing role in flight safety. Preflight planning requires that a pilot consider how crew will help each flight unfold safely.

Flying in New Territory

When planning a flight in a new or unfamiliar area, it is important to insure that balloonists are welcome. If possible, talk to local balloonists who may be familiar with the area, and who may be able to point out local no-fly areas, potential launch and landing sites, and potential landowner problem areas. To locate local balloonists:

- Call the nearest Flight Standards District Office (FSDO) and ask for the name of a balloon pilot examiner or FAA Safety team member.
- Look in the local telephone book under "Balloons—Manned."
- Check for local balloon clubs in the area.
- Check the BFA membership roster online at www.bfa.net.

If there are no balloonists in the local area, talk to other aircraft pilots or local law enforcement offices. Let them know that a balloon flight is being planned and ask for advice.

Equipment

Pack all equipment; have it ready and double checked the night before a flight. New pilots should create and follow a checklist that covers all of the home station preparations for the chase vehicle and equipment. Simple tasks, such as ensuring that radios and cell phones are properly charged, can easily be overlooked. Check to see that the balloon, fan, and vehicle are fueled; vehicle tires are inflated; required documentation is in the balloon; and all necessary maps, radios, and other equipment are loaded in the chase vehicle. *[Figure 3-16]*

Figure 3-16. *It is important to check that all equipment is in proper working condition prior to flight.*

Personal Preparation

The most important single element in preflight preparation is the pilot. A pilot anticipating a flight should do a quick self-analysis to ensure that he or she is mentally and physically capable and competent to perform. It is not uncommon to see pilots and crews drive all night to get to an event, and, without sleep or perhaps suffering from a cold or other ailment, go ahead and fly. Allow enough time and resources to remain healthy, rested, and focused enough to make wise flight preparations and decisions that maximize flight safety.

It is a good policy to do an individual "preflight," using the "I'M SAFE" evaluation checklist as illustrated in Chapter 1, Introduction to Balloon Flight Training, *Figure 1-5*.

Chapter Summary

Flying a hot air balloon requires extensive preflight planning which depends upon a knowledge of meteorology applied to local weather conditions, the ability to navigate using charts and maps, familiarity with the equipment used in ballooning, a good support ground crew, and a dash of common sense. Safe flight experiences for a pilot in ballooning or any other aircraft depend upon continuing education and experience.

Chapter 4

Weather Theory and Reports

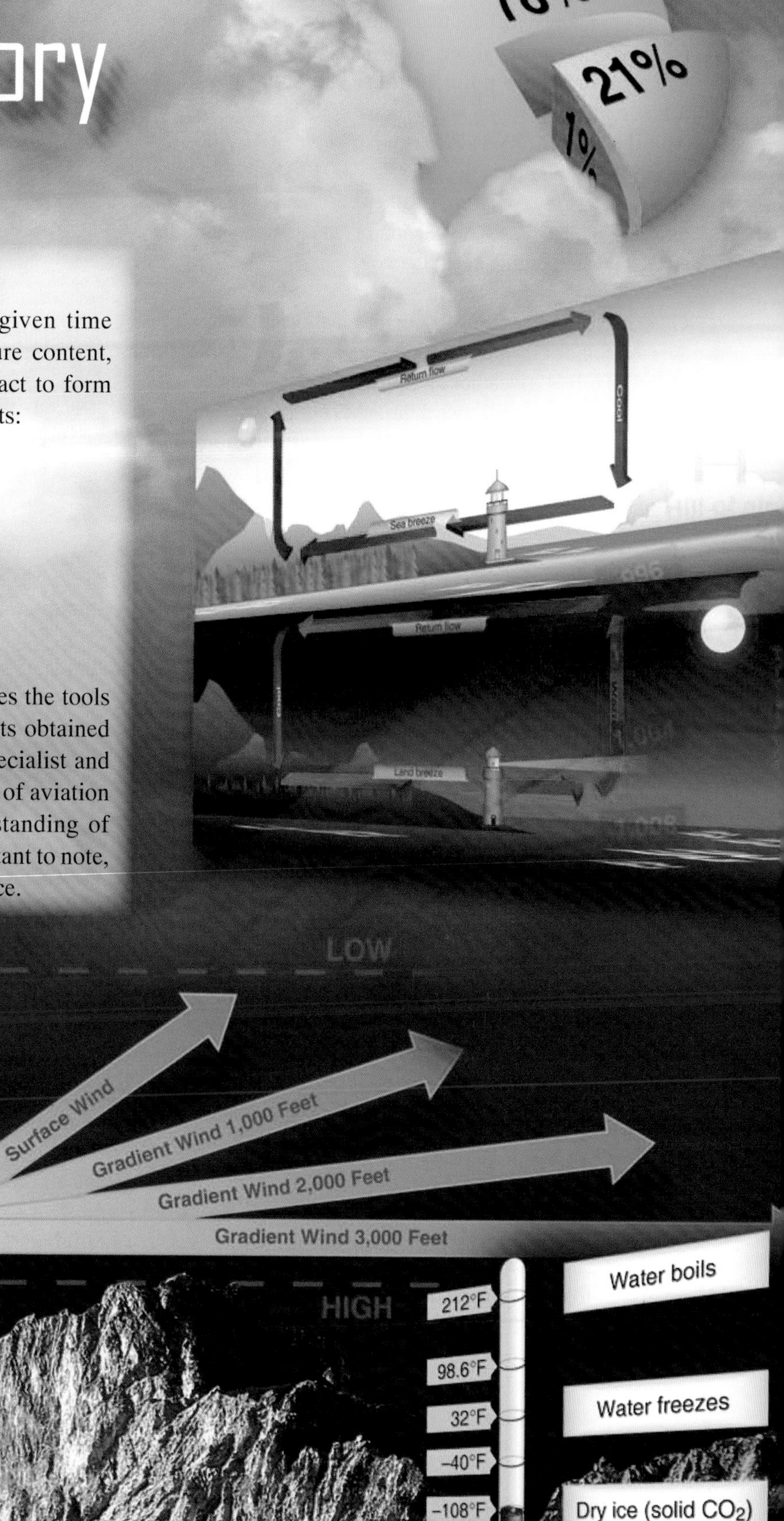

Introduction

Weather is the state of the atmosphere at a given time and place, with respect to temperature, moisture content, turbulence, and cloudiness. These factors interact to form the following five major meteorological elements:

- Atmospheric pressure (high or low),
- Air temperature (heat or cold),
- Wind (calm or storm),
- Clouds (clearness or cloudiness), and
- Precipitation (rain, sleet, snow).

A solid understanding of weather theory provides the tools necessary to understand the reports and forecasts obtained from a Flight Service Station (FSS) weather specialist and other aviation weather services. No other means of aviation relies more heavily on knowledge and understanding of weather for its safety than ballooning. It is important to note, however, that there is no substitute for experience.

There are many excellent texts and online sources available for learning more about weather that are referenced at the appropriate point in this chapter. Much of the following information can be found in the Federal Aviation Administration (FAA) Advisory Circular (AC) 00-06A, Aviation Weather For Pilots and Flight Operations Personnel, and AC 00-45E, Aviation Weather Services, both of which may be found at the FAA's Regulatory and Guidance Library (RGL) located online at rgl.faa.gov.

Other online sources for weather information are also helpful. The National Oceanic and Atmospheric Administration (NOAA) offers a weather tutorial at www.srh.noaa.gov/srh/jetstream/. Developed to meet the needs of educators, weather professionals, and others interested in learning more about weather, it provides a number of concise explanations of weather theory. Additionally, ww2010.atmos.uiuc.edu/(Gh)/home.rxml is a site developed as part of the Weather World 2010 project by the Department of Atmospheric Sciences at the University of Illinois at Urbana-Champaign.

This chapter is designed to give balloon pilots a basic knowledge of weather principles, acquaint them with the weather information available for flight planning, and help them develop sound decision-making skills as they prepare for and execute a safe flight.

The Atmosphere

The atmosphere is the envelope of air that surrounds the Earth. Approximately one-half of the air, by weight, is within the lower 18,000 feet. The remainder of the air is spread over a vertical distance in excess of 1,000 miles. No definite outer atmospheric boundary exists, but the air particles become less numerous with increasing altitude until they gradually overcome Earth's gravity and escape into space. In addition to the rotation of the air with the rotation of the Earth, another type of air movement occurs within the atmosphere. This movement of air around the surface of the Earth is called atmospheric circulation.

Composition

The atmosphere is a blanket of air composed of a mixture of gases that surrounds the Earth and reaches over 560 kilometers (km), 348 miles, from the surface. This blanket of gases provides protection from ultraviolet rays, as well as supporting human, animal, and plant life. Nitrogen accounts for 78 percent of the gases comprising the atmosphere, while oxygen makes up 21 percent. *[Figure 4-1]* Argon, carbon dioxide, and traces of other gases make up the remaining 1 percent. Within this envelope of gases, there are several recognizable layers of the atmosphere as defined by altitude.

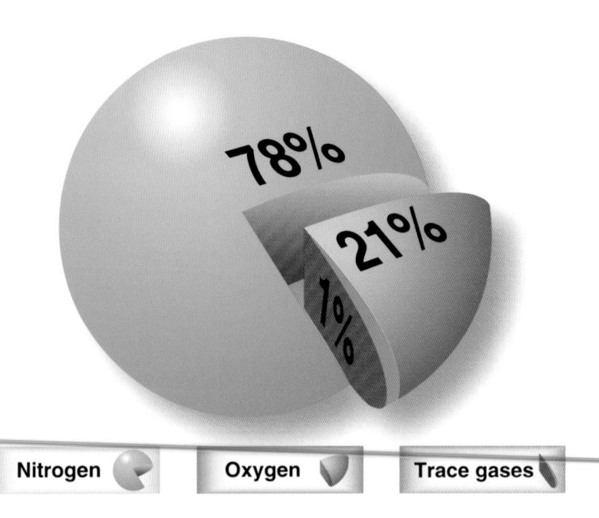

Figure 4-1. *Composition of the atmosphere.*

The first layer, known as the troposphere, extends from sea level up to 20,000 feet (8 km) over the northern and southern poles and up to 48,000 feet (14.5 km) over the equatorial regions. The vast majority of weather, clouds, storms, and temperature variances occur within this first layer of the atmosphere.

At the top of the troposphere is a boundary known as the tropopause, which traps moisture and the associated weather in the troposphere. The altitude of the tropopause varies with latitude and with the season of the year which causes it to take on an elliptical shape, as opposed to round. Location of the tropopause is important because it is commonly associated with the location of the jet stream and possible clear air turbulence.

The Standard Atmosphere

To provide a common reference when discussing weather, the International Standard Atmosphere (ISA) has been established. To arrive at the standard atmosphere, conditions throughout the atmosphere with respect to latitudes, seasons, and altitudes were averaged. The standard reference point is 59 °F or 15 °C, and 29.92 inches of mercury ("Hg) or 1013.2 millibars (mb). Pressure does not decrease linearly with altitude, but for the first 10,000 feet, 1 "Hg for each 1,000 feet approximates the rate of pressure change. There is also a standard temperature lapse rate of 3.5 °F or 2 °C per 1,000 feet of altitude, up to 36,000 feet.

At sea level, the atmosphere exerts pressure on the Earth at a force of 14.7 pounds per square inch (psi). This means a column of air one inch square, extending from the surface up to the upper atmospheric limit, weighs about 14.7 pounds.

A person standing at sea level also experiences the pressure of the atmosphere, but the pressure is a force of pressure over the entire surface of the skin. The actual pressure at a given place and time will differ with altitude, temperature, and density of the air. These conditions also affect balloon performance, especially with regard to useful load and burner performance.

Measurement of Atmospheric Pressure

Constant pressure charts and hurricane pressure reports are written using millibars (mb). Since weather stations are located around the globe, all local barometric pressure readings are converted to a sea level pressure to provide a standard for records and reports. To achieve this, each station converts its barometric pressure by adding approximately 1 "Hg for every 1,000 feet of elevation gain. For example, a station at 5,000 feet above sea level, with a reading of 24.92 "Hg, reports a sea level pressure reading of 29.92 "Hg. Using common sea level pressure readings helps ensure aircraft altimeters are set correctly, based on the current pressure readings. In order to compensate for pressure variations due to different station elevations, all observations are mathematically corrected to mean sea level (MSL). Altimeter settings are obtained by mathematically reducing station pressure to MSL. This enables the pilot to read MSL altitudes on the altimeter.

When charting atmospheric pressures over various areas of the Earth, the meteorologist is primarily interested in the pressure difference per unit of distance—the pressure gradient.

The MSL pressure is plotted in mb at each reporting station on a surface weather map. The isobars outline pressure areas in somewhat the same manner that contour lines outline terrain features on contour maps. Standard procedure on surface weather maps in North America is to draw isobars spaced at four mb of pressure apart, with intermediate two mb spacing when appropriate. Although the isobar patterns are never the same on any two weather maps, they do show patterns of similarity.

By tracking barometric pressure trends across a large area, weather forecasters can more accurately predict movement of pressure systems and the associated weather. For example, tracking a pattern of rising pressure at a single weather station generally indicates the approach of fair weather. Conversely, decreasing or rapidly falling pressure usually indicates approaching bad weather and possibly severe storms.

Temperature

Temperature is a measurement of the amount of heat and expresses a degree of molecular activity. Since different substances have different molecular structures, equal amounts of heat applied to equal volumes of two different substances will result in unequal heating. Every substance has its own unique specific heat. For example, a land surface becomes hotter than a water surface when equal amounts of heat are added to each. The degree of "hotness" or "coldness" of a substance is known as its temperature, and is measured with a thermometer.

The Earth's surface is heated during the day by the sun. This incoming solar radiation is called insolation, while heat radiated from the Earth by outgoing radiation is called terrestrial radiation. The cooling that occurs at night is terrestrial radiation.

Temperature Scales

Two temperature scales are important to the balloon pilot: Fahrenheit (F) and Celsius (C). On the Fahrenheit scale, the freezing point is 32° and the boiling point is 212°, a difference of 180°. On the Celsius scale, the freezing point is 0° and the boiling point is 100°. For many years, the Celsius scale was the choice for technicians and those countries and organizations utilizing the metric system. In recent years, the United States has transitioned to almost exclusive use of the Celsius scale in weather reports, primarily because of the International Civil Aviation Organization (ICAO) convention agreements. *[Figure 4-2]*

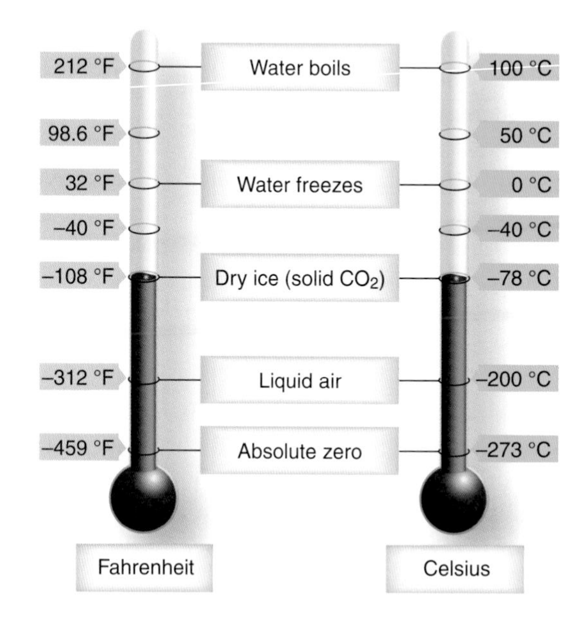

Figure 4-2. *Comparison of Fahrenheit and Celsius temperature scales.*

A quick and easy way to convert Fahrenheit to Celsius is to subtract 30, and divide the number by two. To convert Celsius to Fahrenheit, double the number, and add 30. These formulas give a good approximation for most calculations in ballooning. Conversion charts are also available on the Internet.

Temperature Variations

The amount of solar radiation (insolation) received by any region varies with the time of day, with seasons, and with latitude. These differences in insolation create temperature variations. Temperatures also vary with differences in topographical surface and with altitude. These temperature variations create forces that drive the atmosphere in its motion. Simply stated, heat and, therefore, temperature differences cause weather.

Diurnal variation is the change in temperature from day to night brought about by the daily rotation of the Earth. The Earth receives heat during the day by insolation, but continually loses heat by terrestrial radiation. Warming and cooling depend on an imbalance of insolation and terrestrial radiation. During the day, insolation exceeds terrestrial radiation and the surface becomes warmer. At night, insolation ceases, but terrestrial radiation continues and cools the surface. Cooling continues after sunrise until insolation again exceeds terrestrial radiation. Minimum temperature usually occurs after sunrise, sometimes as much as one hour after. The continued cooling after sunrise is one reason that fog sometimes forms shortly after the sun is above the horizon.

Temperature Variations with Topography

Temperature variations are also induced by water and terrain. Water absorbs and radiates heat energy with less temperature change than does land. Large, deep water bodies tend to minimize temperature changes, while large land masses induce major temperature changes. Wet soil, such as that found in swamps and marshes, is almost as effective as water in suppressing temperature changes. Thick vegetation tends to control temperature changes since it contains some water and also insulates against heat transfer between the ground and the atmosphere. Arid, barren surfaces generate the greatest temperature changes.

These topographical influences are both diurnal and seasonal. For example, the difference between a daily maximum and minimum may be 10° or less over water, near a shore line, or over a swamp or marsh, while a difference of 50° or more is common over rocky or sandy deserts.

Abrupt temperature differences develop along lake and ocean shores. These variations generate pressure differences and local air flows or winds, which may become a consideration in the balloon pilot's study of the air mass.

Prevailing wind is also a factor in temperature control. In an area where prevailing winds are from large water bodies, temperature changes are rather small. Most islands enjoy fairly constant temperatures. On the other hand, temperature changes are more pronounced where prevailing wind is from dry, barren regions.

Temperature Variation with Altitude

Temperature normally decreases with increasing altitude throughout the troposphere. This decrease of temperature with altitude is defined as lapse rate. The standard lapse rate seldom exists. In fact, temperature sometimes increases with height. An increase in temperature with altitude is defined as an inversion.

An inversion often develops near the ground on clear, cool nights when wind is light. The ground radiates and cools much faster than the overlying air. Air in contact with the ground becomes cold, while the temperature a few hundred feet above changes very little. Thus, temperature increases with height. Inversions may also occur at any altitude when conditions are favorable. For example, a current of warm air aloft overrunning cold air near the surface produces an inversion aloft.

Low level inversions, which are usually of most interest to the balloon pilot, generally dissipate as the air mixes with insolation.

Heat Transfer

The heat source for this planet is the sun; energy from the sun is transferred through space and the Earth's atmosphere to the Earth's surface. As this energy warms the Earth's surface and atmosphere, some of it is or becomes heat energy. Heat is transferred into the atmosphere in three ways: radiation, conduction, and convection.

Radiation

Radiation is the transfer of heat by electromagnetic waves. No medium of transfer is required between the radiator and the body being irradiated (receiving the radiation). Heat waves, a form of this electromagnetic energy, may be reflected. In meteorology, the principal reflectors are the Earth's surface, water vapor in the air, and particulate matter in the atmosphere.

Conduction

Conduction is the transfer of heat energy from one substance to another or within a substance. As with electricity, some materials are good conductors while others are poor conductors. Poor conductors are considered to be insulators. Air is one of the poorest conductors of heat in comparison to silver, one of the best conductors. Silver will pass 20,000 times more heat than an equal mass of air across a similar temperature difference during a fixed period of time. Conduction in the atmosphere is considered to be a significant method of heat exchange only at the Earth's surface, where the lowest few centimeters of the atmosphere are actually in contact with the ground or water.

Convection

Convection is the transfer of heat energy in a fluid. This type of heating is most commonly seen in the kitchen when a liquid boils. This type of heat transfer occurs in the atmosphere when the ground is heated by the sun. The warm ground heats the air above it by radiation and conduction, causing the warm air to rise. Meanwhile, the dense cooler air aloft moves in to take the warm ground air's place to be heated.

Heat can be transferred by convection in either a vertical or a horizontal direction. In meteorology, "advection" is the term used for the horizontal transport of heat by wind. It is important to differentiate between the vertical and horizontal paths of convection. In the atmosphere, the amount of heat transferred horizontally over the surface of the Earth by advection is about 1,000 times greater than that transferred by convection.

The Adiabatic Process

The adiabatic process is the change of the temperature of air without transferring heat. In an adiabatic process, compression results in warming, and expansion results in cooling. The adiabatic process takes place in all upward and downward moving air. When air rises into an area of lower pressure, it expands to a larger volume. As the molecules of air expand, the temperature of the air lowers. As a result, when a parcel of air rises, pressure decreases, volume increases, and temperature decreases. When air descends, the opposite is true.

Since air is composed of a mixture of gases subject to heating when compressed and cooling when expanded, air will rise, seeking a level where the pressure of the body of air is equal to the pressure of the air that surrounds it. Whatever the cause of the lifting, the air rises, and the pressure decreases, allowing the "parcel of air" to expand. This continues until it reaches an altitude similar in pressure and density to its own. As it expands, it cools through the adiabatic process and no heat is added or withdrawn from the system in which it operates. As air rises, it is cooled because it is expanding by moving to an altitude where pressure and density is less. This is adiabatic cooling. When the process is reversed and air is forced downward, it is compressed, causing it to heat by a process called adiabatic heating.

Air Masses

An air mass is a large body of air (usually 1,700 kilometers or more across) whose physical properties (temperature and humidity) are horizontally uniform. The weather is a direct result of the continuous alternation of the influences of warm and cold air masses. Warm air masses predominate in the summer, and cold air masses predominate in the winter. However, both cold and warm air, alternately, may prevail almost anywhere in the temperature zone at any season. The basic characteristics of any air mass are temperature and humidity. These properties are relatively uniform throughout the air mass, and it is by measurement of these properties that the various types of air masses are determined.

Characteristics

Air masses acquire the characteristics of the surrounding area, or source region. The characteristics of an air mass consist of the basic properties of moisture and temperature, which include:

- Stability,
- Cloud types,
- Sky coverage,
- Visibility,
- Precipitation,
- Icing, and
- Turbulence.

The terrain surface underlying the air mass is the primary factor in determining air mass characteristics.

A source region is typically an area in which the air remains relatively stagnant for a period of days or longer. During this time of stagnation, the air mass takes on the temperature and moisture characteristics of the source region. Areas of stagnation can be found in polar regions, tropical oceans, and dry deserts. Air masses are classified by region of origination:

- Polar or tropical
- Maritime or continental

A continental polar air mass forms over a polar region and brings cool, dry air with it. Maritime tropical air masses form over warm tropical waters like the Caribbean Sea and bring warm, moist air. As the air mass moves from its source region and passes over land or water, the terrain it passes over modifies its qualities, and thus modifies the nature of the air mass.

An air mass passing over a warmer surface will be warmed from below, and convective currents form, causing the air to rise. This creates an unstable air mass with good surface visibility. Moist, unstable air causes cumulus clouds, localized showers, and turbulence to form. Conversely, an air mass passing over a colder surface does not form convective currents, but instead creates a stable air mass with poor surface visibility. The poor surface visibility is due to the fact that moisture, smoke, dust, and other particles cannot rise out of the air mass and are instead trapped near the surface. A stable air mass can produce low stratus clouds and fog.

Pressure Systems

The differences that occur with heating and cooling the atmosphere in the lower levels also cause density variations. These variations cause small horizontal pressure differences that are only about one ten-thousandth of the magnitude of the normal change of pressure with altitude, but they significantly impact atmospheric circulation and most weather phenomena. *[Figure 4-3]*

A low or cyclone is a pressure system in which the barometric pressure decreases towards the center and the wind flow around the system is counterclockwise in the Northern Hemisphere. Unfavorable flying conditions in the form of low clouds, restricted visibility by precipitation and fog, strong and gusty winds, and turbulence are common in low pressure systems. Thermal low pressure systems caused by intense surface heating and resulting low air density over barren continental areas are relatively dry with few clouds and practically no precipitation. Thermal lows are stationary and predominate over continental areas in the summer. General airflow in a low pressure system, since the atmosphere is attempting to achieve equilibrium, is in (towards the center of the low pressure system), and up. This tendency can affect the overall dynamic of the low pressure system.

A high is a pressure system in which the barometric pressure increases toward the center and the wind flow around the system is clockwise in the Northern Hemisphere. Flying conditions are generally more favorable in highs than in lows because of fewer clouds, light or calm winds, and less concentrated turbulent areas. But, in some situations, visibility may be reduced due to early morning fog, smog, or haze at flight levels. High pressure systems predominate over cold surfaces where the air is dense. General airflow in a high pressure system, in reverse of the low pressure dynamic, is out (away from the center of the pressure system) and down. Again, these airflow tendencies can affect the dynamic of the high pressure system, much like the low.

In the Northern Hemisphere, a general cycle of highs and lows moves through the temperate zones from west to east. The movement of the pressure systems is more rapid in the winter season when the low pressure systems are most intense and the high pressure systems extend farthest to the south. *[Figure 4-4]*

A trough is an elongated area of low pressure, with the lowest pressure along the trough line. The weather in a trough is commonly violent. Also, troughs can be slow moving.

A ridge is an elongated area of high pressure with highest pressure along the ridge line. The weather in a ridge is generally favorable for flying.

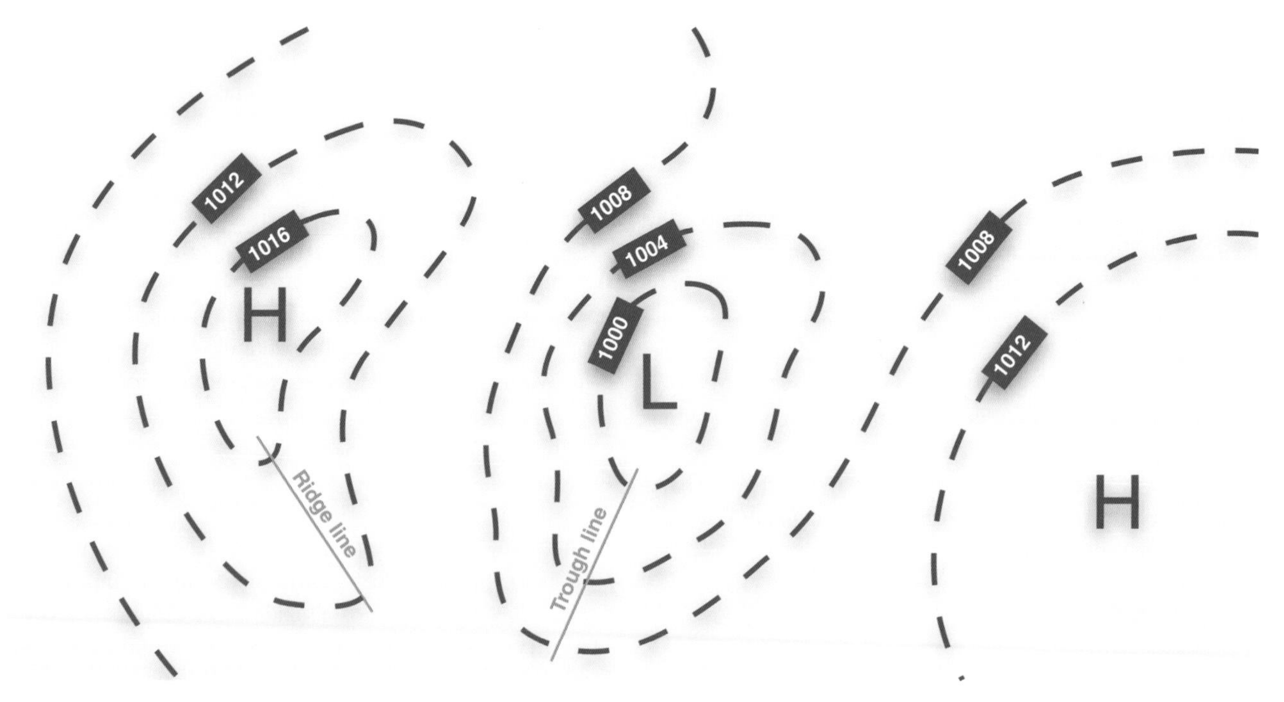

Figure 4-3. *High and low pressure systems.*

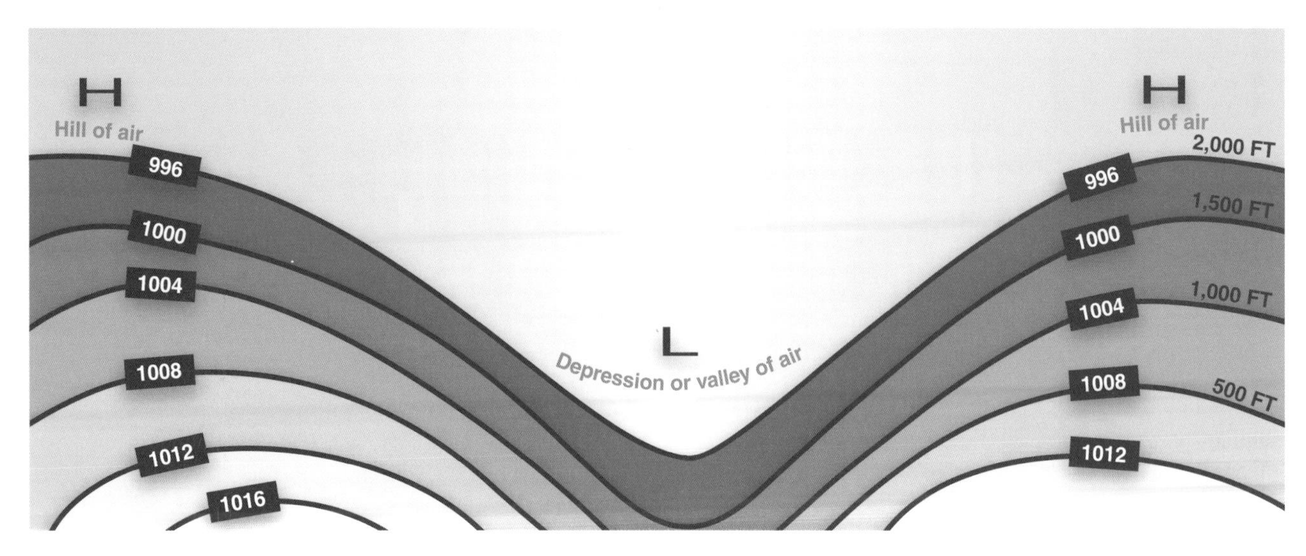

Figure 4-4. *A cross-section of the pressure systems depicted in Figure 4-3.*

Fronts

Fronts are the boundaries between two air masses and are classified as to which type of air mass (cold or warm) is replacing the other. For example, a cold front demarcates the leading edge of a cold air mass displacing a warmer air mass. A warm front is the leading edge of a warmer air mass replacing a colder air mass. Fronts are also transition zones (boundaries) between air masses that have different densities. The density of air is controlled primarily by the temperature of the air. Therefore, fronts in temperate zones usually form between tropical and polar air masses, but they may also form between arctic and polar air masses. A typical surface weather map shows air mass boundary zones at ground level. Designs on the boundary lines indicate the type of front and its direction of movement. On weather maps in local weather stations, fronts may also be indicated by colored lines. A working knowledge of fronts and their accompanying weather hazards is important to pilots.

Types of Fronts

The four major frontal types are:

- Cold,
- Warm,
- Stationary, and
- Occluded.

A front type is determined from the movement of the air masses involved.

Cold Front

A cold front is the leading edge of an advancing mass of cold air. A cold front occurs when a mass of cold, dense, and stable air advances and replaces a body of warmer air. Cold fronts move more rapidly than warm fronts, generally progressing at a rate of 25 to 30 miles per hour (mph). However, extreme cold fronts have been recorded moving at speeds of up to 60 mph. A typical cold front moves in a manner opposite that of a warm front. Because it is so dense, it stays close to the ground and acts like a snowplow, sliding under the warmer air and forcing the less dense air aloft. *[Figure 4-5]* The rapidly ascending air causes the temperature to decrease suddenly, forcing the creation of clouds. The type of clouds that form depends on the stability of the warmer air mass. A cold front in the Northern Hemisphere is normally oriented in a northeast to southwest manner and can extend for several hundred miles, encompassing a large area of land. Prior to the passage of a typical cold front, cirriform or towering cumulus clouds are present, and cumulonimbus clouds are possible. Rain showers and haze are possible due to the rapid development of clouds. The wind from the south-southwest helps to replace the warm temperatures with the relative colder air. A high dew point and falling barometric pressure are indicative of imminent cold front passage.

As the cold front passes, towering cumulus or cumulonimbus clouds continue to dominate the sky. *[Figure 4-6]* Depending on the intensity of the cold front, heavy rain showers form and might be accompanied by lightning, thunder, and/or hail. More severe cold fronts can also produce tornadoes. During cold front passage, the visibility is poor, with winds variable and gusty, and the temperature and dew point drop rapidly. A quickly falling barometric pressure bottoms out during frontal passage, then begins a gradual increase. After frontal passage, the towering cumulus and cumulonimbus clouds begin to dissipate to cumulus clouds, with a corresponding decrease in the precipitation. Good visibility eventually prevails with the winds from the west-northwest. Temperatures remain cooler and the barometric pressure continues to rise.

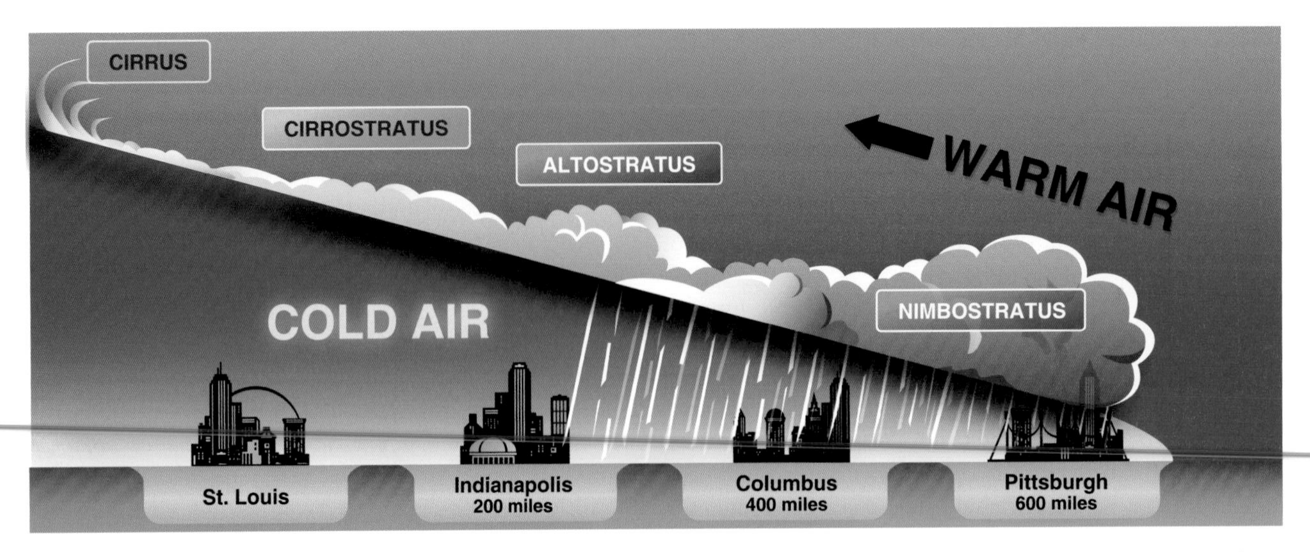

Figure 4-5. *A cold front underrunning warm, moist, stable air. Clouds are stratified and precipitation continuous. Precipitation induces stratus in the cold air.*

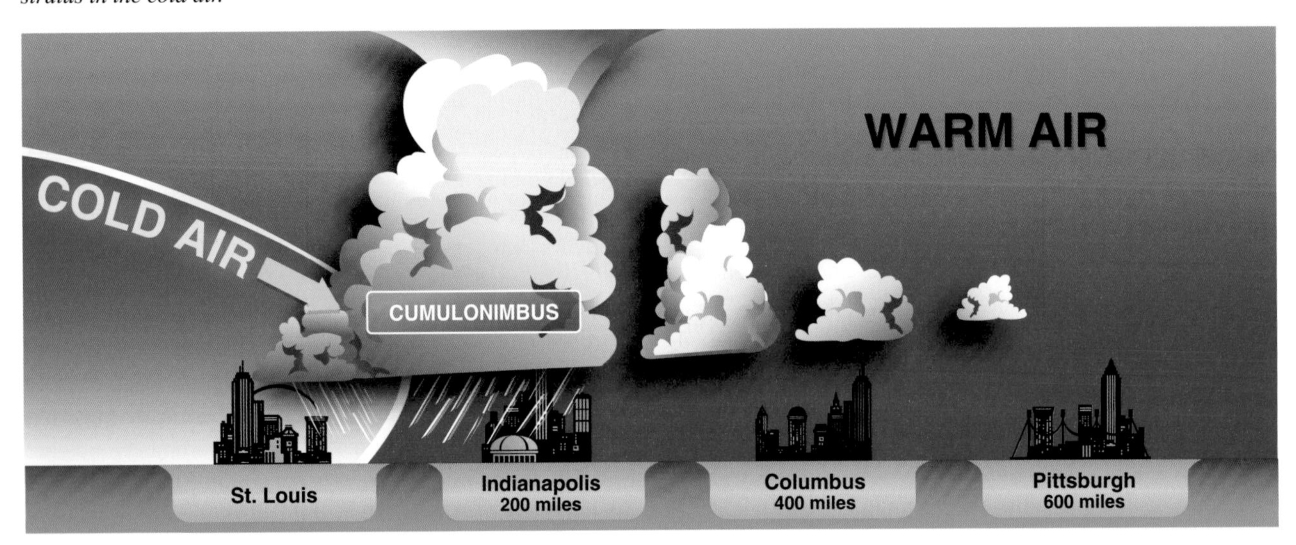

Figure 4-6. *A cold front underrunning warm, moist, unstable air. Clouds are cumuliform with possible showers or thunderstorms near the surface position of the front. Convective clouds often develop in the warm air ahead of the front. The warm, wet ground behind the front generates low-level convection and fair-weather cumulus in the cold air.*

Fast-Moving Cold Front

Fast-moving cold fronts are pushed by intense pressure systems far behind the actual front. The friction between the ground and the cold front retards the movement of the front and creates a steeper frontal surface. This results in a very narrow band of weather concentrated along the leading edge of the front. *[Figure 4-7]* If the warm air being overtaken by the cold front is relatively stable, overcast skies and rain may occur for some distance ahead of the front. If the warm air is unstable, scattered thunderstorms and rain showers may form. A continuous line of thunderstorms, or squall line, may form along or ahead of the front. Squall lines present a serious hazard to pilots as squall type thunderstorms are intense and move quickly. Behind a fast-moving cold front, the skies usually clear rapidly and the front leaves behind gusty, turbulent winds and colder temperatures.

Warm Front

A warm front is actually the trailing edge of a retreating mass of cold air. A warm front occurs when a warm mass of air advances and replaces a body of colder air. Warm fronts move slowly, typically 10 to 25 mph. The slope of the advancing front slides over the top of the cooler air and gradually pushes it out of the area. Warm fronts contain warm air that often has very high humidity. As the warm air is lifted, the temperature drops and condensation occurs. Prior to the passage of a warm front, cirriform or stratiform clouds, along with fog, can be expected to form along the frontal boundary. *[Figure 4-8]* In the summer months, cumulonimbus clouds (thunderstorms) are likely to develop. Light to moderate precipitation is probable, usually in the form of rain, sleet, snow, or drizzle, punctuated by poor visibility. The wind blows from the south-southeast, and the outside temperature

Figure 4-7. *A fast-moving cold front underrunning warm, moist, unstable air. Showers and thunderstorms develop along the surface position of the front.*

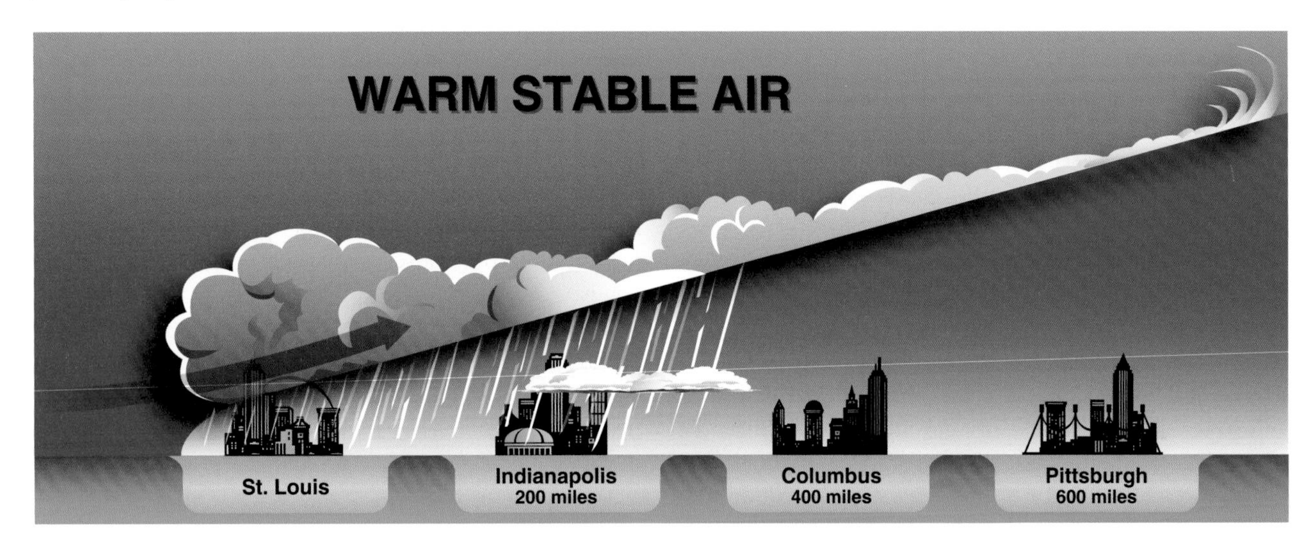

Figure 4-8. *A warm front with overrunning moist, stable air. Clouds are stratiform and widespread over the shallow front. Precipitation is continuous and induces widespread stratus in the cold air.*

is cool or cold, with increasing dew point. Finally, as the warm front approaches, the barometric pressure continues to fall until the front passes completely.

During the passage of a warm front, stratiform clouds are visible and drizzle may be falling. The visibility is generally poor, but improves with variable winds. The temperature rises steadily from the inflow of relatively warmer air. Usually, the dew point remains steady and the pressure levels off. After the passage of a warm front, stratocumulus clouds predominate and rain showers are possible. The visibility eventually improves, but hazy conditions may exist for a short period after passage. The wind generally blows from the south-southwest. With warming temperatures, the dew point rises and then levels off. There is generally a slight rise in barometric pressure, followed by a decrease in barometric pressure.

Stationary Front

When an air mass boundary is neither advancing nor retreating along the surface, the front is called a stationary front. Although there is no movement of the surface position of a true stationary front, an uplift of air may occur along the frontal boundary. If the uplifted air is stable and saturated, stratiform clouds may occur. Intermittent drizzle may occur, and if lifted above the freezing level, icing conditions and frozen precipitation will exist. If the uplifted air is conditionally unstable and saturation occurs, predominately cumuliform clouds will form, possibly generating thunderstorm activity.

Occluded Front

An occluded front occurs when a fast-moving cold front catches up with a slow-moving warm front. As the occluded front approaches, warm front weather prevails, but is immediately followed by cold front weather. There are two types of occluded fronts that can occur, and the temperatures of the colliding frontal systems play a large part in defining the type of front and the resulting weather. A cold front occlusion occurs when a fast-moving cold front is colder than the air ahead of the slow-moving warm front. When this occurs, the cold air replaces the cool air and forces the warm front aloft into the atmosphere. Typically, the cold front occlusion creates a mixture of weather found in both warm and cold fronts, if the air is relatively stable. A warm front occlusion occurs when the air ahead of the warm front is colder than the air of the cold front. When this is the case, the cold front rides up and over the warm front. If the air forced aloft by the warm front occlusion is unstable, the weather will be more severe than the weather found in a cold front occlusion. Embedded thunderstorms, rain, and fog are likely to occur.

Surface Fronts

The air mass boundaries indicated on a surface weather map are called surface fronts. A surface front is the position of a front at the Earth's surface. The weather map shows only the location of fronts on the surface, but these fronts also have vertical extent. For example, the colder, heavier air mass tends to flow under the warmer air mass. The underrunning mass produces the lifting action of warm air over cold air, causing clouds and associated frontal weather.

The vertical boundary between the warm and cold air masses is a frontal surface, and slopes upward over the colder air mass. The frontal surface lifts the warmer air mass and produces frontal cloud systems. The slope of the frontal surface varies with the speed of the moving cold air mass, and the roughness of the underlying terrain. Under normal conditions, the angle of inclination (slope ratio) between the frontal surface and the Earth's surface is greater with cold fronts than with warm fronts. The approximate height of the frontal surface over any station is determined from the analysis of upper air observations.

Frontal passage (FROPA) affects ballooning activities because it can generate precipitation, wind shifts, significant changes in temperature, and many other conditions hazardous to ballooning. Balloon pilots usually do not fly in the face of an approaching front; in fact, many have a rule that they do not fly within 18 to 24 hours prior to frontal passage, particularly if the approaching front has any significant strength associated with it. The FSS often can advise of the time a cold front will pass a given reporting station, which assists in flight planning.

Winds and Currents

Pressure and temperature changes produce two kinds of motion in the atmosphere—vertical movement of ascending and descending currents, and horizontal movement in the form of wind. Both types of motion in the atmosphere are important as they affect the takeoff, landing, and inflight operations. More important, however, is that these motions in the atmosphere, otherwise called atmospheric circulation, cause weather changes.

Understanding wind and current circulation patterns is important for a balloon pilot because balloons are maneuvered solely through interaction with the different layers of wind and current. By using knowledge of the Coriolis force, pressure gradient force, and surface friction, it is possible to predict with a high degree of accuracy the potential track over the countryside and land at a predetermined point. This skill is the mark of a competent, safety conscious balloon pilot.

Atmospheric Circulation

Three forces cause the wind to move as it does: the Coriolis force, the pressure gradient force, and surface friction. All three forces work together at the same time.

As defined earlier, atmospheric circulation is the movement of air around the surface of the Earth caused by the uneven heating of the Earth's surface that upsets the equilibrium of the atmosphere, creates changes in air movement, and affects atmospheric pressure. Because the Earth has a curved surface that rotates on a tilted axis while orbiting the sun, the equatorial regions of the Earth receive a greater amount of heat from the sun than the polar regions. The amount of sun heating the Earth depends upon the time of day, time of year, and the latitude of the specific region. All of these factors affect the length of time and the angle at which sunlight strikes the surface.

In general atmospheric circulation theory, areas of low pressure exist over the equatorial regions, and areas of high pressure exist over the polar regions due to a difference in temperature. Solar heating causes air to become less dense and rise in equatorial areas. The resulting low pressure allows the high pressure air at the poles to flow along the planet's surface toward the equator. As the warm air flows toward the poles, it cools, becoming more dense, and sinks back toward the surface. *[Figure 4-9]* This pattern of air circulation is correct in theory, but the circulation of air is modified by other forces.

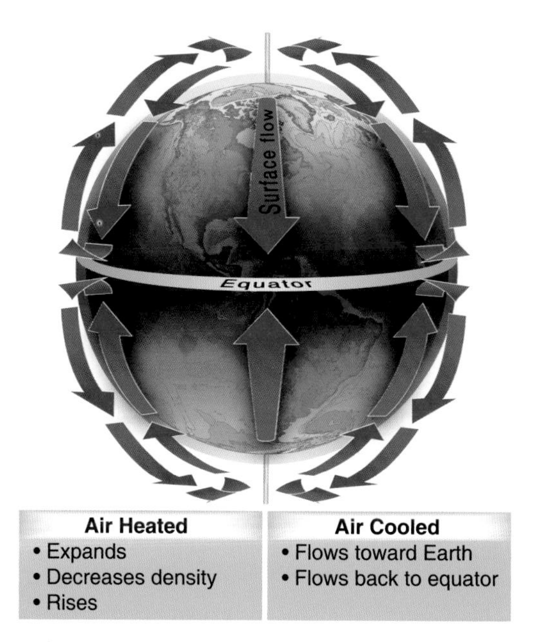

Figure 4-9. *General circulation theory.*

The speed of the Earth's rotation causes the general flow to break up into three distinct cells in each hemisphere. *[Figure 4-10]* In the Northern Hemisphere, the warm air at the equator rises upward from the surface, travels northward, and is deflected eastward by the rotation of the Earth. By the time it has traveled one-third of the distance from the equator to the North Pole, it is no longer moving northward, but eastward. This air cools and sinks in a belt-like area at about 30° latitude, creating an area of high pressure as it sinks toward the surface. Then, it flows southward along the surface back toward the equator. Coriolis force bends the flow to the right, thus creating the northeasterly trade winds that prevail from 30° latitude to the equator. Similar forces create circulation cells that encircle the Earth between 30° and 60° latitude, and between 60° and the poles. This circulation pattern results in the prevailing westerly winds in the conterminous United States.

Figure 4-10. *Three cell circulation pattern caused by the rotation of the Earth.*

Circulation patterns are further complicated by seasonal changes, differences between the surfaces of continents and oceans, and other factors such as frictional forces caused by the topography of the Earth's surface which modify the movement of the air in the atmosphere. For example, within 2,000 feet of the ground, the friction between the surface and the atmosphere slows the moving air. The wind is diverted from its path because the frictional force reduces the Coriolis force. Thus, the wind direction at the surface varies somewhat from the wind direction just a few thousand feet above the Earth.

Coriolis Force

The Coriolis force is not perceptible to humans as they walk around because humans move slowly and travel relatively short distances compared to the size and rotation rate of the Earth. However, the Coriolis force significantly affects bodies that move over great distances, such as an air mass or body of water.

The Coriolis force deflects air to the right in the Northern Hemisphere, causing it to follow a curved path instead of a straight line. The amount of deflection differs depending on the latitude. It is greatest at the poles, and diminishes to zero at the equator. The magnitude of Coriolis force also differs with the speed of the moving body—the faster the speed, the greater the deviation. In the Northern Hemisphere, the rotation of the Earth deflects moving air to the right and changes the general circulation pattern of the air.

Pertinent facts about the Coriolis force:

- The Coriolis force deflection is perpendicular to the flow of air.
- The Coriolis force will deflect air to the right in the Northern Hemisphere, and to the left in the Southern Hemisphere.
- The Coriolis force is strongest at the Poles and decreases to zero at the Equator.
- The Coriolis force is zero with calm winds and increases in magnitude as wind speed increases.
- Coriolis force, in combination with other forces involved, will determine the different circulation patterns over the Earth.

Pressure Gradient

Pressure gradient is the difference in pressure between high and low pressure areas. It is the rate of change in pressure in a direction perpendicular, or across the isobars. Wind speed is directly proportional to the pressure gradient. This means the strongest winds are in the areas where the pressure gradient is the greatest. Since pressure applied to a fluid is exerted equally in all directions throughout the fluid, a pressure gradient exists in the horizontal (along the surface), as well as in the vertical (with altitude) plane in the atmosphere. *[Figure 4-11]*

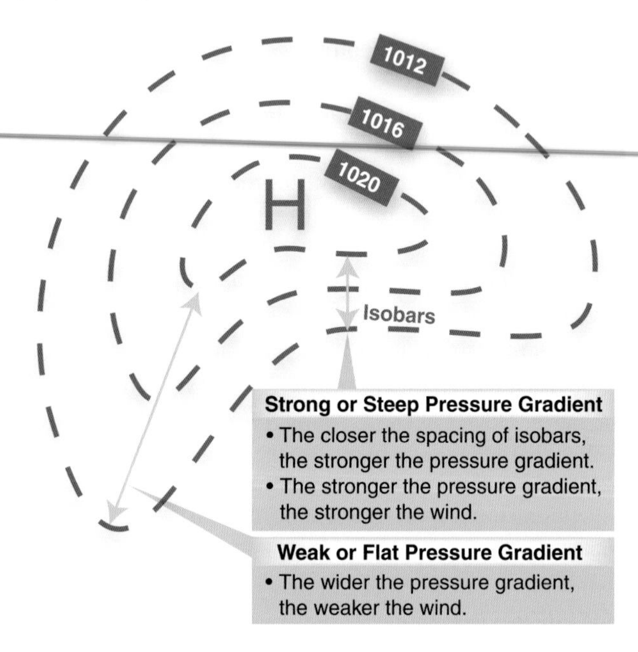

Figure 4-11. *Principles of pressure gradients.*

The horizontal pressure gradient is steep or strong when the isobars determining the pressure gradient are close together. It is flat or weak when the isobars are far apart. If isobars are considered as depicting atmospheric topography, a high pressure system represents a hill of air, and a low pressure system represents a valley of air. The vertical pressure gradient always indicates a decrease in pressure with altitude, but the rate of pressure decrease (gradient) varies directly with changes in air density with altitude. The vertical cross section through a high and a low depicts the surface pressure gradient.

The pressure gradient force is a force that tries to equalize pressure differences. This is the force that causes high pressure to push air toward low pressure. Thus, air would flow from high to low pressure if the pressure gradient force was the only force acting on it.

Surface Friction

Friction is the third component that determines the flow of wind. Because the surface of the Earth is rough, it not only slows the wind down, it also causes the diverging winds from highs and converging winds near lows. Since the Coriolis force varies with the speed of the wind, a reduction in the wind speed by friction means a reduction of the Coriolis force. This results in a momentary disruption of the balance. When the new balance (including friction) is reached, the air flows at an angle across the isobars from high pressure to low pressure. This angle varies from 10° over the ocean to more than 45° over rugged terrain. Frictional effects on the air are greatest near the ground, but the effects are also carried aloft by turbulence. Surface friction is effective in slowing the wind up to an average altitude of 2,000 feet above the ground. Above this level, the effect of friction decreases rapidly and may be considered negligible. Air above 2,000 feet above the ground normally flows parallel to the isobars. *[Figures 4-12* and *4-13]*

Wind Patterns

Since air always seeks out lower pressure, it flows from areas of high pressure into those of low pressure. In the Northern Hemisphere, this flow of air from areas of high to low pressure is deflected to the right and produces a clockwise circulation around an area of high pressure known as anti-cyclonic circulation. The opposite is true of low pressure

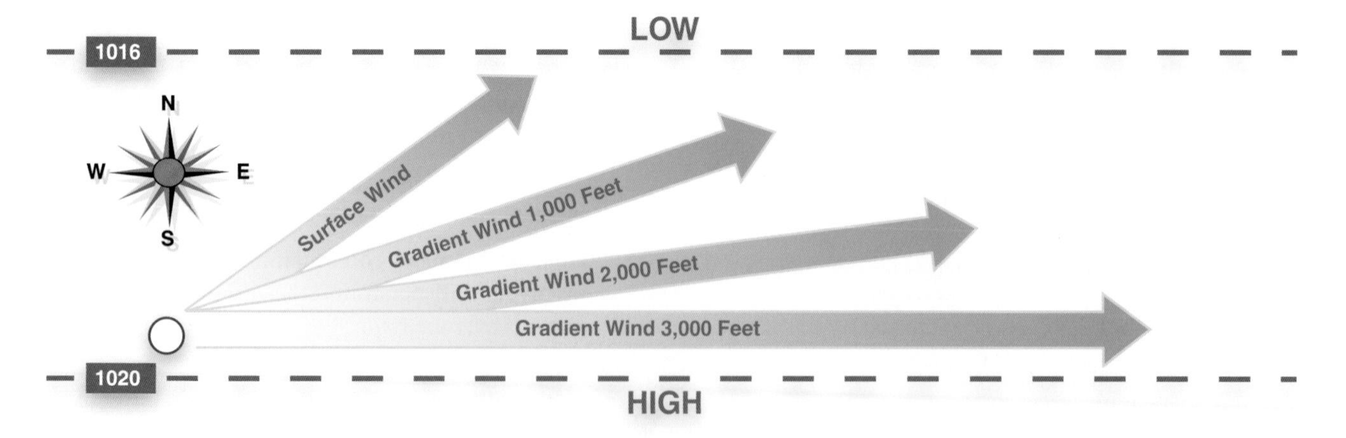

Figure 4-12. *Examples of variations of wind direction with height.*

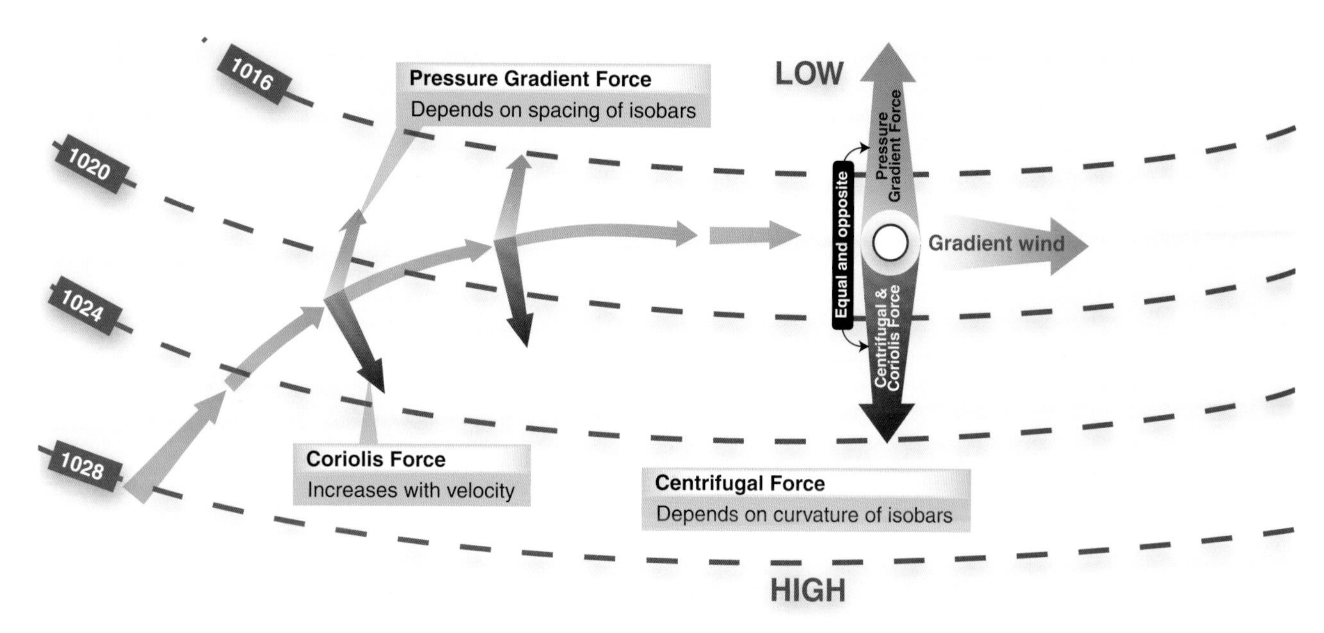

Figure 4-13. *Gradient winds.*

areas; the air flows toward a low and is deflected to create a counter-clockwise or cyclonic circulation.

High pressure systems are generally areas of dry, stable, descending air. Good weather is typically associated with high pressure systems for this reason. Conversely, air flows into a low pressure area to replace rising air. This air tends to be unstable, and usually brings increasing cloudiness and precipitation. Thus, bad weather is commonly associated with areas of low pressure.

Convective Currents

Convection currents refer to the upward moving portion of a convection circulation, such as a thermal or the updraft in cumulus clouds. The uneven heating of the air, due to different surfaces radiating heat in varying amounts, create small areas of local circulation. For example, plowed ground, rocks, sand, and barren land give off a large amount of heat, while water, trees, and other areas of vegetation tend to absorb and retain heat. Convective currents cause the bumpy, turbulent air sometimes experienced when flying at lower altitudes during warmer weather. On a low altitude flight over varying surfaces, updrafts are likely to occur over pavement or barren places, and downdrafts often occur over water or expansive areas of vegetation like a group of trees. Typically, these turbulent conditions can be avoided by flying at higher altitudes.

Convective currents are particularly noticeable in areas with a land mass directly adjacent to a large body of water, such as an ocean, large lake, or other appreciable area of water. During the day, land heats faster than water, so the air over the land becomes warmer and less dense. It rises and is replaced by cooler, denser air flowing in from over the water. This causes an onshore wind, called a sea breeze. Conversely, at night land cools faster than water, as does the corresponding air. In this case, the warmer air over the water rises and is replaced by the cooler, denser air from the land, creating an offshore wind called a land breeze. This reverses the local wind circulation pattern. Convective currents can occur anywhere there is an uneven heating of the Earth's surface. *[Figure 4-14]*

Convection currents close to the ground can affect a pilot's ability to control the balloon. On final approach, for example, the rising air from terrain devoid of vegetation sometimes produces a ballooning effect that can cause a pilot to overshoot the intended landing spot. On the other hand, an approach over a large body of water or an area of thick vegetation tends to create a sinking effect that can cause an unwary pilot to land short of the intended landing spot. This could prove particularly hazardous to a balloon landing in a small, confined area, as the "undershoot" of the approach could potentially put the balloon into the trees or power lines.

The Jet Stream

The jet stream refers to relatively strong winds concentrated in a narrow stream in the atmosphere. These winds are normally horizontal, high altitude winds. The position and orientation of jet streams vary from day to day. General weather patterns (hot/cold, wet/dry) are related closely to the position, strength, and orientation of the jet stream (or jet streams). A jet stream at low levels is known as a low level jet stream. Since it is of interest primarily to high level flight, further discussion is not necessary.

Figure 4-14. *Land-sea breezes.*

Local and Small-Scale Winds

Gradient Winds

Pressure gradients initiate the movement of air and as soon as the air acquires velocity, the Coriolis force deflects it to the right in the Northern Hemisphere. As the speed of the air along the isobars increases, the Coriolis force becomes equal and opposite to the pressure gradient force. After a period of time, the air moves directly parallel to the curved isobars if there is no frictional drag with the surface. The air no longer moves toward lower pressure because the pressure gradient force is completely neutralized by the Coriolis force and the centrifugal force.

Orographic Winds

The term "orographic" has multiple meanings, when placed in the context of weather phenomena. In a general sense, according to the American Meteorological Society, wind flows that are caused, affected, or influenced by mountains may be said to be orographic winds flows. The term has come to mean any winds that are affected by terrain, not just mountains; this definition is probably the most frequently used, when discussing balloon flight.

As a specific term, "orographic lifting" is defined as an ascending air flow caused by mountains. The mechanisms that produce the orographic lifting fall into two broad categories:

1. The upward deflection of horizontal large-scale air flow by the terrain acting as an obstacle or barrier, or
2. The daytime heating of mountain surfaces to produce an anabatic flow (see below) along the slopes and updrafts in the vicinity of mountain peaks.

This definition, while strictly referring only to lifting by mountains, is sometimes extended to include the effects of hills or long sloping terrain. When sufficient moisture is present in the rising air, Orographic fog or clouds may form.

Anabatic Winds

Anabatic winds are those that blow up a steep slope or mountain side. It is sometimes referred to as an upslope flow. These winds typically occur during the daytime in calm, sunny weather. A hill or mountaintop may be warmed by the sun, which in turn heats the air just above it. As that air rises through convection, it creates a low pressure region, into which the air at the bottom of the slope flows, and causes winds.

Katabatic Winds

Katabatic winds are the reverse of anabatic winds; that is, they flow down slope, and most frequently at night. They are created by the effect of the air near the ground losing heat thru radiational cooling at a faster rate than air at a similar altitude over the surrounding land mass.

Clouds

Clouds are weather signposts in the sky. They provide the balloon pilot with visible evidence of the atmospheric motion, water content, and degree of stability. In this sense, clouds are of significant importance to the aeronaut. However, when they become too numerous or widespread, form at low levels, or show extensive vertical development, they present weather hazards to ballooning.

Clouds are visible condensed moisture, consisting of droplets of water or crystals of ice. They are supported and transported by air movements as slow as one-tenth of a mile per hour. Cloud formation is the direct result of saturation producing processes which take place in the atmosphere. A pilot should be able to identify cloud formations that are associated with weather hazards. Knowledge of cloud types will also assist the pilot in interpreting weather conditions from weather reports and existing weather.

Cloud Formation

Clouds are often indicative of future weather. For clouds to form, there must be adequate water vapor and condensation nuclei (miniscule particles of matter like dust, salt, and smoke), as well as a method by which the air can be cooled. When the air cools and reaches its saturation point, the invisible water vapor changes into a visible state. Through the processes of sublimation and condensation, moisture condenses or sublimates onto condensation nuclei. The nuclei are important because they provide a means for the moisture to change from one state to another.

Cloud type is determined by its height, shape, and behavior. They are classified according to the height of their bases as low, middle, or high clouds, as well as clouds with vertical development. The International Cloud Classification is designed to provide a uniform cloud classification system. *[Figure 4-15]* Within this system, cloud types are usually divided into four major groups and further classified in terms of their forms and appearance.

International Cloud Classification
Abbreviations and Weather Map Symbols

Base Altitude	Cloud Type	Abbreviation	Symbol
Bases of high clouds usually above 18,000 feet	Cirrus	Ci	
	Cirrocumulus	Cc	
	Cirrostratus	Cs	
18,000 FT			
Bases of middle clouds range from 6,500 feet to 18,000 feet	Altocumulus	Ac	
	Altostratus	As	
6,500 FT			
Bases of low clouds range from surface to 6,500 feet	*Cumulus	Cu	
	*Cumulonimbus	Cb	
	Nimbostratus	Ns	
	Stratocumulus	Sc	
	Stratus	St	
Surface			

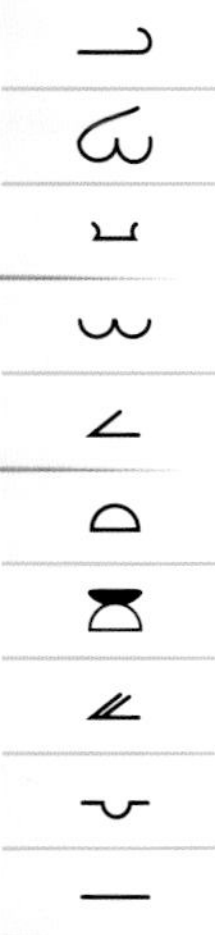

* Cumulus and cumulonimbus are clouds with vertical development. Their bases are usually below 6,500 feet, but may be slightly higher. The tops of the cumulonimbus some times exceed 60,000 feet.

Figure 4-15. *Cloud classification per international agreement.*

The four major groups are:

- Low clouds,
- Middle clouds,
- High clouds, and
- Clouds with vertical development.

Cloud classification can be further broken down into specific cloud types according to the outward appearance and cloud composition. Knowing these terms can help identify visible clouds. The following is a list of cloud classifications:

- Cumulus—heaped or piled clouds
- Stratus—formed in layers
- Cirrus—ringlets, fibrous clouds, also high-level clouds above 20,000 feet
- Castellanus—common base with separate vertical development, castle-like
- Lenticularus—lens shaped, formed over mountains in strong winds
- Nimbus—rain-bearing clouds
- Fracto—ragged or broken
- Alto—meaning high, also middle-level clouds existing at 5,000 to 20,000 feet

Figure 4-16. *Stratus clouds.*

Figure 4-17. *Stratocumulus clouds.*

Low clouds are those that form near the Earth's surface. The low cloud group consists of stratus and stratocumulus clouds. *[Figures 4-16* and *4-17]* Clouds in this family create low ceilings, hamper visibility, and can change rapidly. Because of this, they influence flight planning and can make visual flight rules (VFR) flight impossible. The bases of these clouds can start near the surface, with the top extending to 6,500 feet or more above the terrain. Low clouds are of great importance to the balloon pilot, as they can create low ceilings and poor visibility. The heights of the cloud bases may change rapidly. If low clouds form below 50 feet, they are classified as fog, and may completely blanket landmarks and landing fields.

Middle clouds form around 6,500 feet above ground level (AGL) and extend up to 20,000 feet AGL. They are composed of water, ice crystals, and supercooled water droplets. The middle cloud group consists of altocumulus *[Figure 4-18]*, altostratus, and nimbostratus *[Figure 4-19]* clouds. Altocumulus clouds, which usually form when altostratus clouds are breaking apart, also may contain light turbulence and icing. Altostratus clouds can produce turbulence and may contain moderate icing. The altocumulus has many variations in appearance and formation, whereas the altostratus varies mostly in thickness, from very thin to several thousand feet. Bases of the middle clouds start as low as 6,500 feet and tops can range as high as 20,000 feet above the terrain. These clouds may be composed of ice crystals or water droplets (which may be supercooled). Altocumulus rarely produces precipitation, but altostratus usually indicates the proximity of unfavorable flying weather and precipitation.

Figure 4-18. *Altocumulus clouds.*

High clouds form above 20,000 feet AGL and usually form only in stable air. The high cloud group consists of cirrus, cirrocumulus, and cirrostratus clouds. The mean base level of these three cloud types starts at 18,000 feet or higher above terrain. Cirrus clouds *[Figure 4-20]* may give indications of approaching weather changes. Cirriform clouds are composed of ice crystals, are generally thin, and the outline of the sun or moon may sometimes be seen through them, producing a halo or corona effect. High clouds are generally of no interest to the balloon pilot, other than they may indicate future conditions.

Figure 4-19. *Nimbostratus clouds.*

Figure 4-20. *Cirrus clouds.*

Clouds with extensive vertical development are cumulus clouds that build vertically into towering cumulus or cumulonimbus clouds, often developing into thunderstorms. The bases of these clouds form in the low to middle cloud region, but can extend into high altitude cloud levels. Towering cumulus clouds indicate areas of instability in the atmosphere, and the air around and inside them is turbulent. These clouds generally have their bases below 6,500 feet above the terrain and tops sometimes extend above 60,000 feet. Clouds with extensive vertical development are caused by lifting action, such as convective currents, orographic lift, or frontal lift.

Scattered cumulus or isolated cumulonimbus clouds seldom present a flight problem, since these clouds can usually be circumnavigated without difficulty. However, these clouds may rapidly develop in groups or lines of cumulonimbus. They may also become embedded and hidden in stratiform clouds, resulting in hazardous flight conditions.

Within the high, middle, and low cloud groups are two main subdivisions. These are:

- Clouds formed when localized vertical currents carry moist air upward to the condensation level. These vertical development clouds are characterized by their lumpy or billowy appearance, and are designated cumuliform type clouds, meaning "accumulation" or "heap." Turbulent flying conditions usually exist in, below, around, and above cumuliform clouds.
- Clouds formed when complete layers of air are cooled until condensation takes place. These clouds are stratiform type clouds, meaning "layered out," since they lie mostly in horizontal layers or sheets. Flight in stratiform cloud conditions is usually smooth.

In addition to the two main subdivisions discussed above, is the word nimbus, meaning "rain cloud." These clouds normally produce heavy precipitation, either liquid or solid. For example, a stratiform cloud producing precipitation is referred to as nimbostratus, and a heavy, swelling cumulus cloud that has grown into a thunderstorm is referred to as cumulonimbus.

Cumulonimbus clouds contain large amounts of moisture and unstable air, and usually produce hazardous weather phenomena such as lightning, hail, tornadoes, gusty winds, and wind shear. These extensive vertical clouds can be obscured by other cloud formations and are not always visible from the ground or while in flight. When this happens, these clouds are said to be embedded, hence the term, embedded thunderstorms.

To pilots, the cumulonimbus cloud is perhaps the most dangerous cloud type. It appears individually or in groups and is known as either an air mass or orographic thunderstorm. Heating of the air near the Earth's surface creates an air mass thunderstorm; the upslope motion of air in the mountainous regions causes orographic thunderstorms. Cumulonimbus clouds that form in a continuous line are nonfrontal bands of thunderstorms or squall lines.

Knowledge of principal cloud types and the factors that affect them helps the pilot visualize expected weather conditions, and to recognize potential weather hazards.

Ceilings and Visibilities

Ceilings and visibilities have an important role in the classification of sky conditions, and are critical for the definition of flight restrictions. It is necessary to define these terms to make those distinctions clear for the balloon pilot.

Ceiling

For aviation purposes, a ceiling is the lowest layer of clouds reported as being broken or overcast, or the vertical visibility into an obscuration like fog or haze.

Observations are made using the concept of the "celestial dome," the hemisphere of sky which can be seen from a specific point on the ground. Cloud coverage is reported as the total cloud cover at and below a specific layer, and is reported in one-eighth increments (octals). A ceiling is reported as broken when five-eighths to seven-eighths of the sky is covered with clouds. Overcast means the entire sky is covered with clouds.

Current ceiling information is reported by the aviation routine weather report (METAR) and automated weather stations of various types. Ceilings are reported in height AGL.

Visibility

Closely related to cloud cover and reported ceilings is visibility information. Visibility refers to the greatest horizontal distance at which prominent objects can be viewed with the naked eye. Visibilities reported in standard weather reports are horizontal surface visibilities and are generally considered linear. Predominant visibility is the greatest horizontal distance over which objects can be seen and identified over at least half of the horizon. In the United States, prevailing visibilities are reported in statute miles and portions thereof.

Since prevailing visibility is used for reporting purposes, three miles visibility does not mean that a pilot must have one and one half miles visibility in front of and behind the balloon, but that the predominant visibility in most quadrants must be three miles.

Current visibility is reported in METAR and other aviation weather reports, as well as automated weather stations. Visibility information is available during a preflight weather briefing.

Temperature/Dew Point Relationship

The relationship between dew point and temperature defines the concept of relative humidity. The dew point, given in degrees, is the temperature at which the air can hold no more moisture. When the temperature of the air is reduced to the dew point, the air is completely saturated and moisture begins to condense out of the air in the form of fog, dew, frost, clouds, rain, hail, or snow.

As moist, unstable air rises, clouds often form at the altitude where temperature and dew point reach the same value. When lifted, unsaturated air cools at a rate of 5.4 °F per 1,000 feet and the dew point temperature decreases at a rate of 1 °F per 1,000 feet. This results in a convergence of temperature and dew point at a rate of 4.4 °F. A pilot can determine the height of the cloud base by applying the convergence rate to the reported temperature and dew point in the following manner:

Temperature (T) = 85 °F
Dew point (DP) = 71 °F
Convergence Rate (CR) = 4.4°
T – DP = Temperature Dew Point Spread (TDS)
TDS ÷ CR = X
X × 1,000 feet = height of cloud base AGL

Example:

85 °F – 71 °F = 14 °F
14 °F ÷ 4.4 °F = 3.18
3.18 × 1,000 = 3,180 feet AGL
The height of the cloud base is 3,180 feet AGL.

Explanation:
With an outside air temperature (OAT) of 85 °F at the surface, and dew point at the surface of 71 °F, the spread is 14 °F. Divide the temperature dew point spread by the convergence rate of 4.4 °F, and multiply by 1,000 to determine the approximate height of the cloud base.

This relationship is useful in determining the height of the overlying cloud base when completing preflight preparations.

Fog

Fog is a cloud that begins within 50 feet of the surface. It typically occurs when the temperature of air near the ground is cooled to the air's dew point. At this point, water vapor in the air condenses and becomes visible in the form of fog. Fog is classified according to the manner in which it forms and is dependent upon the current temperature and the amount of water vapor in the air.

Fog is composed of minute droplets of water or ice crystals suspended in the atmosphere with no visible downward motion. It is one of the most common and persistent weather hazards encountered by balloonists. Similar to stratus clouds, the base of fog is at the Earth's surface while the base of a cloud is at least 50 feet above the surface. Fog may be distinguished from haze by its dampness and gray color. It is hazardous during takeoffs and landings, as well as the inflight process, because it restricts surface visibility. Knowledge of fog formation and dissipation processes, as well as types of fog help the balloon pilot plan a flight more accurately.

Fog Formation

Since neither condensation nor sublimation occurs unless the relative humidity is near 100 percent, a high relative humidity is of prime importance in the formation of fog. The natural conditions which bring about a high relative humidity (saturation) are also fog-producing processes, such as the evaporation of additional moisture into the air or cooling of the air to its dew point temperature. A high relative humidity can be estimated, from hourly sequence reports, by determining the spread (difference in degrees) between the temperature and dew point. Fog rarely occurs when the spread is more than 2.2 °C. It is most frequent when the spread is less than 1.1 °C.

A light wind is generally favorable for fog formation. It causes a gentle mixing action, which spreads surface cooling through a deeper layer of air and increases the thickness of the fog.

Although most regions of the Earth have sufficient condensation nuclei to permit fog formation, the amount of smoke particles and sulphur compounds in the vicinity of industrial areas is pronounced. In these regions, persistent fog may occur with above average temperature-dew point spreads.

Fog tends to dissipate when the relative humidity decreases. During this decrease, the water droplets evaporate or ice crystals undergo sublimation, and the moisture is no longer visible. Either strong winds or heating processes may cause the decrease in relative humidity.

Some Fog Types and Characteristics

Radiation Fog

Radiation fog forms after the Earth has radiated back to the atmosphere the heat gained during daylight hours. By early morning, the temperature at the surface may drop more than 11 °C. Since the dew point temperature (moisture content) of the air normally changes only a few degrees during the night, the temperature-dew point spread will decrease as the air is cooled by contact with the cold surface. If the radiational cooling is sufficient, and other conditions are favorable, radiation fog will form. Radiation fog is most likely when the:

- Sky is clear (maximum radiational cooling).
- Moisture content is high (narrow temperature-dew point spread).
- Wind is light (less than 7 knots).

Advection Fog

The movement of warm moist air over a colder surface creates advection fog which is common along coastal regions where the temperature of the land surface and the water surface contrasts. The southeastern area of the United States provides ideal conditions for advection fog formation during the winter months. If air flows (advection) from the Gulf of Mexico or the Atlantic Ocean over the colder continent, this warm air is cooled by contact with the cold ground. If the temperature of the air is lowered to the dew point temperature, fog will form. Advection fog, forming under these conditions, may extend over larger areas of the nation east of the Rockies. It may persist throughout the day or night until replaced by a drier air mass.

If advection fog forms over water, it is often referred to as sea fog. Cold ocean currents, such as those off the coast of California, may cool and saturate moist air coming from the warmer areas of the open sea. Sea fog is often dense offshore, as well as onshore.

As advection fog moves inland during the winter, the colder land surface often causes sufficient contact cooling to keep the air saturated. The fog may then persist during the day or with a wind speed of 10 to 15 knots.

Valley Fog

During the evening hours, cold dense air will drain from areas of higher elevation into low areas of valleys. As the cool air accumulates in the valleys, the air temperature may decrease to the dew point temperature, causing a dense formation of valley fog. While higher elevations may often remain clear throughout the night, the ceiling and visibility become restricted in the valley.

Evaporation Fog

Fog formed by the addition of moisture to the air is called evaporation fog. The major types of evaporation fog are frontal fog and steam fog.

Frontal fog is normally associated with slow-moving winter frontal systems. Frontal fog forms when liquid precipitation, falling from the maritime tropical air above the frontal surface, evaporates in the polar air below the frontal surface. Evaporation from the falling drops may add sufficient water vapor to the cold air to raise the dew point temperature to the temperature of the air. The cold air will then be saturated, and frontal fog will form. Frontal fog is common with active warm fronts during all seasons. It occurs ahead of the surface front in an area approximately 100 miles wide. It is, therefore, frequently mixed with intermittent rain or drizzle. When fog forms ahead of the warm front, it is called prefrontal fog. A similar fog formation may occur in the polar air along a stationary front. Occasionally, a slow-moving winter cold front with light wind may generate fog. This fog forms in the polar air behind the surface front and is known as postfrontal fog.

Steam fog forms when cold stable air flows over a nonfrozen water surface that is several degrees warmer than the air. The intense evaporation of moisture into the cold air saturates the air and produces fog. Conditions favorable for steam fog are common over lakes and rivers in the fall and over the ocean in the winter when an offshore wind is blowing.

Atmospheric Stability and Instability

A stable atmosphere resists upward or downward movement, and small vertical disturbances dampen out and disappear. An unstable atmosphere allows an upward or downward disturbance to grow into a vertical or convective current allowing small vertical air movements to become larger, resulting in turbulent airflow and convective activity. Instability can lead to significant turbulence, extensive vertical clouds, and severe weather.

Rising air expands and cools due to the decrease in air pressure as altitude increases. The opposite is true of descending air; as atmospheric pressure increases, the temperature of descending air increases as it is compressed. This adiabatic process (heating or cooling) takes place in all upward and downward moving air.

When air rises into an area of lower pressure, it expands to a larger volume. As the molecules of air expand, the temperature of the air lowers. As a result, when a parcel of air rises, pressure decreases, volume increases, and temperature decreases. When air descends, the opposite is true.

Since water vapor is lighter than air, moisture decreases air density, causing it to rise. Conversely, as moisture decreases, air becomes denser and tends to sink. Since moist air cools at a slower rate, it is generally less stable than dry air since the moist air must rise higher before its temperature cools to that of the surrounding air. The dry adiabatic lapse rate (unsaturated air) is 3 °C (5.4 °F) per 1,000 feet. The moist adiabatic lapse rate varies from 1.1 °C to 2.8 °C (2 °F to 5 °F) per 1,000 feet. The combination of moisture and temperature determine the stability of the air and the resulting weather. Cool, dry air is very stable and resists vertical movement, which leads to good and generally clear weather. The greatest instability occurs when the air is moist and warm, as it is in the tropical regions in the summer. Typically, thunderstorms appear on a daily basis in these regions due to the instability of the surrounding air.

The normal flow of air tends to be horizontal. If this flow is disturbed, a stable atmosphere will resist any upward or downward displacement. It will tend to return quickly to normal horizontal flow. An unstable atmosphere, on the other hand, will allow these upward and downward disturbances to grow, resulting in rough (turbulent) air. An example is the towering thunderstorm which grows as a result of large and intensive vertical movement or air. It climaxes in lightning, thunder, and heavy precipitation, sometimes including hail.

Atmospheric resistance to vertical motion, called stability, depends upon the vertical distribution of the air's weight at a particular time. The weight varies with air temperature and moisture content. In comparing two parcels of air, warmer air is lighter than colder air, and moist air is lighter than dry air. If air is relatively warmer or moister that its surroundings, it is forced to rise and would be unstable. If the air is colder or dryer than its surroundings, it will sink until it reaches its equilibrium, and would be stable. The atmosphere can be at equilibrium only when light air is above heavier air.

Temperature has a significant effect on the stability or instability of the air mass. Air heated near the Earth's surface on a hot summer day will rise. The speed and vertical extent of its travel depends on the temperature distribution of the atmosphere. Vertical air currents, resulting from the rise of air, can vary from the severe downdraft and compensating downdraft associated with thunderstorms to the closely spaced upward and downward bumps that are felt on warm days when flying at low levels. Since the temperature of air is an indication of its density, a comparison of temperatures from one level to another can approximate the degree of the atmosphere's stability, or how much it will tend to resist vertical motion.

Types of Stability

The five types of atmospheric stability are:

- Absolute stability
- Absolute instability
- Conditional instability
- Neutral stability
- Convective stability

Absolute stability occurs when the actual lapse rate in a layer of air is less than the moist adiabatic lapse rate; that air is absolutely stable regardless of the amount of moisture it contains. A parcel of absolutely stable air which is lifted becomes cooler than the surrounding air and sinks back to its original position as soon as the lifting force is removed.

Similarly, if forced to descend, it becomes warmer than the surrounding air; like a cork in water, it rises to its original position upon removal of the outside force.

Absolute instability exists when the actual lapse rate in a layer of air is greater than the dry adiabatic lapse rate; that air is absolutely unstable regardless of the amount of moisture it contains. A parcel of air lifted even slightly will immediately be warmer than its surroundings, and, as with a hot air balloon, will be forced to rise.

Conditional instability exists when the temperature lapse rate of the air involved lies between the moist and dry adiabatic rates of cooling. Before the displaced air actually becomes unstable, it must be lifted to a point where it is warmer than the surrounding air. When this point has been reached, the relatively warmer air continues to rise freely until, at some higher altitude, its temperature has cooled to the temperature of the surrounding air. In the instability process, numerous variables tend to modify the air. One of the most important of these variables is the process called entrainment. In this process, air adjacent to the cumulus or mature thunderstorm is drawn into the cloud primarily by strong updrafts within the cloud. The entrained air modifies the temperature of the air within the cloud as the two become mixed.

Neutrally stable air is air with the same temperature, and there is no parcel to rise or descend. For example, the surface area in contact with that air is of the same temperature.

The term convective instability refers to a condition in which the air becomes unstable after lifting. From a physical standpoint, it closely resembles a conditionally unstable air mass, but has the mechanical lifting of thermal activity impacting on the overall characteristics of the air.

Effects of Stable and Unstable Air

The degree of stability of the atmosphere helps to determine the type of clouds formed, if any. For example, if very stable air is forced to ascend a mountain slope, clouds will be layer-like with little vertical development and little or no turbulence. Unstable air, if forced to ascend the slope, would cause considerable vertical development and turbulence in the cumulus-type clouds.

If air is subsiding (sinking), the heat of compression frequently causes an inversion of temperature which increases the stability of the subsiding air. When this occurs, as in winter high pressure systems, a surface inversion formed by radiational cooling is sometimes already present. The subsidence-produced inversion, in this case, will intensify the surface inversion, placing a strong "lid" above smoke and haze. Poor visibility in the lower levels of the atmosphere results, especially near industrial areas. Such conditions frequently persist for days, notably in the Great Basin region of the western United States.

Weather Hazards

Turbulence

Turbulence is the irregular motion of the atmosphere as indicated by gusts and lulls in the wind. Since turbulence is associated with many different weather situations, knowledge of its causes and its behavior will help a balloon pilot avoid or minimize its effects.

Turbulence can be divided into four categories according to the specific causes:

- Thermal—caused by localized convective currents due to surface heating or unstable lapse rates and cold air moving over warmer ground or water
- Mechanical—resulting from wind flowing over irregular terrain or obstructions
- Frontal—resulting from the local lifting of warm air by cold air masses, or the abrupt wind shift (shear) associated with most cold fronts
- Wind shear—marked gradient in wind speed and/or direction due to general vibrations in the temperature and pressure fields aloft

Two or more of the above causative factors often work together. In addition, turbulence is produced by manmade phenomena.

Thermal

A thermal is simply the updraft in a small-scale convective current. Convective currents (vertical or horizontal air movements) develop in air, which is heated by contact with a warm surface. This heating from below occurs when either cold air is advected (moved horizontally) over a warmer surface or the ground is strongly heated by solar radiation.

The strength of convective currents depends in part on the extent to which the Earth's surface has been heated, which depends upon the nature of the surface. Barren surfaces, such as sandy or rocky wasteland and plowed fields, are heated more rapidly than surfaces covered in vegetation. Thus, barren surfaces generally cause stronger convection currents. In comparison, water surfaces are heated more slowly.

When air is very dry, convective currents may be present although convective-type clouds (cumulus) are absent. The general upper limits of the convective currents are often marked by the tops of cumulus clouds, which form in them when the air is moist, or by haze lines. However, turbulence may extend beyond this boundary. Varying types of surfaces, and the resultant thermal conditions, can affect a balloon to a considerable extent.

The balloon pilot caught in a thermal will recognize the condition by an increase in altitude without application of heat from the balloon's heater. This ascent can be rapid and may exceed the maximum rate of climb limitations in the balloon's flight manual. Since the air mass is also rising with the balloon, there is no significant pressure against the top of the balloon. Thus, the top cap will not be pushed open (commonly referred to as "floating the top").

Depending on their size, some thermals may have a rotative motion similar to a small low pressure system. This motion draws the balloon in and forces it to fly in an uncontrolled circle. For balloons caught in a thermal, remember the adage "altitude is your friend." First, the pilot should insure there is sufficient altitude to clear potential obstacles. Second, maintain the temperature in the balloon appropriate for level flight. Many pilots attempt to descend immediately, but this may put the balloon, as well as the passengers, at risk of an uncontrolled descent with possible injury. Most thermals have a short lift span. In almost all cases, the thermal will "spit" the balloon out the top after a short time, and the pilot may descend and land as necessary.

Figure 4-21. *Surface obstructions cause eddies and other irregular air movements.*

Mechanical

When the air near the surface of the Earth flows over obstructions, such as irregular terrain, (bluffs, hills, mountains) and buildings, the normal horizontal wind flow is disturbed. As a result, it is transformed into eddies or other irregular air movements. *Figure 4-21* shows how the buildings or other obstructions near a launch site or landing field can cause turbulence.

The strength and magnitude of mechanical turbulence depends on:

- The speed of the wind,
- The nature of the obstruction,
- The stability of the air, and
- The angle at which the wind moves over the obstacle.

Stability seems to be the most important factor in determining the strength and vertical extent of the mechanical turbulence.

Frontal

Frontal turbulence is cause by the lifting of warm air by a frontal surface, leading to instability and/or the mixing or shear between the warm and cold air masses. The vertical currents in the warm air are strongest when the warm air is moist and unstable. The most severe cases of frontal turbulence are generally associated with fast moving cold fronts. In these cases, mixing between the two air masses, as well as the differences in wind speed and/or direction add to the intensity of the turbulence.

Wind Shear

Wind shear is a relatively steep gradient in wind velocity along a given line of direction (either vertical or horizontal) and produces churning motions (eddies) which result in turbulence. The greater the change of wind speed and/or direction in the given direction, the greater the shear and associated turbulence.

Clear-air turbulence (CAT), or sudden severe turbulence that occurs in cloudless regions, is associated with wind shear, particularly between the core of a jet stream and the surrounding air. CAT is not limited to the vicinity of the jet stream and may occur in isolated regions of the atmosphere. For example, the turbulence in a mountain wave can also be classified as CAT because the identifying clouds in the wave do not necessarily have to occur for the turbulence to be present.

Sometimes during a climb or descent, a balloon encounters a narrow zone of wind shear with its accompanying turbulence at the top of a temperature inversion. These inversions occur anywhere from just above the surface to the tropopause.

Strong inversions near the ground are an extreme form of wind shear that adversely affect balloon takeoffs and landings. For example, a pocket of calm, cold air forms in a valley as a result of nighttime cooling, but the warmer air moving over it has not been affected appreciably. Due to the difference between the two bodies of air, a narrow layer of very turbulent air may form. A balloon climbing or descending through this zone will usually encounter considerable turbulence, as well as changes in lift.

Low Level Wind Shear

Wind shear is a sudden, drastic change in windspeed and/or direction over a very small area. Wind shear can subject a balloon to violent updrafts and downdrafts, as well as abrupt changes to the horizontal movement of the balloon. While wind shear can occur at any altitude, low level wind shear is especially hazardous due to the proximity of a balloon to the ground. Directional wind changes of 180° and speed changes of 50 knots or more are associated with low level wind shear. Low level wind shear is commonly associated with passing frontal systems, thunderstorms, and temperature inversions with strong upper level winds (greater than 25 knots).

Wind shear is hazardous to a balloon for several reasons. The rapid changes in wind direction and velocity changes the wind's relation to the balloon disrupting the normal flight attitude and performance of the balloon.

Obstructions and Wind

As mentioned earlier, obstructions on the ground affect the flow of wind and can be an unseen danger, causing yet another atmospheric hazard for pilots. For example, ground topography and large buildings can break up the flow of the wind and create wind gusts that change rapidly in direction and speed. Obstructions range from manmade structures, such as hangars, to large natural obstructions, such as mountains, bluffs, or canyons. A safe pilot is vigilant when flying in or out of launch or landing sites that have large buildings or natural obstructions located near them.

The intensity of the turbulence associated with ground obstructions depends on the size of the obstacle and the primary velocity of the wind. This can affect the takeoff and landing performance of a balloon, and can present a very serious hazard. During the landing phase of flight, a balloon may "drop in" due to the turbulent air and be too low to clear obstacles during the approach. Disrupted airflow often extends horizontally as much as ten times the height of the object, if the winds are in the eight to ten knot range. Balloon pilots should be aware of this, and make adjustments accordingly when attempting to launch next to an obstruction, or when landing just past one.

This same condition is even more noticeable when flying in mountainous regions. While the wind flows smoothly up the windward side of the mountain and the upward currents help to carry an aircraft over the peak of the mountain, the wind on the leeward side does not act in a similar manner. As the air flows down the leeward side of the mountain, the air follows the contour of the terrain and is increasingly turbulent. This tends to push an aircraft into the side of a mountain. The stronger the wind, the greater the downward pressure and turbulence become.

Due to the effect terrain has on the wind in valleys or canyons, downdrafts may be severe. Thus, a prudent balloonist is well advised to seek out another balloon pilot with mountain flying experience and get a mountain "checkout" before conducting a flight in or near mountainous terrain.

Mountain Wave

A mountain wave is the wavelike effect, characterized by updrafts and downdrafts, that occurs above and behind a mountain range when rapidly flowing air encounters the mountain range's steep front. The characteristics of a typical mountain wave are represented in *Figure 4-22* which illustrates how air flows with relative smoothness in its lifting component as the wave current moves along the windward side of the mountain. Wind speed gradually increases, reaching a maximum near the summit. On passing the crest, the flow breaks into a much more complicated pattern, with downdrafts predominating.

An indication of the possible intensities in the mountain wave is reflected in verified records of sustained downdrafts and updrafts in excess of 3,000 feet per minute (fpm). Turbulence in varying degrees can be expected, with particularly severe turbulence in the lower levels. Proceeding downwind 5 to 10 miles from the summit, the airflow begins to ascend as part of a definite wave pattern. Additional waves, generally less intense than the primary wave, may form downwind. This event is much like the series of ripples that form downstream from a rock submerged in a swiftly flowing river. The distance between successive waves (wavelength) usually ranges from 2 to 10 miles, depending on existing wind speed and atmospheric stability, although waves up to 20 miles apart have been reported.

Figure 4-22. *Characteristics of a mountain wave.*

Characteristic cloud forms peculiar to wave action provide the best means of identification. Lenticular clouds, formed by mountain waves, are smooth in contour. *[Figure 4-23]* These clouds may occur alone or in layers at heights above 20,000 feet MSL, and be quite ragged when the airflow at that level is turbulent. The roll cloud forms at a lower level, generally near the height of the mountain ridge. The cap cloud must always be avoided in flight because of turbulence, concealed mountain peaks, and strong downdrafts on the lee slope. The lenticulars, like the roll and cap clouds, are stationary. They are constantly forming on the windward side and dissipating on the lee side of the mountain wave.

Rotors

Rotors or eddies can also be found embedded in mountain waves. Formation of rotors can also occur as a result of down slope winds. Their formation usually occurs where wind speeds change in a wave or where friction slows the wind near the ground. These rotors are often experienced as gusts or wind shear. Clouds may also form within a rotor.

Figure 4-23. *Multiple lenticular clouds over Mount Shasta, California.*

Research on mountain waves and rotors or eddies continues, but there is no doubt that pilots need to be aware of these phenomena and take appropriate precautions. Although mountain wave activity is normally forecast, many local factors may affect the formation of rotors and eddies. When planning a flight, a pilot should take note of the winds and terrain to assess the likelihood of waves and rotors. There may be telltale signs in flight, including disturbances on water or wheat fields and the formation of clouds, provided there is sufficient humidity to allow cloud formation.

Thunderstorms

A thunderstorm is a local storm produced by a cumulonimbus cloud and accompanied by lightning and thunder. Thunderstorms and cumulonimbus clouds contain many of the most severe atmospheric hazards for the balloon pilot. They are almost always accompanied by strong gusty winds, severe turbulence, heavy rain showers, and lightning. These hazards may extend well away from the central core of the thunderstorm mass, sometimes as much as 30 miles or more.

Ceilings and visibility in the precipitation areas under the thunderstorms are normally poor. Because of the heavy precipitation, the ceiling reported is at best an estimate of where the pilot may break out into visual contact with the surface. The weather observer determines the vertical visibility into the precipitation, which may be significantly different from the slant-range of the pilot.

Potentially hazardous turbulence, as well as many other hazards associated with thunderstorms, make a thunderstorm a dangerous weather formation. The safe balloon pilot avoids any conditions that may expose him or her to potential thunderstorm activity. He or she exercises discretion and good judgment when potential thunderstorm conditions exist.

Structure of Thunderstorms

Convective Cells

The fundamental structural element of the thunderstorm is the unit of convective circulation known as a convective cell. A mature thunderstorm contains one or more of these cells in different stages of development, each varying in diameter from one to five miles. By radar analysis and measurement of drafts, it has been determined that each cell is generally independent of surrounding cells in the same storm. Each thunderstorm progresses through a life cycle from 1 to 3 hours, depending upon the number of cells contained and their stage of development. In the initial stage (cumulus), the cloud consists of a single cell. As the development progresses, however, new cells may form as older cells dissipate.

Stages in Thunderstorm Development

The life cycle of each thunderstorm cell consists of three distinct stages:

- Cumulus,
- Mature, and
- Dissipating.

Cumulus Stage

Although most cumulus clouds do not become thunderstorms, the initial stage of a thunderstorm is always a cumulus cloud. *[Figure 4-24]* The chief distinguishing feature of the cumulus or building stage is an updraft that prevails throughout the entire cell. This updraft may vary from a few feet per second to as much as 6,000 fpm (65 knots) in mature cells. As an updraft continues through the vertical extent of the cell, water droplets grow in size and raindrops are formed.

Figure 4-24. *Congestive cumulus cloud.*

Mature Stage

The beginning of surface rain and adjacent updrafts and downdrafts initiates the mature stage. *[Figure 4-25]* By this time, the average cell has attained a height of 25,000 feet. As the drops begin to fall, the surrounding air begins a downward motion because of frictional drag. This descending air will be colder that its surroundings and its rate of downward motion is accelerated, forming the downdraft. The downdraft reaches maximum speed a short time after rain begins to fall in the cloud. Downdrafts occur at all levels within the storm, and their speed ranges from a few feet per minute to about 2,500 fpm (25 knots). Significant downdrafts never extend from the top of the cell because moisture is not sufficient in the upper levels for raindrops to form. At these high levels, only ice crystals, snowflakes, and supercooled water are present. Therefore, their rate of fall is insufficient to cause appreciable downdrafts. The mature cell generally extends far above 25,000 feet—in some cases up to 70,000 feet. In the middle levels, around 14,000 feet, strong updrafts and downdrafts are adjacent to each other. A shear action exists between these drafts and produces strong and frequent gusts.

Figure 4-25. *Mature thunderstorm.*

Dissipating or Anvil Stage

Throughout the life span of the mature cell, more and more air aloft is entrained by the falling raindrops. Consequently, the downdraft spreads out to take the place of the weakening updrafts. As this process progresses, the entire lower portion of the cell becomes an area of downdraft. Since updrafts are necessary to produce condensation and release latent heat energy, the entire structure begins to dissipate.

[Figure 4-26] The strong winds aloft carry the upper section of the cloud into the familiar anvil form (cumulonimbus cloud). However, the appearance of the anvil does not always indicate that the thunderstorm is dissipating.

Figure 4-26. *Dissipating thunderstorm.*

A significant thunderstorm hazard is the rapid change in wind direction and wind speed immediately prior to storm passage at the surface. These strong winds are the result of the horizontal spreading of the storm's downdraft current as they approach the surface of the earth. This initial wind surge, as observed at the surface, is known as a first gust. The speed of this first gust may exceed 75 knots and vary 180° in direction from the previously prevailing surface winds. First-gust speeds average about 15 knots over prevailing velocities, and average an approximately 40° change in the direction of the wind. First gusts usually precede the heavy precipitation, and strong gusts may continue for approximately 5 to 10 minutes with each thunderstorm cell. First gusts are not limited to the area ahead of the storm's movement. They may be found in all sectors, including the area back of the storm's movement.

During the passage of a thunderstorm, rapid and marked surface variations generally occur. These variations usually occur in a particular sequence characterized by:

- An abrupt fall in pressure as the storm approaches.
- An abrupt rise in pressure associated with rain showers as the storm moves on and the rain ceases.

All thunderstorms are similar in physical makeup, but for purposes of identification they are divided into two general groups: frontal and air mass. This division gives the balloon pilot an indication of the method by which the storms are formed and the distribution of the clouds over the area. The specific nomenclature of these thunderstorms depends upon the manner in which the lifting action occurs.

Frontal Thunderstorms

Thunderstorms may occur within the cloud system of any front: warm, cold, stationary, or occluded. Frontal thunderstorms are caused by the lifting of warm, moist, conditionally unstable air over a frontal surface. Thunderstorms may also occur many miles ahead of a rapidly moving cold front, and are called prefrontal or squall line thunderstorms.

Warm front thunderstorms are caused when warm, moist, conditionally unstable air is forced aloft over a colder, denser shelf of retreating air. Because the frontal surface is shallow, the air is lifted gradually. The lifting condensation level is normally reached long before the level of free convection, thus producing stratiform clouds. The level of free convection is normally reached in isolated areas along the frontal surface. This is the area where the greatest amount of water vapor is present in the warm air being lifted. Therefore, warm front thunderstorms are generally scattered. When the level of free convection is reached, warm front thunderstorms may form. These thunderstorms may prove particularly hazardous, as they are frequently obscured by the surrounding stratiform clouds.

Cold front thunderstorms are caused by the forward motion of a wedge of cold air under a mass of warm, moist, conditionally unstable air (cold front), increasing the possibility for thunderstorms to develop. The slope of a typical cold frontal surface is relatively steep, so the lifting condensation level and the level of free convection are usually near the same altitude. Cold front thunderstorms are typically positioned along the frontal surface in what appears to be a continuous line. These storms are easily recognized from the air, because they are partly visible from the front and rear of the storm line. However, if the slope of the frontal surface is shallow, the lifting action is not sufficient to produce thunderstorms in lines (line squalls). With a shallow front, the thunderstorms form behind the surface front and are widely scattered. Such storms may be concealed by the surrounding cloud layers.

Lines of thunderstorms frequently develop ahead of rapidly moving cold fronts. These are known as prefrontal squall lines, and frequently form parallel to the cold front. Prefrontal squall line thunderstorms are usually more violent than cold front thunderstorms. They are most active between noon and midnight. The cold front cloud system usually weakens during the period of the greatest prefrontal squall line activity because the warm air displaced by the frontal surface has lost its moisture and energy in the prefrontal thunderstorms. In the United States, tornadoes are frequently associated with strong prefrontal squall lines. Pre-frontal squall line thunderstorms are indicated on a surface weather map by an alternate dash-dot-dash line (display).

The distribution of stationary front thunderstorms is controlled by the slope of the frontal surface. Steeply sloped stationary fronts tend to have lines of storms, whereas shallow stationary fronts tend to have widely scattered storms.

Occluded front thunderstorms are associated with the two types of occluded fronts (warm front and cold front occlusions), and are usually cold front thunderstorms that have been moved into the area of warm frontal weather by the occlusion process. They are found along the upper front, and are normally strongest for a distance of 50 to 100 miles north of the peak of the warm sector.

Air Mass Thunderstorms

The two types of air mass thunderstorms are locally convective and orographic. Both types form within air masses, and are randomly distributed throughout the air mass.

Convective thunderstorms are often caused by solar heating of the land, which provides heat to the air, thereby resulting in thermal convection. Relatively cool air flowing over a warmer water surface may also produce sufficient convection to cause thunderstorms. The land-type convective thunderstorms normally form during the afternoon hours, after the Earth has gained maximum heating from the sun. If cool, moist, conditionally unstable air is passing over this land area, heating from below will cause convective currents, thereby resulting in towering cumulus or thunderstorm activity. Dissipation usually occurs during the early evening hours, as the land begins to lose its heat to the atmosphere. Although convective thunderstorms form as individual cells, they may become so numerous over a particular geographical area that continued flight cannot be maintained.

Thunderstorms over the ocean are most common during the night and early morning. They frequently occur offshore when a land breeze is blowing toward the water. The cool land breeze is heated by the warmer water surface, which results in sufficient convection to produce thunderstorms. After sunrise, heating of the land surface reverses the airflow (sea breeze). The thunderstorms then dissipate over the water, but they may re-form over the warmer land surface. As an example, the air mass weather that exists in Florida combines both types of convective thunderstorms. Circulation around a semipermanent high pressure system off the southeastern United States (Bermuda high) carries moist ocean air over the warm land surface of the Florida Peninsula. At night, thunderstorms off the Florida Coast are caused by the warm water of the Gulf Stream heating the surface air, while the upper air is cooling by radiation to space. This heating from below produces thermal convection over the water. When the sun rises, the heat balance necessary to maintain storm formation over the water is destroyed. By day, the storms appear to move inward over the land areas, but actually dissipate off the coast and re-form over the hot landmass. The heated land surface sets up an unstable lapse rate over the Peninsula and causes storm development to continue until nocturnal cooling occurs. Usually, convective type storms are randomly distributed and easily recognized.

Orographic thunderstorms will form on the windward side of a mountain if conditionally unstable air is lifted above the level of free convection. The storm activity is usually scattered along the individual peaks of the mountains. Occasionally, however, this activity may form a long unbroken line of storms similar to a squall line. The storms persist as long as the circulation causes upslope motion. From the windward side of the mountains, identification of orographic storms may sometimes be difficult because the storm clouds are obscured by other clouds (usually stratiform). Almost without exception, orographic thunderstorms enshroud mountain peaks or hills.

Minimum Factors

The minimum factors essential to the formation of a thunderstorm are conditionally unstable air with relatively high moisture content and some type of lifting action. Lifting of warm air will not necessarily cause free convection. The air may be lifted to a point where the moisture condenses and clouds form. These cloud layers, however, will be stable if the level of free convection has not been reached by the lifting. Conversely, it is possible for dry heated air to rise convectively without the formation of clouds. In this condition, turbulence might be experienced in perfectly clear weather. Cumulonimbus cloud formations require a combination of conditionally unstable air, some type of lifting actions, and high moisture content. Once a cloud has formed, the latent heat of condensation released by the change of state from vapor to liquid tends to make the air more unstable.

Some type of external lifting action is necessary to bring the warm surface air to the point where it will continue to rise freely (the level of free convection). For example, an air mass may be lifted by thermal convection, terrain, fronts, or convergence.

Summary

By no means is the information contained here a complete discussion of all the weather information and factors affecting balloon flight. There are many resources available, both through government and private agencies, which may be of value to the pilot in planning a flight. A pilot should take the time to explore the internet, read weather books, and gain a complete understanding of the myriad of weather information and products that are available.

The second section of this chapter will expose the reader to some of the weather reporting products available, both through FSS briefings and Internet searches. With the knowledge gained from the first half of this chapter, the pilot will be able to make a good interpretation of the reports, and determine how present and future conditions will affect the decision to fly.

How to Obtain Weather Information

An integral part of flight preparation for any pilot is checking the weather conditions expected to occur during the flight. FAA regulations place the responsibility for flight planning on the pilot. To effectively plan a flight, a pilot needs to understand what weather information is available, how to obtain it, and how it can be applied to a flight.

While weather forecasts are not 100 percent accurate, meteorologists, through careful scientific study and computer modeling, have the ability to predict the weather patterns, trends, and characteristics with increasing accuracy. Through a complex system of weather services, government agencies, and independent weather observers, pilots and other aviation professionals receive the benefit of this vast knowledge base in the form of up-to-date weather reports and forecasts. These reports and forecasts enable pilots to make informed decisions regarding weather and flight safety.

Sources for Weather Information

There are many sources available for today's pilot when gathering information about weather prior to a flight. A review of pertinent weather reports and information is required by FAA regulations and the following sources provide excellent weather information. For the balloon pilot, experience and study helps him or her determine the preferred sources for weather information.

FAA Flight Service Stations (FSS)

The FAA FSS is the primary source for preflight weather information. FSS can be contacted by calling 1-800-WXBRIEF. It also logs pilot contacts to provide background information in the event of an accident or incident, as well as substantiating workload statistics. It is one of the two sources of an official weather briefing.

Although there are currently 44 FSS operations, the FAA plans to automate the service and reduce the physical locations eventually to three. Due to this, occasional extended wait times can occur, particularly during periods of marginal or severe weather. To facilitate the process, inform the briefer prior to the start of the briefing of the information needed.

Internet Sources

A wealth of internet sources exist for the balloon pilot seeking information about current weather conditions.

- *Direct User Access Terminal Service (DUATS) (www.duats.com*—probably the most well-known of the many internet based weather sites, DUATS provides an on-line "official" briefing service at no cost to the aviation community. The site is a contracted function of the FAA, and is available to anyone who registers. Besides the normal briefing that can be received from the FSS, there are a number of graphic products available that enhance the user's understanding of current and forecast weather.
- *National Oceanic and Atmospheric Administration (NOAA) (www.noaa.gov)*—NOAA operates the National Weather Service (NWS) at www.nws.noaa.gov which provides access to weather information collected on a continuing basis throughout the country.

  NOAA also offers pilots an interactive page at www.weather.gov/view/largemap.php that is activated on a state-by-state basis and provides hourly weather observations at selected points. *[Figure 4-27]* This weather information helps a pilot determine trends in winds and pressure systems, which are prime indicators of general conditions for balloon pilots.

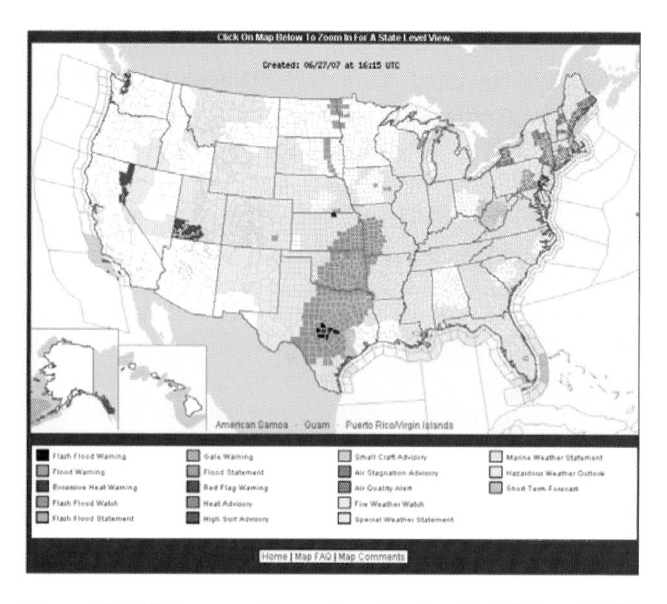

Figure 4-27. *This map can be used to gather hourly AWOS and ASOS type weather information from various sites in each state.*

- *National Center for Atmospheric Research and the University Corporation for Atmospheric Research (www.ucar.edu)*—a collaborative effort of research centers, universities, and weather offices around the United States, this site provides numerous real-time and forecast weather products and graphics.
- *Aviation Digital Data Services (ADDS) (www.aviationweather.gov)*—the Aviation Digital Data Service (ADDS) makes text, digital and graphical forecasts, analyses, and observations of aviation-related weather variables available to the aviation community. The ADDS graphics and charts are often easier to interpret than those of the AWC website.
- *Helicopter Emergency Medical Services (HEMS) Weather Display (www.weather.aero/hems)*—the ADDS development team created an experimental tool designed to show weather conditions for short-distance and low-altitude flights common for the helicopter emergency medical services (HEMS) community at the request of the FAA. *[Figure 4-28]* This interactive site allows the user to determine ceiling and visibility and winds (at 500 foot increments) for an area as small as $5k^2$. While not specifically targeted nor designed for the balloon pilot's use, the information obtained from this site is helpful to the pilot planning a flight at some distance from a normal weather reporting facility.

Interpreting Weather Charts and Reports

A weather chart is any chart or map that presents data and analysis that describe the state of the atmosphere over an extended region at a given time. Weather charts provide a picture of the overall movement of major weather systems and fronts and are used in flight planning.

Three useful weather charts for balloon pilots are: surface analysis, weather depiction, and radar summary charts. These three charts present current weather information and provide "big picture" information for weather systems across the United States. The composite moisture stability chart, constant pressure analysis charts, and significant weather prognostic charts provide additional information for flight planning.

Knowledge of all these weather charts, reports, and forecasts may not be necessary for the pilot planning a local flight, but an understanding of large scale weather patterns and systems bring a greater understanding of how those systems affect local weather. It is important to gain an understanding of the primary charts used, and develop interpolation skills to be able to perform safe, adequate flight planning.

Surface Analysis Chart

The surface analysis chart is computer-generated, covers the contiguous 48 states and adjacent areas, and is transmitted every 3 hours with an analysis of the current surface weather. It shows the areas of high and low pressure, fronts, temperatures, dew points, wind directions and speeds, local weather, and visual obstructions. *[Figure 4-29]*

Surface weather observations for reporting points across the United States are also depicted on this chart. Each of these reporting points is illustrated by a station model.

- Type of observation—a round model indicates an official weather observer made the observation. A square model indicates the observation is from an automated station. Stations located offshore give data from ships, buoys, or offshore platforms.
- Sky cover—the station model depicts total sky cover and will be shown as clear, scattered, broken, overcast, or obscured/partially obscured.
- Clouds—cloud types are represented by specific symbols. Low cloud symbols are placed beneath the station model, while middle and high cloud symbols are placed directly above the station model. Typically, only one type of cloud will be depicted with the station model.
- Sea level pressure—sea level pressure given in three digits to the nearest tenth of a mb. For 1,000 mb or greater, prefix a ten to the three digits. For less than 1,000 mb, prefix a nine to the three digits.
- Pressure change/tendency—pressure change in tenths of mb over the past 3 hours. This is depicted directly below the sea level pressure.
- Precipitation—a record of the precipitation that has fallen over the last 6 hours to the nearest hundredth of an inch.
- Dew point—dew point is given in degrees Fahrenheit.
- Present weather—over 100 different weather symbols are used to describe the current weather.
- Temperature—temperature is given in degrees Fahrenheit.
- Wind—true direction of wind is given by the wind pointer line, indicating the direction from which the wind is coming. A short barb is equal to five knots of wind, a long barb is equal to ten knots of wind, and a pennant is equal to 50 knots.

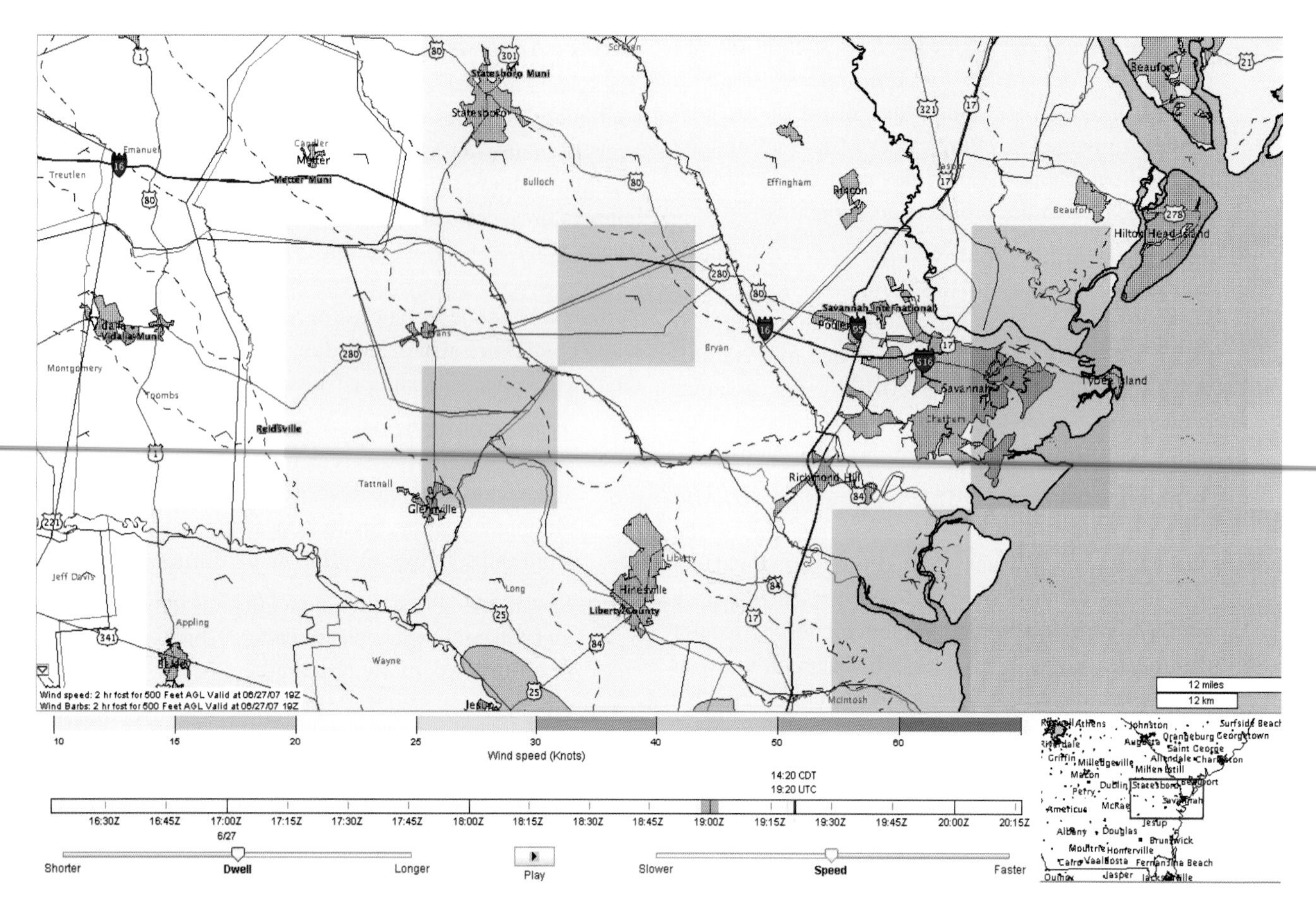

Figure 4-28. *Example of the information found on the Helicopter Emergency Medical Services (HEMS) weather display site.*

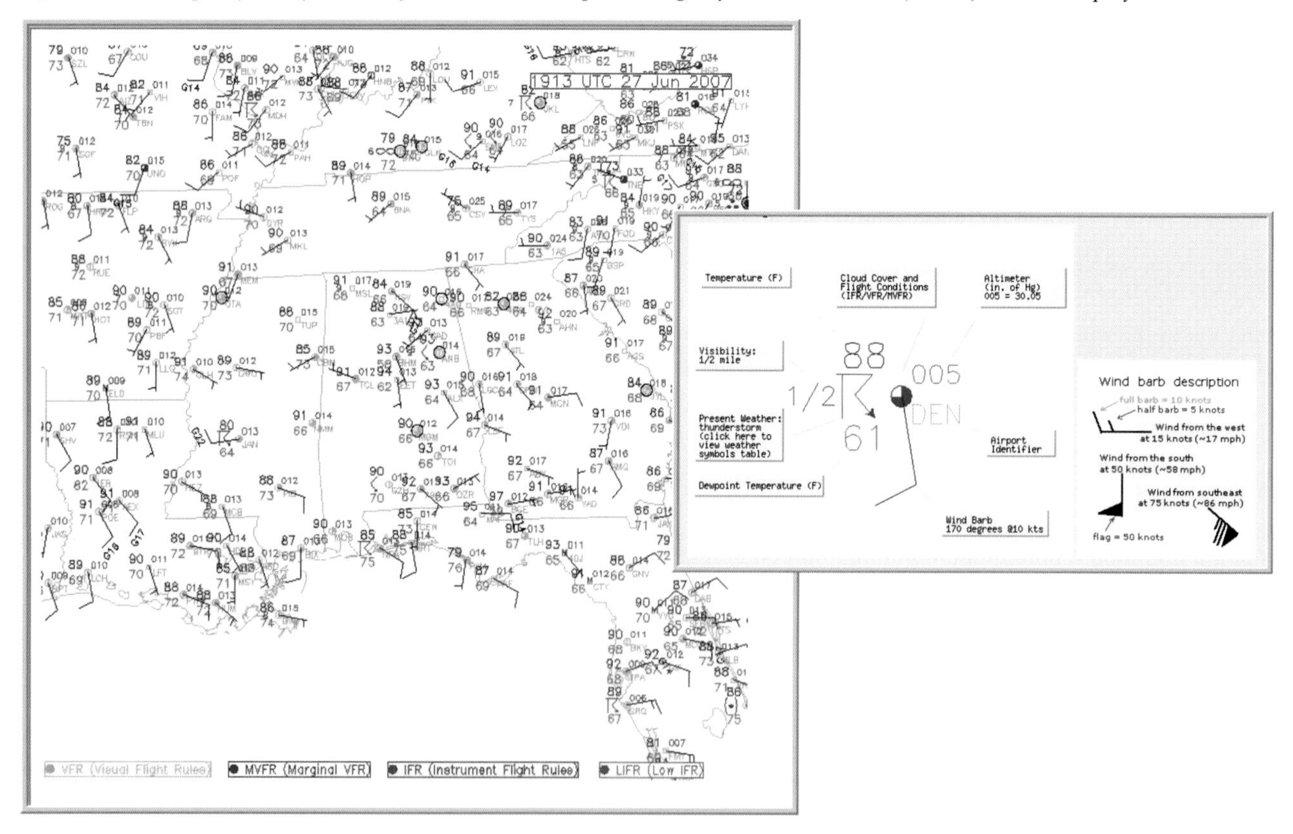

Figure 4-29. *Example of a section of a surface analysis chart with station model legend (inset).*

Weather Depiction Chart

A weather depiction chart details surface conditions as derived from METAR and other surface observations. It is prepared and transmitted by computer every 3 hours beginning at 0100 Zulu time (0100Z), and is valid at the time of the plotted data. Designed to be used for flight planning, it gives an overall picture of the weather across the United States. *[Figure 4-30]*

The weather depiction typically displays major fronts or areas of high and low pressure. It also provides a graphic display of Instrument Flight Rules (IFR), VFR, and marginal VFR (MVFR) weather. Areas of IFR conditions (ceilings less than 1,000 feet and visibility less than 3 miles) are shown by a hatched area outlined by a smooth line. MVFR regions (ceilings 1,000 to 3,000 feet, visibility 3 to 5 miles) are shown by a non-hatched area outlined by a smooth line. Areas of VFR (no ceiling or ceiling greater than 3,000 feet and visibility greater than 5 miles) are not outlined. Weather depiction charts show a modified station model that provides sky conditions in the form of total sky cover, cloud height or ceiling, weather, and obstructions to visibility, but does not include winds or pressure readings like the surface analysis chart. A bracket ( ]) symbol to the right of the station indicates the observation was made by an automated station. A detailed explanation of a station model is depicted in the previous discussion of surface analysis charts.

Radar Summary Chart (SD)

A radar summary chart *[Figure 4-31]* is a computer-generated graphical display of a collection of automated radar weather reports (SDs). The chart is published hourly, 35 minutes past the hour. It displays areas of precipitation as well as information regarding the characteristics of the precipitation. An SD chart includes:

- No information—if information is not reported, the chart will read "NA." If no echoes are detected, the chart will read "NE."
- Precipitation intensity contours—intensity can be described as one of six levels and is shown on the chart by three contour intervals.
- Height of tops—the heights of the echo tops are given in hundreds of feet MSL.
- Movement of cells—individual cell movement is indicated by an arrow pointing in the direction of movement. The speed of movement in knots is the number at the top of the arrow head. "LM" indicates little movement.

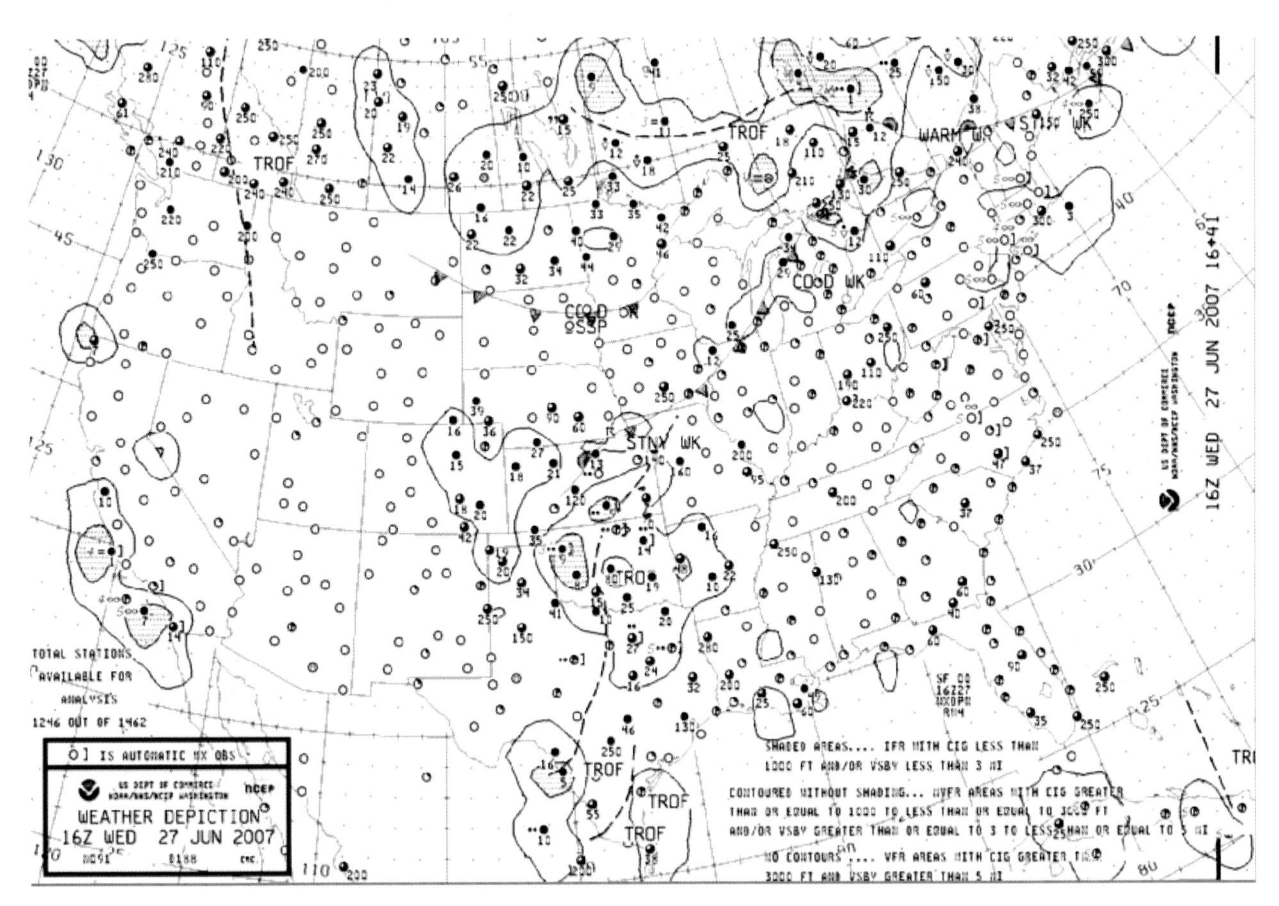

Figure 4-30. *Weather depiction chart.*

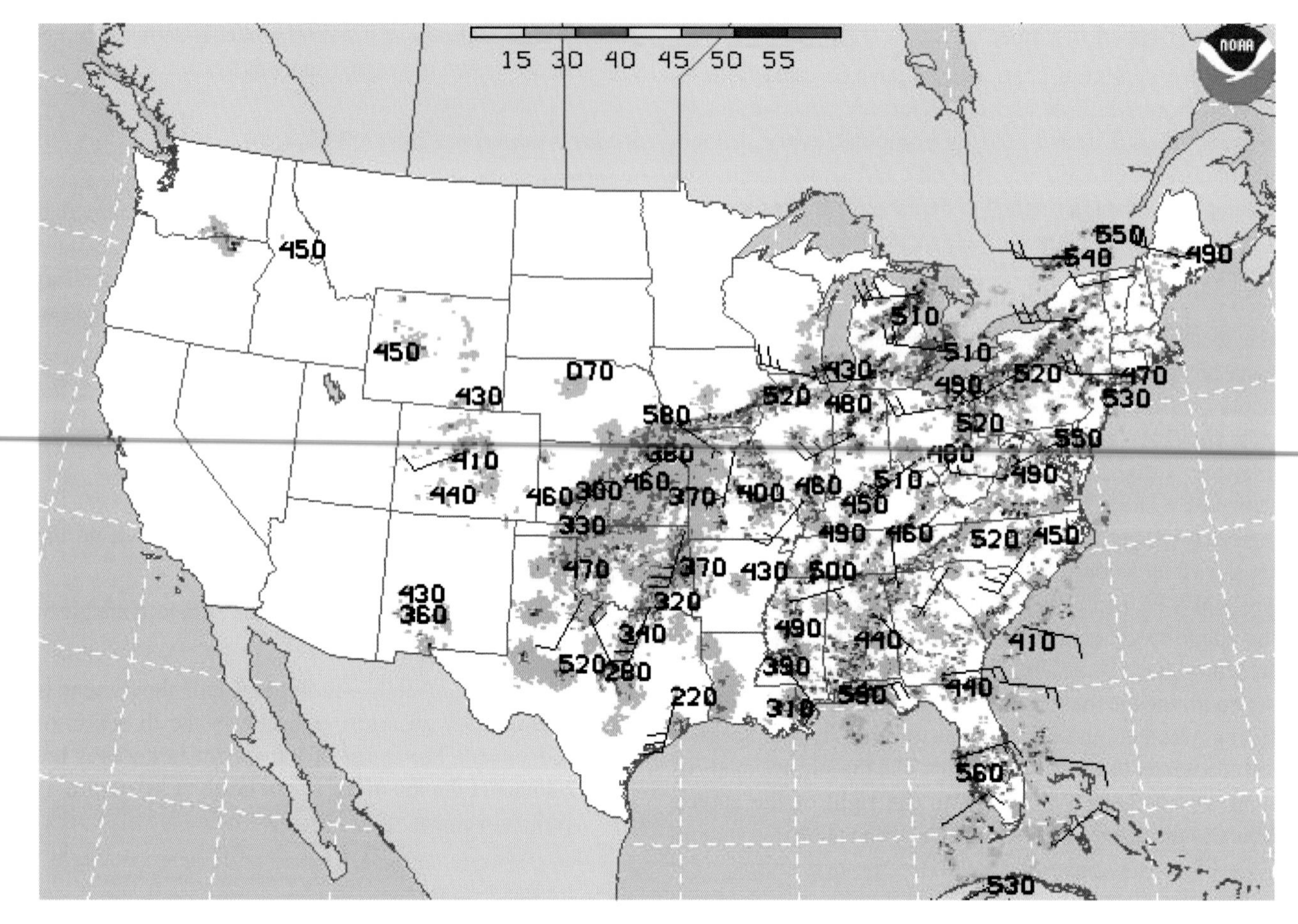

Figure 4-31. *Radar summary chart.*

- Type of precipitation—the type of precipitation is marked on the chart using specific symbols.
- Echo configuration—echoes are shown as areas, cells, or lines.
- Weather watches—severe weather watch areas for tornadoes and severe thunderstorms are depicted by boxes outlined with heavy dashed lines.

A valuable tool for preflight planning, the SD chart has several limitations. Since it depicts only areas of current precipitation, it will not show areas of clouds and fog with no appreciable precipitation, or the height of the tops and bases of the clouds. SD charts should be used in conjunction with current METAR and weather forecasts.

Composite Moisture Stability Chart

The composite moisture stability chart is a chart composed of four panels depicting stability, precipitable water, freezing level, and average relative humidity conditions. This computer-generated chart contains data obtained from upper-air observations, is updated twice a day, and shows the relative stability of the air mass and the potential for thunderstorms or thermal activity.

Stability/Lifted Index Chart

A subdisplay of the composition moisture stability chart is the stability or lifted index (LI) chart, a valuable tool for determining the stability of the atmosphere. The stability or LI chart is the upper left hand panel of the composite moisture stability chart. Two indexes represent the moisture and stability of the air: the K index (KI) and the LI, with these numbers composing a fraction. KI (denominator of the fraction) provides moisture and stability information. Values range from high positive values to low negative values. A high positive KI indicates moist, unstable air. KI values are considered high when at or above +20, and low when less than +20. *[Figure 4-32]*

The LI (numerator of the fraction) is a common measure of atmospheric stability. It is calculated by hypothetically lifting a parcel of air to the 500 mb level, approximately 18,000 feet MSL, and analyzing its stability at that level. A positive LI indicates that a particular parcel of air is stable and resists further upward motion. Large positive values (+8 or greater) would indicate very stable air. Conversely, a negative LI means that a lifted surface parcel of air is unstable, and more likely to rise. Large negative values (–6 or more) indicate very unstable air.

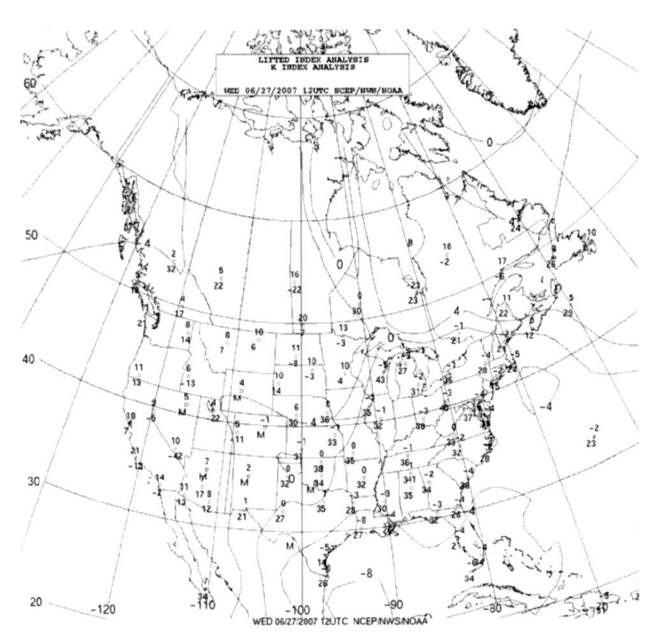

Figure 4-32. *Lifted index chart.*

The KI and LI can be used together to determine the moisture and stability characteristics of a particular air mass. The air masses may be classified as moist and stable, moist and unstable, dry and stable, or dry and unstable. This determination allows the balloon pilot to make an informed decision regarding the likelihood of thermal and potential thunderstorms, and if a safe flight can be conducted.

Constant Pressure Analysis Charts

A constant pressure analysis chart or isobaric chart is a weather map representing conditions on a surface of equal atmospheric pressure. *[Figure 4-33]* For example, a 500 mb chart will display conditions at the level of the atmosphere at which the atmospheric pressure is 500 mb. The height above sea level at which the pressure is that particular value may vary from one location to another at any given time, and also varies with time at any one location, so it does not represent a surface of constant altitude/height.

Constant pressure charts provide the pilot with a clearer picture of how the atmosphere behaves at different altitudes and pressures. For example, a low pressure system that seems insignificant based on surface observations may prove to be a major factor in the weather at five or ten thousand feet.

Constant pressure charts are prepared for selected values of pressure and present weather information at various altitudes. Standard charting values are at 850 mb (approximately 5,000 feet MSL), 700 mb (approximately 10,000 feet MSL), 500 mb (approximately 18,000 feet MSL), as well as higher and lower altitudes. Charts with higher pressure altitudes present

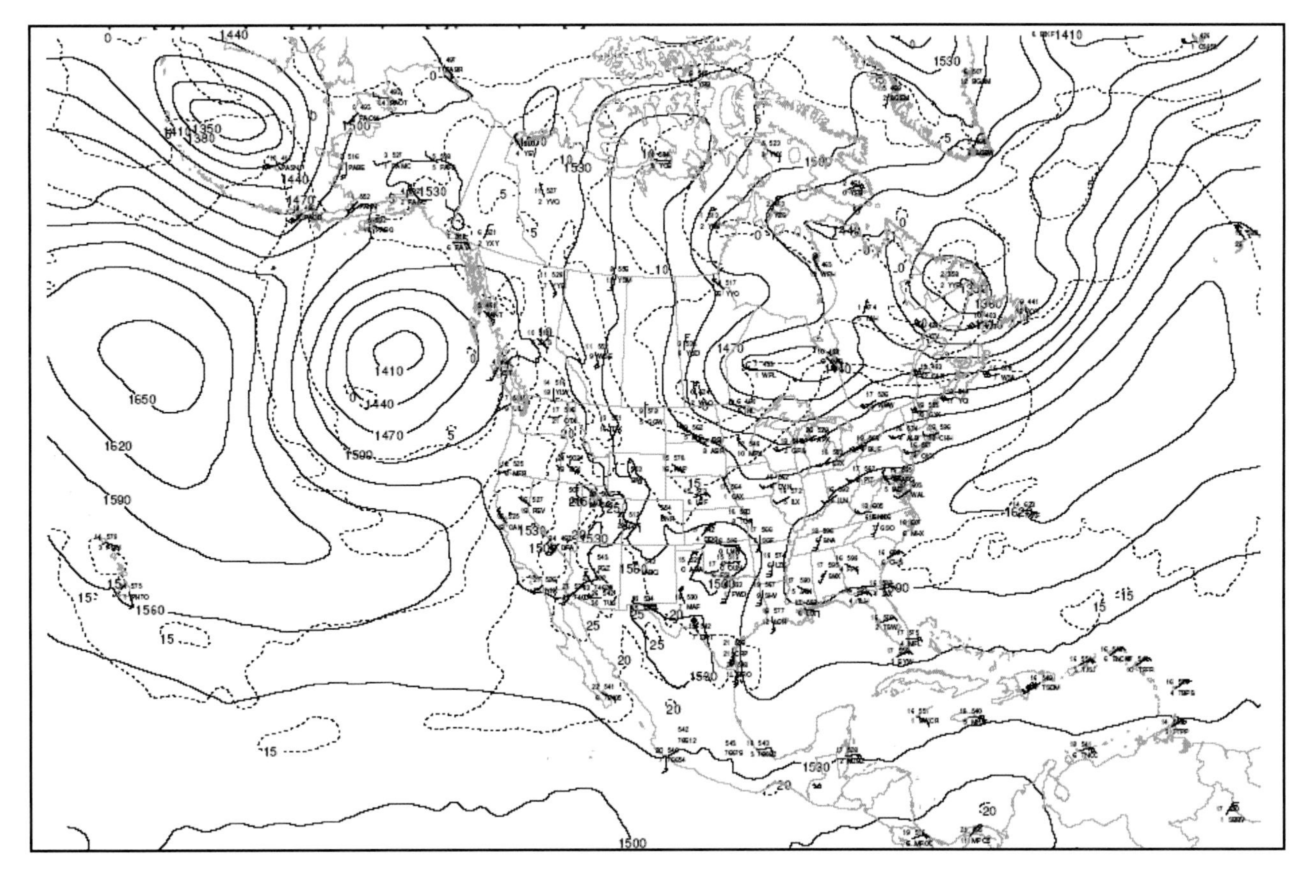

Figure 4-33. *Constant pressure analysis chart.*

information at lower altitudes, and charts with lower pressure altitudes present information at higher altitudes.

Symbology on the constant pressure analysis chart is the same as that of the surface analysis chart. This chart depicts the information at a specific pressure altitude. When compared with the surface analysis of the same time frame, a three-dimensional concept of the atmosphere can be conceptualized, and the pilot can gain a greater understanding of the atmospheric dynamics involved in weather patterns.

Significant Weather Prognostic Charts

Significant weather prognostic charts *[Figure 4-34]* display the observed or forecast significant weather phenomena at different flight levels that may affect the operation of aircraft. They are available for low-level significant weather from the surface to FL 240 (24,000 feet), also referred to as the 400 mb level, and high-level significant weather from FL 250 to FL 600 (25,000 to 60,000 feet). This discussion involves the low-level significant weather prognostic chart.

The low-level chart comes in two forms: the 12- and 24-hour forecast chart, and the 36- and 48-hour surface only forecast chart. The first chart is a four-panel chart that includes 12- and 24-hour forecasts for significant weather and surface weather. Charts are issued four times a day at 0000Z, 0600Z, 1200Z, and 1800Z. The valid time for the chart is printed on the lower left corner of each panel. The upper two panels show forecast significant weather, which may include nonconvective turbulence, freezing levels, and IFR or MVFR weather. Areas of moderate or greater turbulence are enclosed in dashed lines. Numbers within these areas give the height of the turbulence in hundreds of feet MSL. Figures below the line show the anticipated base, while figures above the line show the top of the zone of turbulence. Also shown on this panel are areas of VFR, IFR, and MVFR. IFR areas are enclosed by solid lines, MVFR areas are enclosed by scalloped lines, and the remaining, unenclosed area is designated VFR. Zigzag lines and the letters "SFC" (surface) indicate freezing levels in that area are at the surface. Freezing level height contours for the highest freezing level are drawn at 4,000 foot intervals with dashed lines.

Additional Weather Information

Skew-T Plots

Most weather information is derived from radiosondes, or weather balloons, that are released from over 100 stations in the United States twice daily (00Z and 12Z). The observations of temperature and humidity at various pressure altitudes are transmitted back to the releasing station; the radiosondes are also tracked by radar in order to determine wind speed and direction. This information is plotted to create a diagram known as a Skew-T/Log-P plot, commonly referred to as a Skew-T. These plots can be found at many different online

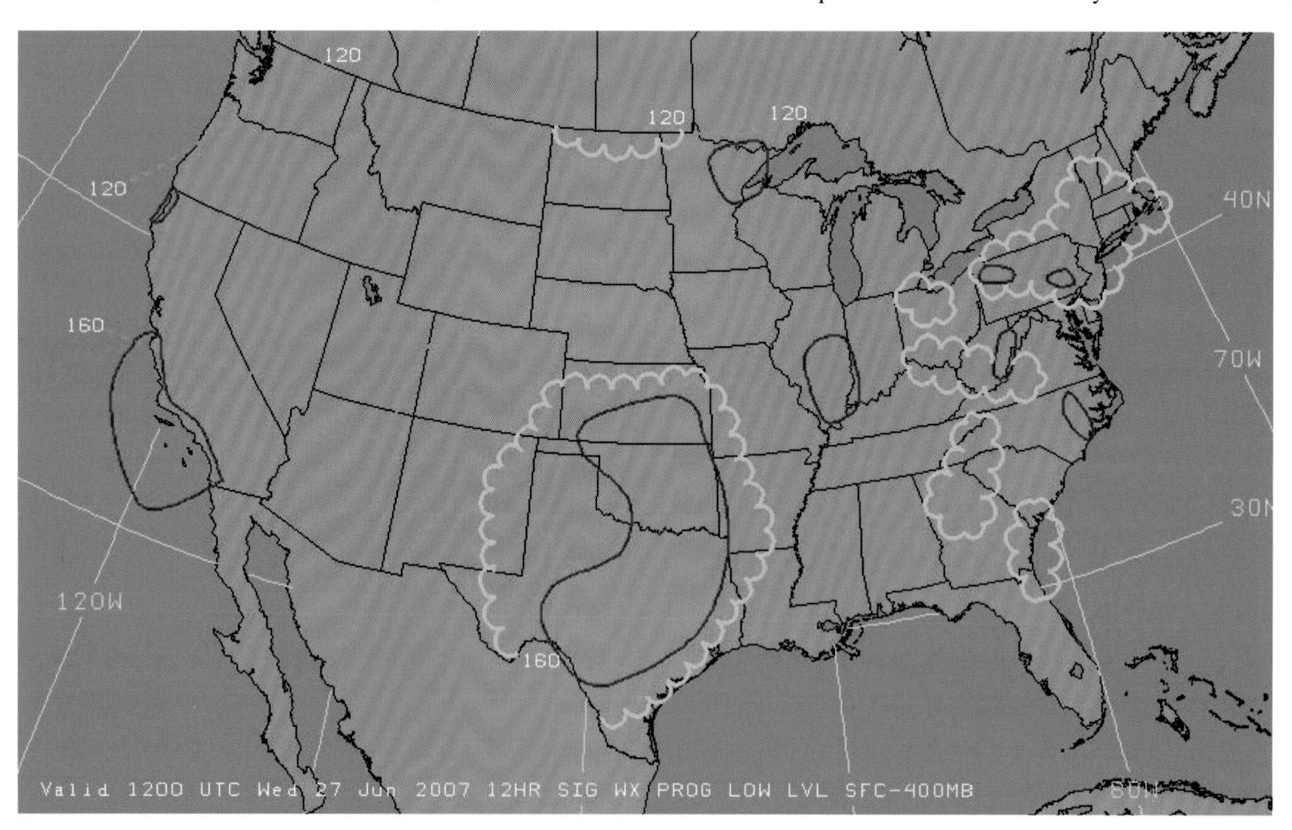

Figure 4-34. *Significant weather prognostic chart.*

weather sites; perhaps the easiest to use is rucsoundings.noaa.gov. The appropriate plot for any of the reporting sites can be found by typing in the closest reporting station identifier (usually an airport), and allowing the graphic to load.

There is a wealth of information that may be derived from the Skew-T plot (or "sounding," as it may be referred to), but this discussion will be limited to those features of immediate interest to the average balloon pilot.

Some of the information that may be derived from the Skew-T, using the example in *Figure 4-35:*

- The two lines running vertically through the center of the graphic (red and blue) show the temperature and dew point at ascending pressure altitudes. The temperature is always plotted to the right of the dew point because temperature is almost always greater than the dew point temperature.
- The right side margin shows wind speed and direction, using the standard "barbed" system common in weather reporting. The scale (in this case, 0 to 40 knots) can be changed, but this setting provides the best resolution.
- The left margin of the chart indicates the pressure altitude for a particular reading. Pressure altitude readings correspond generally to certain altitudes. For example, the reading at 850 mb equates roughly to an altitude of 5,000 feet above MSL. That may not appear to be useful; however, this is where the dynamics of the application come into play. If the computer cursor is moved over the graphic part of the diagram, indicators as to the specific information for that altitude may be seen (not depicted in *Figure 4-35*). A pilot may be able to get information for varying altitudes as close at 125 feet apart, depending on the resolution of the original sounding information.

There is a tutorial available to fully explain the data interpretation of the Skew T plot at www.met.tamu.edu/class/ATMO151/tut/sound/soundmain.html.

Velocity Azimuth Display (VAD) Winds

Velocity Azimuth display (VAD) winds are derived from the output of the 160 WSR-88 radar sites located throughout the United States. These radar systems are used by weather professionals to produce many different products, including the weather radar depictions found on many of the web sites previously discussed, as well as various television station weather reporting. *[Figure 4-36]*

The WSR-88 radar systems can be configured to produce radar returns from dust and other particulate matter that may be in the air. These radar returns can be processed to indicate wind direction and speed at different altitudes. VAD winds are generally reported in 1,000 foot increments, although at

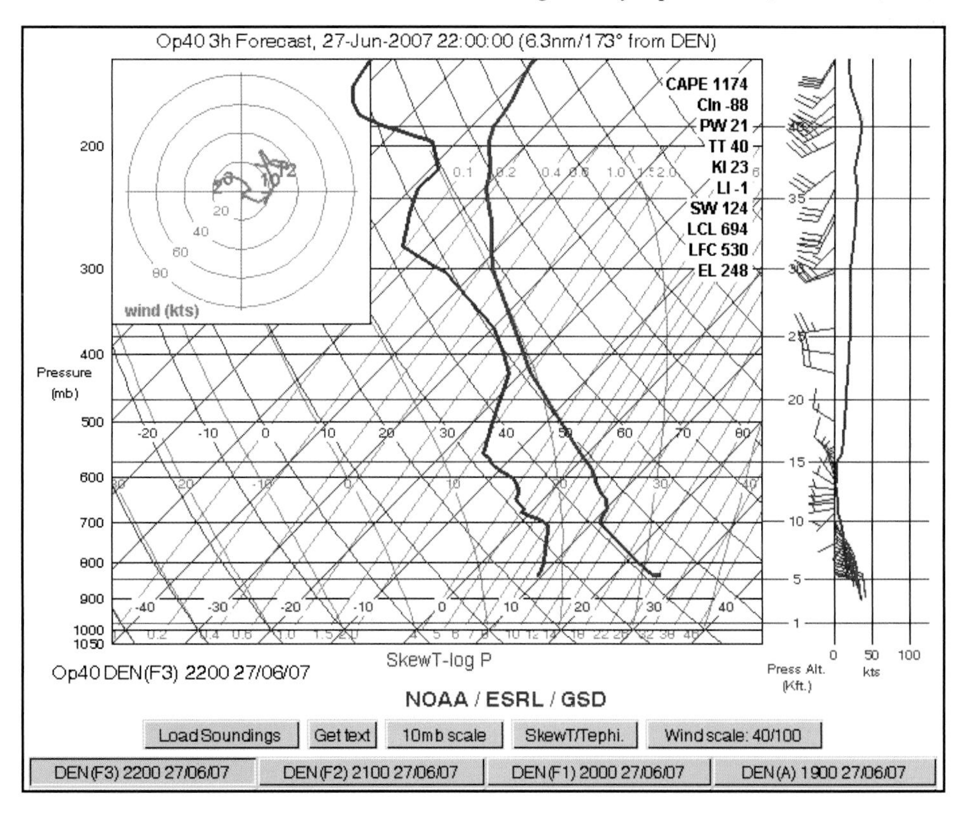

Figure 4-35. *Example of a Skew-T plot.*

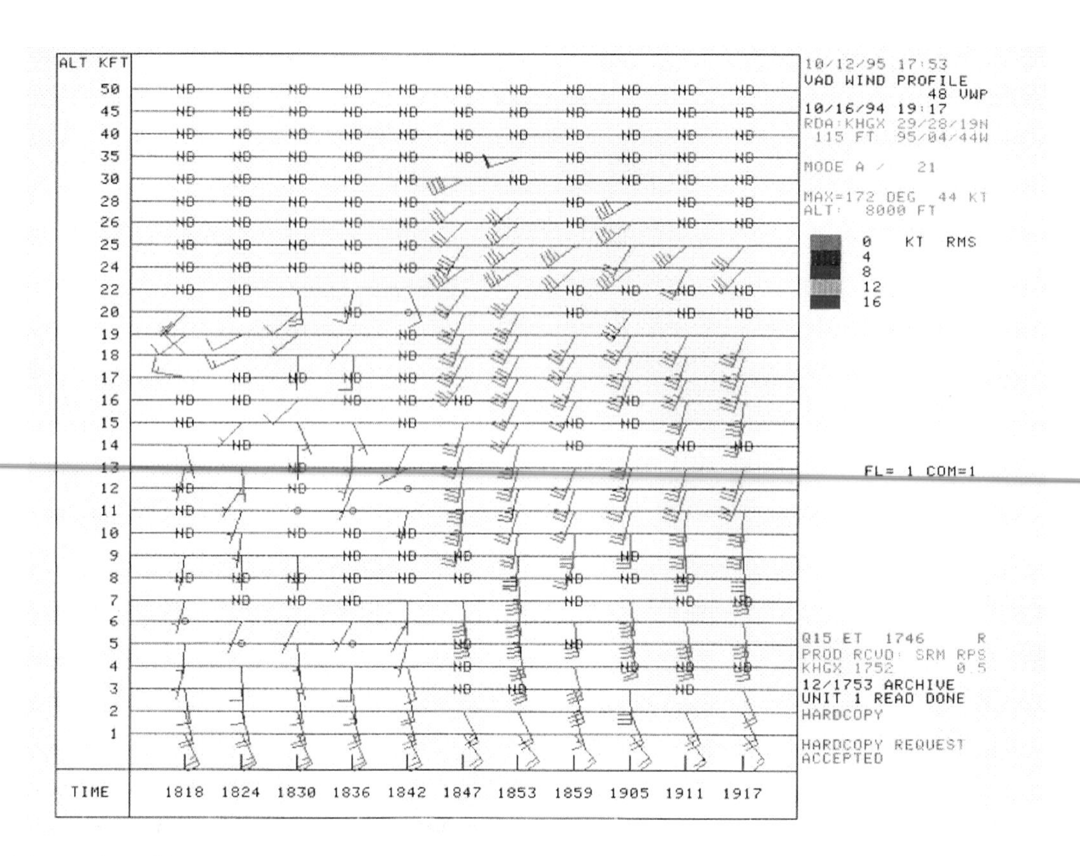

Figure 4-36. *VAD wind graphic.*

times reports may be as small as 150 feet between altitudes. Standard "wind barb" depictions are used to represent the wind direction and speed at different altitudes.

Plymouth State University in Plymouth, New Hampshire operates a Weather Center with a website for locating VAD winds. This user-friendly website offers pilots a convenient selection of reporting stations at http://vortex.plymouth.edu/lnids_conus.html.

Aviation Forecasts

Observed weather condition reports are often used in the creation of forecasts for the same area. A variety of different forecast products are produced and designed to be used in the preflight planning stage. Pilots need to be familiar with the following printed forecasts: wind and temperature aloft forecasts (FD reports), the terminal aerodrome forecast (TAF), aviation area forecast (FA), and inflight weather advisories (SIGMET, AIRMET).

Wind and Temperature Aloft Forecast (FD Report)

Wind and temperature aloft forecasts provide wind and temperature forecasts for specific locations in the contiguous United States, plus network locations in Hawaii and Alaska. The forecasts are made twice a day based on the radiosonde upper air observations taken at 0000Z and 1200Z. Up through 12,000 feet are true altitudes; at and above 18,000 feet are pressure altitudes. Wind direction is always in reference to true north, and wind speed is given in knots. The temperature is given in degrees Celsius; no winds are forecast when a given level is within 1,500 feet of the station elevation. Similarly, temperatures are not forecast for any station within 2,500 feet of the station elevation. If the wind speed is forecast to be greater than 100 knots but less than 199 knots, the computer adds 50 to the direction and subtracts 100 from the speed. A sample FD report is shown in *Figure 4-37*.

To decode this type of data group, the reverse process must be accomplished. For example, when the data appears as "731960," subtract 50 from the 73 and add 100 to the 19; the wind would be 230° at 119 knots with a temperature of –60°C. If the wind speed is forecast to be 200 knots or greater, the wind group is coded as 99 knots. When the data appears as "7799," subtract 50 from 77 and add 100 to 99; the wind is 270° at 199 knots or greater. When the forecast wind speed is calm or less than 5 knots, the data group is coded "9900," which means light and variable.

Aviation Routine Weather Report (METAR)

METAR is an observation of current surface weather reported in a standard international format. While the METAR code has been adopted worldwide, each country is allowed to make

000
FDUM02 KWBC 251405
DATA BASED ON 251200Z
VALID 251800Z FOR USE 1700-2100Z. TEMPS NEG ABV 24000

FT	3000	6000	9000	12000	18000	24000	30000	34000	39000
BHM	9900	9900+15	9900+09	3205+03	2911-08	2814-19	262435	263146	253655
HSV	9900	2706+15	9900+09	3110+02	3012-08	2814-19	272235	262145	251555
MGM	9900	9900+15	9900+09	9900+05	2814-08	2617-18	264433	256244	256655
MOB	1608	2507+15	3009+09	2806+06	2610-07	2721-18	244132	254342	255055
FSM	9900	1807+15	9900+13	9900+07	0808-06	0610-18	061035	020845	051055
LIT	0605	3005+15	2209+10	9900+06	0206-07	0309-18	990035	990045	990055
LCH	1008	9900+15	9900+12	9900+06	9900-09	2316-17	243132	233542	244554
MSY	1508	9900+15	9900+10	9900+06	2608-07	2523-17	253332	244242	254954
SHV	9900	1112+15	1111+10	0907+06	2505-07	9900-18	231134	243244	243055
JAN	9900	1905+15	9900+09	0705+05	2912-07	2908-19	262734	265144	265656
GAG		1812+19	1709+12	1507+06	0509-04	0609-17	041935	051743	053155
OKC	1505	0811+15	1508+13	9900+07	0913-05	0913-17	051434	052344	053056
TUL	1405	0807+15	9900+13	9900+08	0909-05	0608-18	031335	032145	042856
BNA	9900	2613+14	2815+08	3016+03	3118-09	3021-20	292636	282746	282755
MEM	9900	9900+15	2505+10	2905+05	3509-07	3411-19	291035	271445	260554
TRI		2419+13	2420+07	2419+02	2525-10	2532-21	264737	255946	245054
TYS	2414	2416+13	2515+08	2516+03	2519-09	2626-20	273436	264146	264256
ABI		9900+16	1109+11	0810+06	1113-06	0712-17	071534	072044	062354
AMA		1810	1910+13	0808+06	0615-04	0717-16	051934	062044	062755
BRO	1312	1513+15	1514+11	1612+06	2109-06	2224-16	232630	232541	232354
CLL	1307	9900+15	1206+11	1512+06	1506-08	2311-17	232432	223543	224455
CRP	1313	1414+15	1516+10	1411+05	1805-07	2115-17	212331	212641	213454
DAL	9900	9900+14	1408+13	1009+07	1407-07	1106-17	150734	170744	160654
DRT	1112	9900+15	0812+10	0811+05	9900-06	9900-17	240833	230943	211554
ELP		1008	0913+14	0815+07	0414-04	0911-14	021232	011742	362154
HOU	1307	9900+16	1315+10	1311+04	2209-07	2215-17	222632	222842	223854
INK		0912+19	1013+12	0711+06	0415-04	0521-15	042733	032443	021854
LBB		1805+19	9900+12	1007+06	0713-04	0618-16	052134	062044	052455
LRD	1317	1420+15	1515+10	1505+05	9900-06	2711-16	252032	242942	213554
MRF			0512+12	0412+06	0318-04	0315-15	032532	012443	362054
PSX	1011	1009+15	1211+10	1413+05	2014-06	2121-17	212232	213042	213654
SAT	1418	1318+15	1415+10	1409+05	9900-07	2209-17	221932	222542	213854
SPS	1307	0608+16	1308+13	0908+08	1009-05	0818-17	081934	071944	052555
T01	1106	0806+15	1007+10	1207+05	2012-07	2321-17	222631	222841	213454
T06	1410	9900+15	9900+10	9900+05	2011-07	2518-17	242831	233241	234354
T07	1209	9900+16	9900+10	9900+05	2213-07	2420-17	252532	253241	255054
4J3	1607	1905+16	9900+11	9900+05	2313-07	2522-17	252732	262842	273955
H51	1112	1212+15	1312+11	1511+06	1607-06	2216-16	222631	223041	212954
H52	1313	1712+16	1914+11	2014+06	2705-06	2413-16	242630	233641	234153
H61	9900	9900+16	2007+11	2113+06	2216-06	2418-16	261831	283841	274454

Figure 4-37. *Sample FD report.*

modifications to the code. This discussion of METAR will cover elements used in the United States.

Example:

METAR KGGG 161753Z AUTO 14021G26 3/4SM +TSRA BR BKN008 OVC012CB 18/17 A2970 RMK PRESFR

A typical METAR report contains the following information in sequential order:

1. Type of report—the first of two types of METAR reports is the routine METAR report that is transmitted every hour. The second is the aviation selected special weather report (SPECI). This is a special report that can be given at any time to update the METAR for rapidly changing weather conditions, aircraft mishaps, or other critical information.
2. Station identifier—each station is identified by a four-letter code as established by the International Civil Aviation Organization (ICAO). In the 48 contiguous states, a unique three-letter identifier is preceded by the letter "K." For example, Gregg County Airport in Longview, Texas, is identified by the letters "KGGG," K being the country designation and GGG being the airport identifier. In other regions of the world, including Alaska and Hawaii, the first two letters of the four-letter ICAO identifier indicate the region, country, or state. Alaska identifiers always begin with the letters "PA" and Hawaii identifiers always begin with the letters "PH." A list of station identifiers can be found at an FSS or NWS office or online at www.eurocontrol.int/icaoref/.
3. Date and time of report—are depicted in a six-digit group (e.g., 161753Z). The first two digits of the six-digit group are the date. The last four digits are the time of the METAR, which is always given in Coordinated Universal Time (UTC). A "Z" is appended to the end of the time to denote the time is given in Zulu time (UTC) as opposed to local time.
4. Modifier—denotes that the METAR came from an automated source or that the report was corrected. If the notation "AUTO" is listed in the METAR, the report came from an automated source. It also lists "AO1" or "AO2" in the remarks section to indicate the type of precipitation sensors employed at the automated station. When the modifier "COR" is used, it identifies a corrected report sent out to replace an earlier report that contained an error (for example: METAR KGGG 161753Z COR).
5. Wind—reported with five digits (e.g., 14021) unless the speed is greater than 99 knots, which requires six digits. The first three digits indicate the direction the wind is blowing in tens of degrees. If the wind is variable, it is reported as "VRB." The last two digits indicate the speed of the wind in knots (KT) unless the wind is greater than 99 knots, which requires three digits. If the winds are gusting, the letter "G" follows the wind speed. After the letter "G," the peak gust recorded is provided (e.g., 21G26). If the wind varies more than 60° and the wind speed is greater than 6 knots, a separate group of numbers, separated by a "V," will indicate the extremes of the wind directions.
6. Visibility—the prevailing visibility (e.g., 3/4 SM) is reported in statute miles as denoted by the letters "SM." It is reported in both miles and fractions of miles. At times, runway visual range (RVR) is reported following the prevailing visibility. RVR is the distance a pilot can see down the runway in a moving aircraft. When RVR is reported, it is shown with an R, then the runway number followed by a slant, then the visual range in feet. For example, when the RVR is reported as R17L/1400FT, it translates to a visual range of 1,400 feet on runway 17 left.

7. Weather—broken down into two different categories: qualifiers and weather phenomenon (e.g., +TSRA BR). First, the qualifiers of intensity, proximity, and the descriptor of the weather will be given. The intensity may be light (–), moderate ( ), or heavy (+). Proximity depicts only weather phenomena that are in the airport vicinity. The notation "VC" indicates a specific weather phenomenon is in the vicinity of 5 to 10 miles from the airport. Descriptors are used to describe certain types of precipitation and obscurations. Weather phenomena may be reported as being precipitation, obscurations, and other phenomena such as squalls or funnel clouds. Descriptions of weather phenomena as they begin or end, and hailstone size are also listed in the remarks sections of the report.
8. Sky condition—always reported in the sequence of amount, height, and type or indefinite ceiling/height (vertical visibility) (e.g., BKN008 OVC012CB). The heights of the cloud bases are reported with a three-digit number in hundreds of feet above the ground. Clouds above 12,000 feet are not detected or reported by an automated station. The types of clouds, specifically towering cumulus (TCU) or cumulonimbus (CB) clouds, are reported with their height. Contractions are used to describe the amount of cloud coverage and obscuring phenomena. The amount of sky coverage is reported in eighths of the sky from horizon to horizon.
9. Temperature and dew point—always given in degrees Celsius (e.g., 18/17). Temperatures below 0 °C are preceded by the letter "M" to indicate minus.
10. Altimeter setting—reported as inches of mercury in a four-digit number group, and is always preceded by the letter "A" (e.g., A2970). Rising or falling pressure may also be denoted in the remarks sections as "PRESRR" or "PRESFR" respectively.
11. Remarks—comments may or may not appear in this section of the METAR. The information contained in this section may include wind data, variable visibility, beginning and ending times of particular phenomenon, pressure information, and various other information deemed necessary. An example of a remark regarding weather phenomenon that does not fit in any other category would be: OCNL LTGICCG. This translates as occasional lightning in the clouds, and from cloud to ground. Automated stations also use the remarks section to indicate the equipment needs maintenance. The remarks section always begins with the letters "RMK."

Example:

METAR BTR 161753Z 14021G26 3/4SM -RA BR BKN008 OVC012 18/17 A2970 RMK PRESFR

Explanation:

Type of Report	Routine METAR
Location:	Baton Rouge, Louisiana
Date:	16th day of the month
Time:	1753 Zulu
Modifier:	None shown
Wind Information:	140° at 21 kts gusting to 26 kts
Visibility:	3/4 statute mile
Weather:	light rain and mist
Sky Conditions:	Broken 800 ft, overcast 1,200
Temperature:	18 °C, dew point 17 °C
Altimeter:	29.70 "Hg
Remarks:	Barometric pressure is falling.

Terminal Aerodrome Forecasts (TAF)

A terminal aerodrome forecast (TAF) is a report established for the five statute mile radius around an airport. TAF reports are usually given for larger airports. Each TAF is valid for a 24-hour time period, and is updated four times a day at 0000Z, 0600Z, 1200Z, and 1800Z. TAF utilizes the same descriptors and abbreviations as used in the METAR report.

The terminal forecast includes the following information in sequential order:

1. Type of report—TAF can be either a routine forecast (TAF) or an amended forecast (TAF AMD).
2. ICAO station identifier—The station identifier is the same as that used in a METAR.
3. Date and Time of Origin—reported in a six-digit code. The first two indicate the date, the last four indicate the time. Time is always given in UTC as denoted by the "Z" following the number group.
4. Valid period date and time—the valid forecast time period is reported in a six-digit number group. The first two numbers indicate the date, followed by the two-digit beginning time for the valid period, followed by the two digit ending time.
5. Forecast Wind—the wind direction and speed forecast are reported in a five-digit number group. The first three digits indicate the direction of the wind in reference to true north. The last two digits state the windspeed in knots, as denoted by the letters "KT."

As in the METAR, winds greater than 99 knots are given in three digits.

6. Forecast visibility—reported in statute miles, in whole numbers or fractions. If the forecast visibility is greater than 6 miles, it will be coded as "P6SM."
7. Forecast significant weather—weather phenomenon is coded in the TAF reports in the same format as the METAR. If no significant weather is expected during the forecast time period, the denotation "NSW" will be included in the "becoming" or "temporary" weather groups.
8. Forecast sky condition—reported in the same manner as the METAR. Only cumulonimbus (CB) clouds are forecast in this portion of the TAF report as opposed to CBs and towering cumulus in the METAR.
9. Forecast change group—for any significant weather change forecast to occur during the TAF time period, the expected conditions and time period are included in this group. This information may be shown as From (FM), Becoming (BECMG), and Temporary (TEMPO). "From" is used when a rapid and significant change, usually within an hour, is expected. "Becoming" is used when a gradual change in the weather is expected over a period of no more than 2 hours. "Temporary" is used for temporary fluctuations of weather, expected to last for less than an hour.
10. Probability forecast—a given percentage that describes the probability of thunderstorms and precipitation occurring in the coming hours. This forecast is not used for the first 6 hours of the 24-hour forecast.

Example:

TAF KPIR 111130Z 111212 15012KT P6SM BKN090 TEMPO 1214 5SM BR

FM1500 16015G25KT P6SM SCT040 BKN250

FM0000 14012KT P6SM BKN080 OVC150 PROB40 0004 3SM TSRA BKN030CB

FM0400 1408KT P6SM SCT040 OVC080 TEMPO0408 3SM TSRA OVC030CB

BECMG 0810 32007KT=

Explanation:

Routine TAF for Pierre, South Dakota on the 11th day of the month, at 1130Z valid for 24 hours from 1200Z on the 11th to 1200Z on the 12th wind from 150° at 12 knots visibility greater than 6 statute miles, broken clouds at 9,000 feet, temporarily, between 1200Z and 1400Z, visibility 5 statute miles in mist.

From 1500Z, winds from 160° at 15 knots, gusting to 25 knots, visibility greater than 6 statute miles, clouds scattered at 4,000 feet and broken at 25,000 feet.

From 0000Z wind from 140° at 12 knots; visibility greater than 6 statute miles, clouds broken at 8,000 feet, overcast at 15,000 feet between 0000Z and 0400Z, there is 40 percent probability of visibility 3 statute miles thunderstorm with moderate rain showers clouds broken at 3,000 feet with cumulonimbus clouds.

From 0400Z winds from 140°at 8 knots visibility greater than 6 miles clouds at 4,000 scattered and overcast at 8,000 temporarily between 0400Z and 0800Z visibility 3 miles, thunderstorms with moderate rain showers clouds overcast at 3,000 feet with cumulonimbus clouds.

Becoming between 0800Z and 1000Z wind from 320° at 7 knots end of report (=).

Area Forecasts (FA)

The aviation area forecast (FA) gives a picture of clouds, general weather conditions, and visual meteorological conditions (VMC) expected over a large area encompassing several states. There are six areas for which area forecasts are published in the contiguous 48 states. Area forecasts are issued three times a day and are valid for 18 hours. This type of forecast gives information vital to en route operations as well as forecast information for smaller airports that do not have terminal forecasts.

Inflight Weather Advisories

Inflight weather advisories are forecasts provided to en route aircraft, and detail potentially hazardous weather. These advisories are also available to pilots prior to departure for flight planning purposes.

Airman's Meteorological Information (AIRMET)

AIRMETs (report type designator WA) are issued every 6 hours with intermediate updates issued as needed for a particular area forecast region. The information contained in an AIRMET is of operational interest to all aircraft, but the weather section concerns phenomena considered potentially hazardous to light aircraft and aircraft with limited operational capabilities. An AIRMET includes forecast of moderate icing, moderate turbulence, sustained surface winds of 30 knots or greater, widespread areas of ceilings less than 1,000 feet and/or visibilities less than 3 miles, and extensive mountain obscurities. Each AIRMET bulletin has a fixed alphanumeric designator, and is numbered sequentially for easy identification, beginning with the first issuance of the day. SIERRA is the AIRMET code used to denote IFR and mountain obscuration; TANGO is used to denote turbulence, strong surface winds, and low-level wind shear; and ZULU is used to denote icing and freezing levels.

Example:

DFWT WA 241650
AIRMET TANGO UPDT 3 FOR TURBC... STG SFC WINDS AND LLWS VALID UNTIL 242000 AIRMET TURBC... OK TX...UPDT FROM OKC TO DFW TO SAT TO MAF TO CDS TO OKC OCNL MDT TURBC BLO 60 DUE TO STG AND GUSTY LOW LVL WINDS. CONDS CONTG BYD 2000Z

Significant Meteorological Information (SIGMET)

SIGMETs (report type designator WS) are inflight advisories concerning nonconvective weather that is potentially hazardous to all aircraft. They report weather forecasts that include severe icing not associated with thunderstorms, severe or extreme turbulence or clear air turbulence (CAT) not associated with thunderstorms, dust storms, or sandstorms that lower surface or inflight visibilities to below 3 miles, and volcanic ash. SIGMETs are unscheduled forecasts that are valid for 4 hours, but if the SIGMET relates to hurricanes, it is valid for 6 hours.

A SIGMET is issued under an alphabetic identifier, from November through Yankee, excluding Sierra and Tango. The first issuance of a SIGMET is designated as a UWS, or Urgent Weather SIGMET. Re-issued SIGMETs for the same weather phenomenon are sequentially numbered until the weather phenomenon ends.

Example:

SFOR WS 100130 SIGMET ROME02 VALID UNTIL 100530 OR WA FROM SEA TO PDT TO EUG TO SEA OCNL MOGR CAT BTN 280 AND 350 EXPCD DUE TO JTSTR. CONDS BGNG AFT 0200Z CONTG BYD 0530Z.

Convective Significant Meteorological Information (WST)

A Convective SIGMET (WST) is an inflight weather advisory issued for hazardous convective weather that affects the safety of every flight. Convective SIGMETs are issued for severe thunderstorms with surface winds greater than 50 knots, hail at the surface greater than or equal to ¾ inch in diameter, or tornadoes. They are also issued to advise pilots of embedded thunderstorms, lines of thunderstorms, or thunderstorms with heavy or greater precipitation that affect 40 percent or more of a 3,000 square foot or greater region.

Convective SIGMETs are issued for each area of the contiguous 48 states but not Alaska or Hawaii. Convective SIGMETs are issued for the eastern (E), western (W), and central (C) United States. Each report is issued at 55 minutes past the hour, but special reports can be issued during the interim for any reason. Each forecast is valid for 2 hours. They are numbered sequentially each day from 1-99, beginning at 00 Zulu time. If no hazardous weather exists, the Convective SIGMET will still be issued; however, it will state "CONVECTIVE SIGMET.... NONE."

Example:

MKCC WST 221855 CONVECTIVE SIGMET 21C VALID UNTIL 2055 KS OK TX
VCNTY GLD-CDS LINE NO SGFNT TSTMS RPRTD LINE TSTMS DVLPG BY 1955Z WILL MOV EWD 30-35 KT THRU 2055Z HAIL TO 2 IN PSBL

Pilot Weather Reports (PIREPs)

Pilot weather reports (PIREPs) provide valuable information regarding the conditions as they actually exist in the air, and cannot be gathered from any other source. Pilots can confirm the height of bases and tops of clouds, locations of wind shear and turbulence, and the location of inflight icing. If the ceiling is below 5,000 feet, or visibility is at or below 5 miles, ATC facilities are required to solicit PIREPs from pilots in the area. When unexpected weather conditions are encountered, pilots are encouraged to make a report to an FSS or ATC. When a pilot weather report is filed, the ATC facility or FSS will add it to the distribution system to brief other pilots and provide inflight advisories.

PIREPs are easy to file, and a standard reporting form outlines the manner in which they should be filed. *Figure 11-4* shows the elements of a PIREP form. Item numbers 1 through 5 are required information when making a report, as well as at least one weather phenomenon encountered. PIREPs are normally transmitted as an individual report, but may be appended to a surface report. Pilot reports are easily decoded and most contractions used in the reports are self-explanatory.

Example:

UA/OV GGG 090025/ M 1450/ FL 060/ TP C182/ SK 080 OVC/ WX FV 04R/ TA 05/ WV 270030/ TB LGT/ RM HVY RAIN

Explanation:

Type:	Routine pilot report
Location:	25 NM out on the 090° radial, Gregg County VOR
Time:	1450 Zulu
Altitude or Flight Level:	6,000 feet
Aircraft Type:	Cessna 182
Sky Cover:	8,000 overcast

Visibility/Weather:	4 miles in rain
Temperature:	5° Celsius
Wind:	270° at 30 knots
Turbulence:	Light
Icing:	None reported
Remarks:	Rain is heavy

Chapter Summary

A thorough understanding of the weather is a "make or break" item for the balloon pilot; without a complete picture of the weather, a pilot may make an ill-advised decision to launch that may result in injury, damage to the balloon, or worse. It is imperative that a pilot use as many resources as he can, understanding the variables potentially affecting flight, and making an informed decision to conduct a safe flight.

Perhaps the most valuable point to be made is that the balloon pilot must use and exercise common sense. When flying a balloon, the most desirable conditions are good visibility, light winds, and no precipitation. Anything other than that scenario should be reason to pause and consider the possible outcome of a launch. There is never an absolute requirement to fly—there is always the possibility of making the decision to try again another day.

Chapter 5

The National Airspace System

Introduction

The National Airspace System (NAS) is the network of all components regarding airspace in the United States. This comprehensive label includes air navigation facilities, equipment, services, airports or landing areas, aeronautical charts, information/services, rules, regulations, procedures, technical information, manpower, and material. Many of these system components are shared jointly with the military. To conform to international aviation standards, the United States adopted the primary elements of the classification system developed by the International Civil Aviation Organization (ICAO). This chapter provides a general discussion of airspace classification. Detailed information on the classification of airspace, operating procedures, and restrictions is found in the Aeronautical Information Manual (AIM).

This handbook departs from the conventional norm, in that the airspace discussions are presented in reverse order, in the belief that it is much easier to learn the airspace from least complicated to most complicated; also, the information presented for basic visual flight rules (VFR) weather minimums is only that necessary for balloon operations. *[Figure 5-1]* Other information, specifically for instrument flight rules (IFR) flying or night flight, has been omitted, as it has little or no value to the balloon pilot. For a complete discussion of these issues, the balloon pilot should refer to the AIM, or FAA-H-8083-25, Pilot's Handbook of Aeronautical Knowledge.

Airspace Classification

The two categories of airspace are regulatory and non-regulatory. Within these two categories, there are four types: uncontrolled, controlled, special use, and other airspace. *Figure 5-2* presents a profile view of the dimensions of various classes of airspace. It will be helpful to refer to this figure as this chapter is studied, as well as remembering that almost all sport balloon operations are conducted in Class E or G airspace.

Uncontrolled Airspace

Class G Airspace

Class G or uncontrolled airspace is the portion of the airspace that has not been designated as Class A, B, C, D, or E. Class G airspace extends from the surface to the base of the overlying Class E airspace. Although air traffic control (ATC) has no authority or responsibility to control air traffic within Class G airspace, pilots should remember there are VFR minimums which apply while operating in it. Class G airspace is essentially uncontrolled by ATC except when associated with a temporary control tower. *[Figures 5-2 and 5-3]*

In the eastern half of the United States, most Class G airspace is overlaid with Class E airspace, beginning at either 700 or 1,200 feet above ground level (AGL). In the western half of the United States, much of the Class G airspace goes up to 14,500 feet before the Class E airspace begins. The pilot is advised to consult the appropriate sectional chart to ensure that he or she is aware of the airspace limits prior to flight in an unfamiliar area.

Most balloon operations are conducted in Class G airspace. There are no communications requirements to operate in Class G airspace. If operations are conducted at an altitude of 1,200

Basic VFR Weather Minimums			
Airspace		Flight visibility	Distance from clouds
Class G	1,200 feet or less above the surface (regardless of MSL altitude).	1 statute mile	clear of clouds
	More than 1,200 feet above the surface but less than 10,000 feet MSL.	1 statute mile	1,000 feet above 500 feet below 2,000 feet horizontal
Class E	At or above 10,000 feet MSL	5 statute miles	1,000 feet above 1,000 feet below 1 statute mile horizontal
	Less than 10,000 feet MSL	3 statute miles	1,000 feet above 500 feet below 2,000 feet horizontal
Class D		3 statute miles	1,000 feet above 500 feet below 2,000 feet horizontal
Class C		3 statute miles	1,000 feet above 500 feet below 2,000 feet horizontal
Class B		3 statute miles	Clear of clouds

Figure 5-1. *Basic weather minimums for balloon operations in the different classes of airspace.*

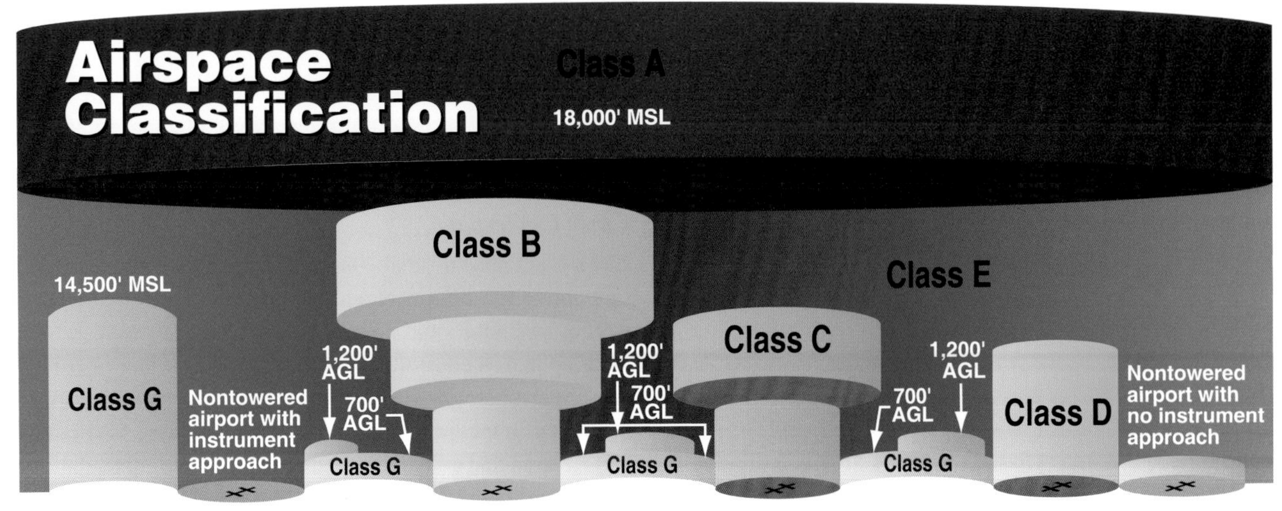

Figure 5-2. *Dimensions of various classes of airspace.*

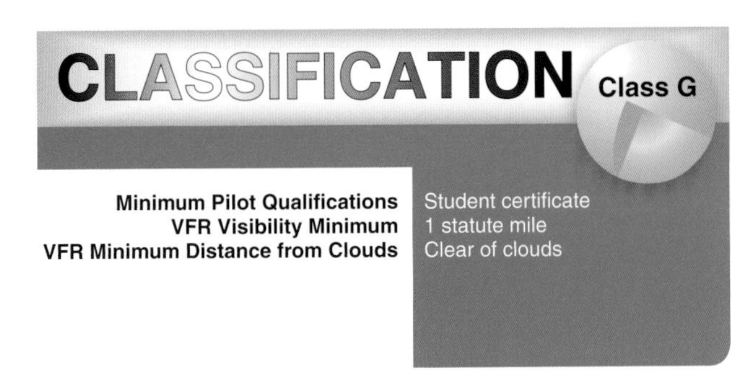

Figure 5-3. *Class G airspace.*

feet or less AGL, the pilot must remain clear of clouds, and there must be one statute mile of visibility. If the operations are conducted more than 1,200 feet AGL, but less than 10,000 feet mean sea level (MSL), visibility requirements remain at one statute mile, but cloud clearances are 1,000 feet above, 500 feet below, and 2,000 feet horizontally from any cloud(s). A popular mnemonic tool used to remember basic cloud clearances is "C152," a popular fixed-wing training aircraft. In this case, the mnemonic recalls, "Clouds 1,000, 500, and 2,000."

Controlled Airspace

Controlled airspace is a generic term that covers the different classifications of airspace and defined dimensions within which ATC service is provided in accordance with the airspace classification. Controlled airspace consists of:

- Class E
- Class D
- Class C
- Class B
- Class A

Class E Airspace

Generally, if the airspace is not Class A, B, C, or D, and is controlled airspace, then it is Class E airspace. Class E airspace extends upward from either the surface or a designated altitude to the overlying or adjacent controlled airspace. *[Figures 5-2 and 5-4]* Also in this class are federal airways and airspace beginning at either 700 or 1,200 feet AGL used to transition to and from the terminal or en route environment. Unless designated at a lower altitude, Class E airspace begins at 14,500 MSL over the United States, including that airspace overlying the waters within 12 nautical miles (NM) of the coast of the 48 contiguous states and Alaska, up to but not including 18,000 feet MSL, and the airspace above FL 600.

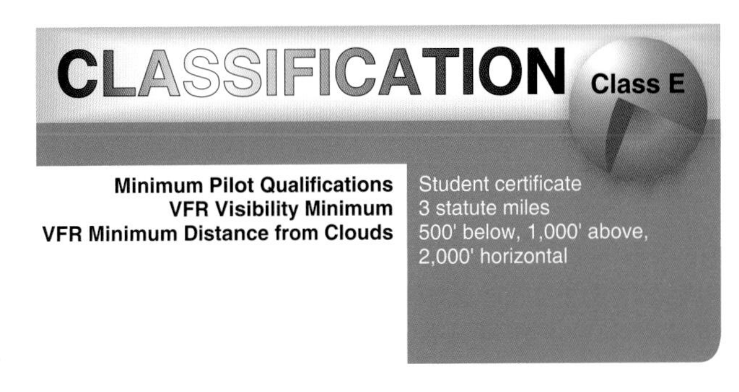

Figure 5-4. *Class E airspace.*

There are no specific communications requirements associated with Class E airspace; however, some Class E airspace locations are designed to provide approaches for instrument approaches, and a pilot would be prudent to ensure that appropriate communications are established when operating near those areas.

Visibility and cloud clearances for operations conducted in Class E airspace are similar to those of the Class G airspace. If balloon operations are being conducted below 10,000 feet MSL, visibility is three statute miles and the basic VFR cloud clearance requirements of 1,000 feet above, 500 feet below, and 2,000 feet horizontal (remember the C152 mnemonic). Operations above 10,000 feet MSL require visibility of five statute miles and cloud clearance of 1,000 feet above, 1,000 feet below, and one statute mile horizontally.

Class D Airspace

Class D is that airspace from the surface to 2,500 feet AGL (but charted in MSL) surrounding those airports that have an operational control tower. *[Figures 5-2 and 5-5]* The configuration of each Class D airspace area is individually tailored. When instrument procedures are published, the airspace is normally designed to contain the procedures.

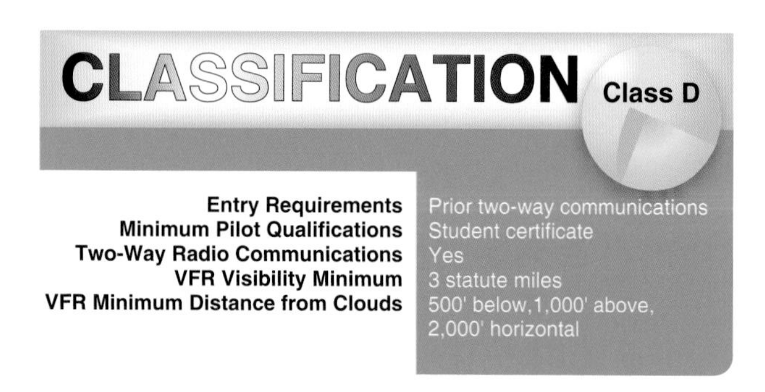

CLASSIFICATION Class D

Entry Requirements	Prior two-way communications
Minimum Pilot Qualifications	Student certificate
Two-Way Radio Communications	Yes
VFR Visibility Minimum	3 statute miles
VFR Minimum Distance from Clouds	500' below,1,000' above, 2,000' horizontal

Figure 5-5. *Class D airspace.*

Unless otherwise authorized, each aircraft must establish two-way radio communications with the ATC facility providing air traffic services prior to entering the airspace and thereafter maintain those communications while in the airspace. Balloon pilots approaching Class D airspace should call the ATC facility at a point where, if entry is denied, a safe landing can still be executed. It is important to understand that if the controller responds to the initial radio call without using the balloon's call sign, radio communications have not been established, and the balloon may not enter the Class D airspace. Many airports associated with Class D airspace do not operate on a 24-hour-a-day basis. When not in operation, the airspace will normally revert to Class E or G airspace, with no communications requirements. In either case, balloon pilots are reminded that, in the absence of an emergency situation, there is no absolute requirement to land at the airport associated with the Class D airspace. In most cases, the first radio call will be made at a distance of five statute miles from the airport.

Balloon pilots operating in the vicinity of an airport should exercise discretion. While balloons indeed have the right of way over all powered aircraft as stated in Title 14 of the Code of Federal Regulations (14 CFR) part 91, section 91.113, a pilot would cause less disruption by landing well short of the airport. This is nothing more than common courtesy and common sense. The visibility requirement for Class D airspace is three statute miles; cloud clearances are the standard VFR 1,000/500/2,000.

Class C Airspace

Class C airspace normally extends from the surface to 4,000 feet above the airport elevation surrounding those airports having an operational control tower, that are serviced by a radar approach control, and with a certain number of IFR or passenger operations. *[Figures 5-2 and 5-6]* This airspace is charted in feet MSL, and is generally of a five NM radius surface area that extends from the surface to 4,000 feet above the airport elevation, and a 10 NM radius area that extends from 1,200 feet to 4,000 feet above the airport elevation. There is also a noncharted outer area with a 20 NM radius, which extends from the surface to 4,000 feet above the primary airport, and this area may include one or more satellite airports. *[Figure 5-7]*

CLASSIFICATION Class C

Entry Requirements	Prior two-way communications
Minimum Pilot Qualifications	Student certificate
Two-Way Radio Communications	Yes
VFR Visibility Minimum	3 statute miles
VFR Minimum Distance from Clouds	500' below,1,000' above, 2,000' horizontal

Figure 5-6. *Class C airspace.*

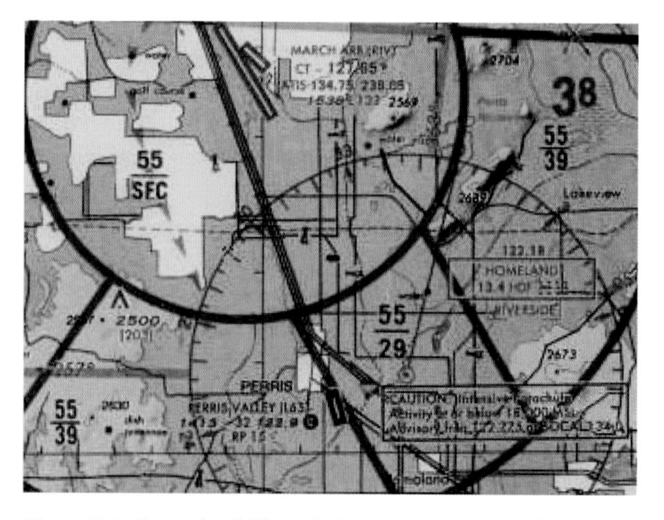

Figure 5-7. *Example of Class C airspace on a sectional.*

Balloon operations in Class C airspace, while technically feasible, are usually not advisable. To fly an aircraft in Class C airspace requires not only a two-way radio to establish and maintain contact with ATC, but also an encoding transponder. Very few balloons carry a transponder in normal operations. As a practical matter, under certain circumstances and with prior coordination, it is possible to enter or depart Class C airspace without an operational transponder, but the practice is generally discouraged. It is possible to fly in the vicinity of a Class C airport; for example, under the lower lateral limits of the outer layer, without needing to penetrate the inner circle. Again, a balloon pilot is well advised to remain clear of the controlled airspace, as a matter of common courtesy and good public relations.

Cloud clearances in Class C airspace are the same as Class D airspace: visibility of three statute miles and a distance from clouds of 1,000/500/2,000.

Class B Airspace

Class B airspace is generally airspace from the surface to 10,000 feet MSL surrounding the nation's busiest airports in terms of airport operations. *[Figures 5-2 and 5-8]* The configuration of each Class B airspace area is individually tailored, consists of a surface area and two or more layers (some Class B airspace areas resemble upside-down wedding cakes). An ATC clearance is required for all aircraft to operate in the area, and all aircraft that are so cleared receive separation services within the airspace.

Figure 5-8. *Class B airspace.*

As a general rule, balloons do not freely operate within Class B airspace. Equipment requirements are the same as for Class C airspace; however, due to air traffic congestion, the balloon pilot requesting entry to Class B airspace will likely be denied entry, as ballooning operations inside the Class B airspace constitute a potential traffic conflict. Most traffic transiting the Class B airspace is being flown under an IFR flight plan, or, in the very least, provide a radar signature which allows ATC to provide traffic separation. Due to their construction, balloons do not provide a radar return much larger than a small flock of birds (and are frequently mistaken as such). Should it become necessary for operational reasons to fly through Class B airspace, that flight should be coordinated at least one hour prior, as provided for by 14 CFR section 91.215. It is permissible, and perfectly legal, to operate a balloon under the lateral limits of the Class B airspace. Balloon pilots operating in the vicinity of Class B airspace are encouraged to be familiar with the airspace, including visual ground reference points when conducting flight under the lateral limits to avoid inadvertently encroaching on Class B airspace. Since aircraft operating in Class B airspace are under radar control, there is a slight difference in the cloud clearance requirements. Visibility remains three statute miles, but the only cloud clearance requirement is that the pilot remain clear of clouds.

Class A Airspace

Class A airspace is generally the airspace from 18,000 feet MSL up to and including FL 600, including the airspace overlying the waters within 12 NM of the coast of the 48 contiguous states and Alaska. Unless otherwise authorized, all operations in Class A airspace are conducted under IFR. For that reason, balloons do not utilize the Class A airspace, unless operated on a waiver or are conducting some special operation, such as a record attempt.

Special Use Airspace

Special Use Airspace (or Special Area of Operation) is the designation for airspace in which certain activities must be confined, or where limitations may be imposed on aircraft operations that are not part of those activities. Certain special use airspace areas can create limitations on the mixed use of airspace. The special use airspace depicted on sectional charts includes the area name or number, effective altitude, time and weather conditions of operation, the controlling agency, and the chart panel location. On National Aeronautical Charting Group (NACG) sectional charts, this information is available on one of the end panels.

Special use airspace usually consists of:

- Prohibited areas
- Restricted areas
- Warning areas
- Military operation areas (MOAs)
- Alert areas
- Controlled firing areas

Prohibited Areas

Prohibited areas contain airspace of defined dimensions within which the flight of aircraft is prohibited. Such areas are established for security or other reasons associated with the national welfare. These areas are published in the Federal Register and are depicted on aeronautical charts. The area is charted as a "P" followed by a number (e.g., "P-49"). *[Figure 5-9]* Examples of prohibited areas include Camp David and the National Mall in Washington, D.C., where the White House and Congressional buildings are located.

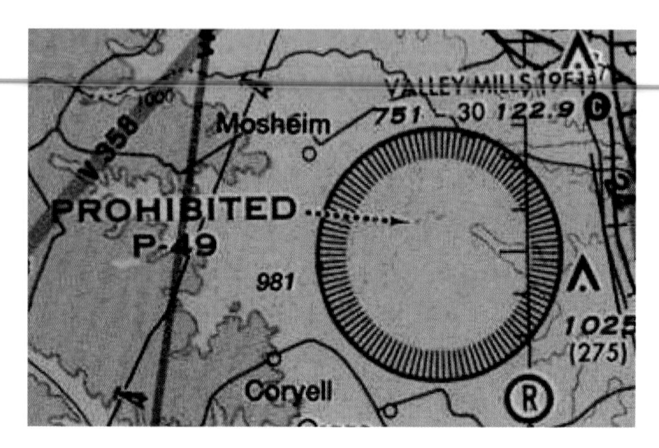

Figure 5-9. *An example of a prohibited area is Crawford, Texas.*

Restricted Areas

Restricted areas are areas where operations are hazardous to nonparticipating aircraft and contain airspace within which the flight of aircraft, while not wholly prohibited, is subject to restrictions. Activities within these areas must be confined because of their nature, or limitations may be imposed upon aircraft operations that are not a part of those activities, or both. Restricted areas denote the existence of unusual, often invisible, hazards to aircraft (e.g., artillery firing, aerial gunnery, or guided missiles). Penetration of restricted areas without authorization from the using or controlling agency may be extremely hazardous to the aircraft and its occupants. ATC facilities apply the following procedures:

1. If the restricted area is not active and has been released to the Federal Aviation Administration (FAA), the ATC facility will allow the aircraft to operate in the restricted airspace without issuing specific clearance for it to do so.
2. If the restricted area is active and has not been released to the FAA, the ATC facility will issue a clearance which will ensure the aircraft avoids the restricted airspace.

Restricted areas are charted with an "R" followed by a number (e.g., "R-4401") and are depicted on the sectional charts. *[Figure 5-10]* Restricted area information can be obtained on the back of the chart. *[Figure 5-11]*

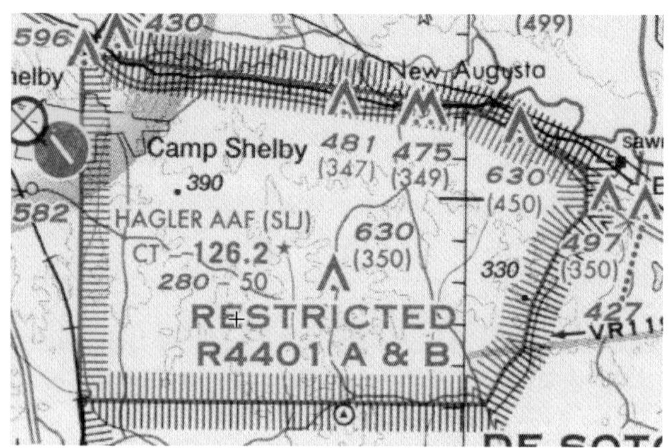

Figure 5-10. *Restricted areas on a sectional chart.*

SPECIAL USE AIRSPACE ON NEW ORLEANS SECTIONAL CHART

Unless otherwise noted altitudes are MSL and in feet. Time is local.
"TO" an altitude means "To and including."
FL – Flight Level
NO A/G – No air to ground communications.
Contact nearest FSS for information.

† Other times by NOTAM
NOTAM – Use of this term in Restricted Areas indicates FAA and DoD NOTAM systems. Use of this term in all other Special Use areas indicates the DoD NOTAM system.

U.S. P–PROHIBITED, R–RESTRICTED, A–ALERT, W–WARNING, MOA–MILITARY OPERATIONS AREA

NUMBER	ALTITUDE	TIME OF USE	CONTROLLING AGENCY/ CONTACT FACILITY	FREQUENCIES
R-2103A	TO BUT NOT INCL 10,000	CONTINUOUS	CAIRNS APP CON	
R-2103B	10,000 TO 15,000	BY NOTAM 6 HRS IN ADVANCE	JACKSONVILLE CNTR	
R-2905A, B	TO 10,000	INTERMITTENT AS ANNOUNCED BY NOTAM	TYNDALL APP CON	
R-2908	TO 12,000	NOV-DEC 0800-1600 MON-FRI † 24 HRS IN ADVANCE	PENSACOLA TRACON	
R-2914A	UNLIMITED EXCL AIRSPACE WITHIN R-2917	CONTINUOUS	JACKSONVILLE CNTR	
R-2914B	8500 TO UNLIMITED	CONTINUOUS	JACKSONVILLE CNTR	
R-2915A, B	UNLIMITED	CONTINUOUS	JACKSONVILLE CNTR	
R-2915C	8500 TO UNLIMITED	CONTINUOUS	JACKSONVILLE CNTR	
R-2917	TO BUT NOT INCL FL 230	CONTINUOUS	EGLIN APP CON	
R-2918	UNLIMITED	CONTINUOUS	JACKSONVILLE CNTR	
R-4401A	TO 4000	BY NOTAM 24 HOURS IN ADVANCE	HOUSTON CNTR	
R-4401B	4000 TO 18,000	BY NOTAM 24 HOURS IN ADVANCE	HOUSTON CNTR	
		† 24 HRS IN ADVANCE		
W-54A	TO FL 400	0700-2400 DAILY†	HOUSTON CNTR	
W-54B	TO BUT NOT INCL FL 240	0700-2400 DAILY†	HOUSTON CNTR	
W-59A	5000 TO FL 500	0900-2100†	HOUSTON CNTR	
W-92	TO FL 400	0700-2400	HOUSTON CNTR	
W-151A, B, C&D	UNLIMITED	0600-0130 DAILY†	JACKSONVILLE CNTR	
W-151E, F	UNLIMITED	INTERMITTENT	JACKSONVILLE CNTR	
W-155A, B	TO FL 600	SR-0100 MON-FRI†	JACKSONVILLE CNTR	
W-155C	TO FL 600	SR-0100 DAILY†	JACKSONVILLE CNTR	
W-453	TO [illegible] 500	INTERMITTENT SR-SS†	HOUSTON CNTR	

Figure 5-11. *Specific information about a restricted area can be found on the back of the sectional chart.*

Warning Areas

Warning areas consist of airspace which may contain hazards to nonparticipating aircraft in international airspace. The activities may be much the same as those for a restricted area. Warning areas are established beyond the three mile limit and are depicted on aeronautical charts. Warning Areas are also defined on the backside of the appropriate sectional chart. *[Figure 5-11]*

Military Operation Areas

Military operation areas (MOA) consist of airspace of defined vertical and lateral limits established for the purpose of separating certain military training activity from IFR traffic. There is no restriction against a pilot operating VFR in these areas; however, a pilot should be alert since training activities may include acrobatic and abrupt maneuvers. MOAs are depicted by name and with defined boundaries on sectional, VFR terminal area, and en route low altitude charts and are not numbered (e.g., "Camden Ridge MOA"). *[Figure 5-12]* However, the MOA is also further defined on the back of the sectional chart with times of operation, altitudes affected, and the controlling agency. *[Figure 5-13]*

Alert Areas

Alert areas are depicted on aeronautical charts with an "A" followed by a number (e.g., "A-211") to inform nonparticipating pilots of areas that may contain a high volume of pilot training or an unusual type of aerial activity. Pilots should exercise caution in alert areas. All activity within an alert area shall be conducted in accordance with regulations, without waiver, and pilots of participating aircraft, as well as pilots transiting the area, shall be equally responsible for collision avoidance. *[Figure 5-14]*

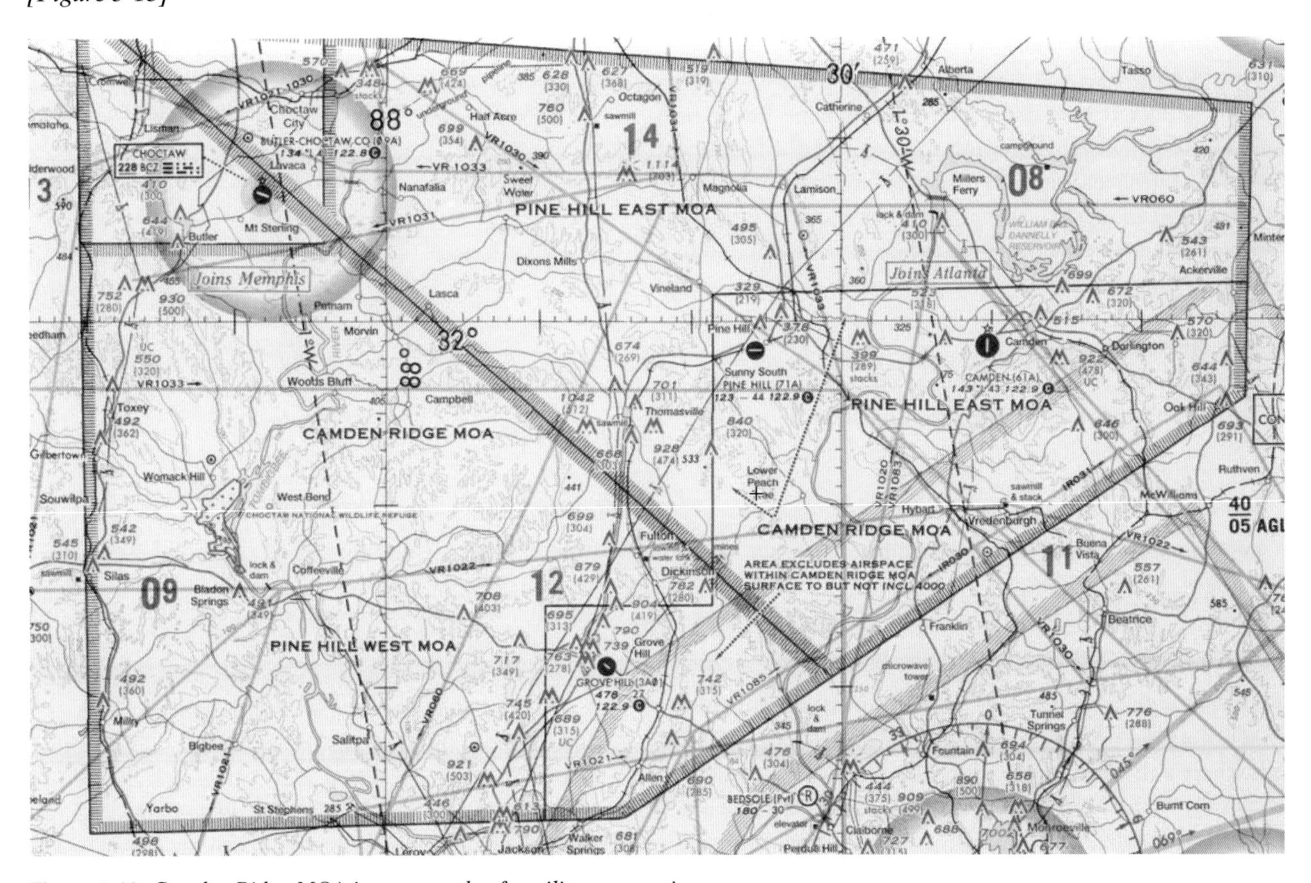

Figure 5-12. *Camden Ridge MOA is an example of a military operations area.*

MOA NAME	ALTITUDE*	TIME OF USE †	CONTROLLING AGENCY/ CONTACT FACILITY	FREQUENCIES
CAMDEN RIDGE	500 AGL TO BUT NOT INCL 10,000	0700-2300 (APPROXIMATELY FIVE HRS PER DAY)	ATLANTA CNTR	
DE SOTO 1	500 AGL TO 10,000	0830-1730 MON-FRI	HOUSTON CNTR	
DE SOTO 2	100 AGL TO 5000	0830-1730 MON-FRI	HOUSTON CNTR	

Figure 5-13. *Additional information found on the sectional chart for Camden Ridge MOA.*

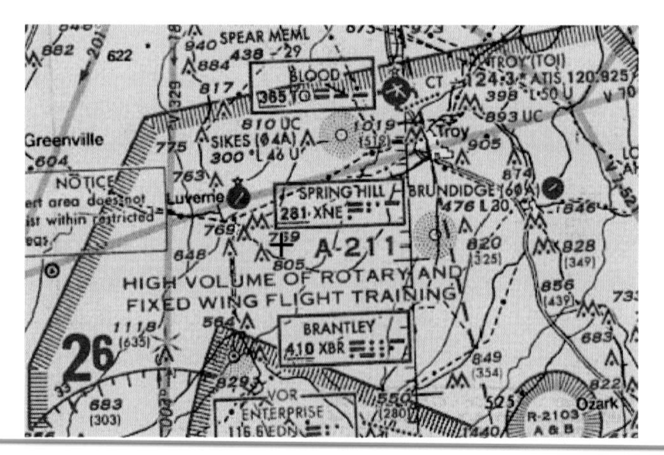

Figure 5-14. *Alert area (A-211).*

Controlled Firing Areas

Controlled firing areas contain military activities, which, if not conducted in a controlled environment, could be hazardous to nonparticipating aircraft. The difference between controlled firing areas and other special use airspace is that activities must be suspended when a spotter aircraft, radar, or ground lookout position indicates an aircraft might be approaching the area.

Other Airspace Areas

"Other airspace areas" is a general term referring to the majority of the remaining airspace. It includes:

- Airport advisory areas
- Military training routes (MTR)
- Temporary flight restrictions (TFR)
- National security areas

Airport Advisory Areas

An airport advisory area is an area within 10 statute miles (SM) of an airport where a control tower is not operating, but where a flight service station (FSS) is located. At these locations, the FSS provides advisory service to arriving and departing aircraft.

Military Training Routes (MTRs)

Military Training Routes (MTRs) are routes used by military aircraft to maintain proficiency in tactical flying. These routes are usually established below 10,000 feet MSL for operations at speeds in excess of 250 knots. Some route segments may be defined at higher altitudes for purposes of route continuity. Routes are identified as IFR (IR), and VFR (VR), followed by a number. MTRs with no segment above 1,500 feet AGL are identified by four number characters (e.g., IR1206, VR1207). MTRs that include one or more segments above 1,500 feet AGL are identified by three number characters (e.g., IR206, VR207). IFR Low Altitude En Route Charts depict all IR routes and all VR routes that accommodate operations above 1,500 feet AGL. IR routes are conducted in accordance with IFR regardless of weather conditions.

MTR are usually indicated with a blue line on the sectional chart. A balloon pilot flying in the area of numerous VRs or IRs should question the briefer during the weather brief to find out if any of the routes are in use, and a possible time frame for opening and closing. While it is true that the balloon pilot has the right of way, the balloon will generally come out worse in a midair conflict with a fast-moving military aircraft. MTRs, such as the example depicted in *Figure 5-15,* are also further defined on the back of the sectional charts.

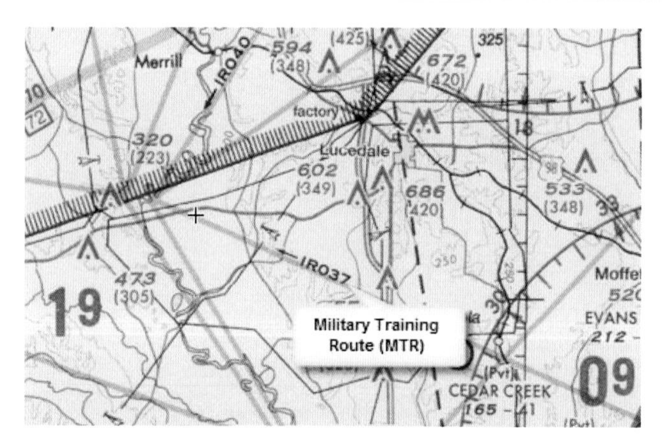

Figure 5-15. *Military training routes (MTR) chart symbols.*

Temporary Flight Restrictions (TFRs)

Temporary flight restrictions (TFRs) are put into effect when traffic in the airspace would endanger or hamper air or ground activities in the designated area. For example, a forest fire, chemical accident, flood, or disaster-relief effort could warrant a TFR, which would be issued as a Notice to Airmen (NOTAM). The NOTAM begins with the phrase "FLIGHT RESTRICTIONS" followed by the location of the temporary restriction, effective time period, area defined in statute miles, and altitudes affected. The NOTAM also contains the FAA coordination facility and telephone number, the reason for the restriction, and any other information deemed appropriate. The pilot should check the NOTAMs as part of flight planning.

The reasons for establishing a temporary restriction are:

- Protect persons and property in the air or on the surface from an existing or imminent hazard;
- Provide a safe environment for the operation of disaster relief aircraft;
- Prevent an unsafe congestion of sightseeing aircraft above an incident or event, which may generate a high degree of public interest;

- Protect declared national disasters for humanitarian reasons;
- Protect the President, Vice President, or other public figures; and
- Provide a safe environment for space agency operations.

Since the events of September 11, the use of TFRs has become much more common. There have been a number of incidents of aircraft incursions into TFRs, which have resulted in pilots undergoing security investigations, and certificate suspensions. It is a pilot's responsibility to be aware of TFRs in their proposed area of flight. One way to check is to visit the FAA's web site, at www.tfr.faa.gov, and verify that there is not a TFR in the area. *[Figure 5-16]*

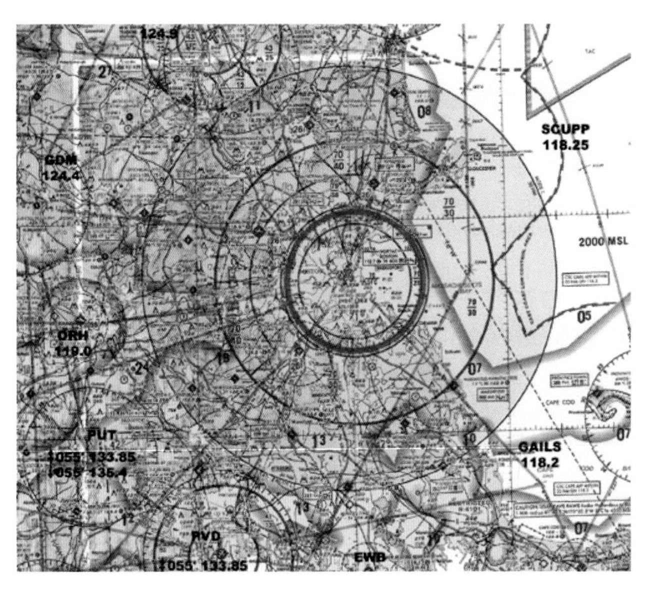

Figure 5-16. *A temporary flight restriction (TFR) imposed at Boston's Logan Airport to general aviation and restricting operations at surrounding airports.*

Another TFR issue that the balloon pilot needs to be aware of are the restrictions imposed under 14 CFR part 91, section 91.145. In this regulation, the FAA has codified restrictions against the overflight of major outdoor events, such as the World Series, the Rose Bowl, and NASCAR events. Several years ago, it was relatively common to see hot air balloons promoting products or services being flown over these events, providing exposure for the commercial sponsor. Now, there are significant distance and altitude restrictions for such events, and any operations inside the TFR area must be conducted under the provisions of a waiver. Should such an operation be contemplated, the balloon pilot should consult with the local Flight Service District Office (FSDO) well in advance of the event.

National Security Areas

National security areas (NSAs) consist of airspace with defined vertical and lateral dimensions established at locations where there is a requirement for increased security and safety of ground facilities. Flight in NSAs may be temporarily prohibited by regulation under the provisions of 14 CFR part 99, and prohibitions will be disseminated via NOTAM.

Radio Communications

When utilizing the NAS, there are many occasions when it is necessary to communicate with a control tower, a controlling agency, or another aircraft. Balloon pilots should have a working knowledge of correct radio procedures in order to properly exercise their privileges, and not interfere with other aviation traffic.

Aviation very-high-frequency (VHF) radios may be used for communications between the balloon pilot and towers, pilot to FSS, air to air (pilot to pilot), air to ground in some circumstances, and to get information from weather facilities and automated stations. VHF radios are not to be used as a "telephone service" or in a Citizen's Band-type of operation; this use is specifically prohibited by the Federal Communications Commission (FCC), and may result in monetary fines.

There is no license requirement for a pilot operating in the United States; however, a pilot who may be operating outside the country is required to hold a restricted radiotelephone permit issued by the FCC. There is also no station license requirement for a handheld radio used in a balloon, as it is not permanently mounted. A station license may be required when a balloon is operated outside the country; it is best to check with the appropriate local authorities before contemplating such an operation.

The single most important aspect of radio communication is clarity. Keeping one's transmissions brief, and using correct terminology and phraseology when operating an aircraft radio is imperative. All frequencies are shared with others; a pilot's idle chatter to a chase crew could inadvertently block another pilot's emergency transmission at an airfield many miles away. A good pilot will minimize radio requirements, and think before speaking.

Phonetic Alphabet

The ICAO, of which the United States is a member, has adopted a phonetic alphabet, which should be used in radio communications as necessary. When communicating with air traffic control facilities, pilots should use this alphabet to identify their aircraft. *[Figure 5-15]*

Character	Telephony	Phonic (Pronunciation)
A	Alfa	(AL-FAH)
B	Bravo	(BRAH-VOH)
C	Charlie	(CHAR-LEE) or (SHAR-LEE)
D	Delta	(DELL-TAH)
E	Echo	(ECK-OH)
F	Foxtrot	(FOKS-TROT)
G	Golf	(GOLF)
H	Hotel	(HOH-TEL)
I	India	(IN-DEE-AH)
J	Juliett	(JEW-LEE-ETT)
K	Kilo	(KEY-LOH)
L	Lima	(LEE-MAH)
M	Mike	(MIKE)
N	November	(NO-VEM-BER)
O	Oscar	(OSS-CAH)
P	Papa	(PAH-PAH)
Q	Quebec	(KEH-BECK)
R	Romeo	(ROW-ME-OH)
S	Sierra	(SEE-AIR-RAH)
T	Tango	(TANG-GO)
U	Uniform	(YOU-NEE-FORM) or (OO-NEE-FORM)
V	Victor	(VIK-TAH)
W	Whiskey	(WISS-KEY)
X	Xray	(ECKS-RAY)
Y	Yankee	(YANG-KEY)
Z	Zulu	(ZOO-LOO)
1	One	(WUN)
2	Two	(TOO)
3	Three	(TREE)
4	Four	(FOW-ER)
5	Five	(FIFE)
6	Six	(SIX)
7	Seven	(SEV-EN)
8	Eight	(AIT)
9	Nine	(NIN-ER)
0	Zero	(ZEE-RO)

Figure 5-17. *Phonetic alphabet.*

Radio Procedures

The Pilot/Controller Glossary (found in the AIM) contains correct language for communications between a pilot at ATC facilities. Good phraseology enhances safety and is the mark of a conscientious pilot. Jargon, slang, and "CB" codes should not be used in radio communications.

A pilot should listen to the radio before transmitting. Do not clutter the frequency with inappropriate requests or irrelevant information. Most pilots will think before keying their radio, and keep the transmission as brief as possible. The pilot should be occupied with control of the balloon and should not be spending time on the radio.

When Departing a Nontowered Airport

A balloon pilot departing a nontowered airport should make a call on the field's UNICOM or CTAF (Common Traffic Advisory Frequency), both of which are generally shown on the appropriate sectional chart. The pilot should state his or her location and intentions, as a "self-announce" advisory, so that other air traffic in the area is aware of the operation.

Example: "Williamson traffic, balloon 3584 Golf departing the grass near hanger 12 to the east."

When Approaching a Nontowered Airport

If an approach to a nontowered airport is conducted, with the intent to land, the pilot should again "self-announce" on the appropriate UNICOM frequency or CTAF. The current location should be stated, as well as the pilot's intentions.

Example: "Air Acres traffic, balloon 3584 Golf approaching from the west for landing at the departure end of the runway."

When Transiting a Nontowered Facility

If flight in the vicinity of a nontowered airport is likely, with no intent to land at the airfield, it is appropriate to let other aircraft in the area know the intentions, particularly if there is the possibility of interfering with the traffic pattern.

Example: "Enterprise traffic, balloon 7510 Delta at 700 feet, transiting from east to west along the north boundary of the field, no landing."

When Approaching Controlled Airspace

When approaching controlled airspace, and it appears that entry into the airspace is likely (or desired), the balloon pilot should contact the appropriate approach control with enough time to execute a landing, should entry be denied. Remember that a balloon usually does not have the option of turning around.

Light Gun Signals		
Color and type of signal	**Aircraft on the ground**	**Aircraft in flight**
Steady green	Cleared for takeoff	Cleared to land
Steady red	Stop	Give way to other aircraft and continue circling
Flashing red	Taxi clear of the runway in use	Airport unsafe, do not land
Alternating red and green	Exercise extreme caution	Exercise extreme caution

Figure 5-18. *Light gun signals.*

Example: "Huntsville Approach, balloon 903 Hotel Delta, at 3,000."

The approach control facility (in this example, Huntsville Approach) will either reply by saying, "Aircraft calling Huntsville Approach, standby", in which case the balloon pilot does not have a clearance to enter the airspace; or, approach will reply, saying "Balloon 903 Hotel Delta, Huntsville, standby." In this case, the aircraft call sign was used, and the pilot may enter the airspace. Further communication may continue until the balloon has either landed or cleared the airspace.

In the event that radio communication has failed, or that no radio is available in the balloon, and continued flight into the immediate vicinity of an towered airport facility cannot be avoided, control towers have a "light gun" available. This device, which looks like an oversized flare gun, allows tower personnel to "shoot" a directed beam of light at an approaching aircraft, and may signal field availability options to the approaching balloon. *[Figure 5-16]*

Many balloon pilots fly for years without using a radio of any kind. Using an aircraft radio is not a difficult task, and, with practice, can provide a more enjoyable flight. Most ATCs express an interest in balloons, and go out of their way to assist balloon pilots in accomplishing their flight safely.

Chapter Summary

At first glance, the NAS appears to be a complex arena in which to operate such a simple aircraft. This chapter simplifies the airspace to the reader, and makes it readily apparent that it is possible to operate a hot air balloon safely, without causing conflict.

Simple courtesy and common sense goes a long way in airspace operations. A complete and thorough understanding of the airspace, combined with good decision-making, will allow the pilot to do what he or she wishes, with recognition of the needs of other users of the sky.

Chapter 6

Layout To Launch

Introduction

This chapter introduces inflation and launch of the balloon. It also provides useful information on checklists, crew and crew management, false lift, and landowner relations.

Preflight Operations

The preflight, as an aeronautical term, is generally agreed to be the airworthiness check of an aircraft immediately before flight. In the broadest sense, preflight is everything accomplished in preparation for a flight. In this chapter, preflight operations occur at the balloon launch site, up to and including the preflight inspection.

Checklists

The value of using a checklist is well known to the airlines and the military. Regulations require air carrier pilots and military pilots to use checklists. Also, Federal Aviation Administration (FAA) practical tests require pilot certificate applicants to use checklists. Checklists are effective and contribute to safe flying because routine and familiarity breed complacency. Like military and airline pilots, balloonists who fly every day need a checklist to ensure nothing is omitted. For example, professional balloon ride operators are subject to distractions and interruptions during their preflight, layout, assembly, and inspection. Infrequent balloon flyers, which include most balloonists, need checklists because long periods of inactivity create memory lapses. A typical balloonist may make only 25 to 30 flights per year or less. A checklist does not replace proficiency, but it helps. *[Figure 6-1]*

Figure 6-1. *Using a checklist.*

A checklist can also save time. By arranging the layout, assembly, and inspection in a logical order, and accomplishing more than one task at a time, duplication and wasted time is minimized. For instance, a properly arranged preflight checklist includes many tasks that are performed while the fan is running, so people are not just standing around waiting for the envelope to inflate. Also, a checklist eliminates needless walking. Students and new pilots need checklists because they are forming habit patterns, and need prompting to reinforce training and confirm good habits.

There are two theories regarding checklists. One popular procedure is the "call-and-response" method. One person reads the checklist, and the pilot responds that the item is in the proper configuration. This is rather time consuming, and probably not appropriate for ballooning activities. At the other extreme are large groupings of components and items to check with a casual glance to confirm that those items look correct. Perhaps the best methodology suited for ballooning is smaller groupings (i.e., basket, burner and fuel system, instruments, envelope security) and using the checklist as an outline to ensure that each item is checked. It may be appropriate to create a habit pattern of physically touching particular items to ensure security and proper operation. Pilots are cautioned that a checklist is not necessarily a "to-do" list. Flight training should emphasize proper procedures and habits, with a checklist used to confirm that tasks have been completed.

Preflight Inspection Checklist

Title 14 of the Code of Federal Regulations (14 CFR) and the practical test standards (PTS) for balloon pilot certification require the pilot to inspect the balloon by systematically utilizing an appropriate checklist prior to each flight. Most balloon manufacturers include a preflight inspection checklist in the flight manual. This may be used as the basis for a personalized preflight checklist. Each balloon manufacturer lists maximum allowable damage with which a balloon may fly and still be considered airworthy. Balloon pilots should be familiar with the manufacturer's maximum allowable damage rules for their balloon and abide by them.

Using a written checklist, the pilot should make certain that the balloon is correctly laid out for inflation, all control lines are attached, the fuel system is operating correctly, maximum allowable damage limits are not exceeded, and all components are ready for flight. The pilot is responsible for all aspects of flight, including preflight operations.

The best checklist is one each pilot develops for his or her individual balloon. A good source for checklist items is the manufacturer of the balloon. Also, checklists from other pilots and manufacturers may be incorporated into a personal checklist. Remember, a checklist is a living document that changes when modifications or additions are appropriate.

Emergency Checklist

Carefully study and memorize emergency checklists. Do not try to read a checklist during an emergency; that is for an aircraft with two- or three-person crews and at high altitude. During an emergency, take prompt action to resolve the problem, and when the situation permits, refer to the balloon's flight manual to ensure all necessary items have been accomplished. The single most important action in an emergency is to continue to fly the balloon and regain control of the situation. Appendix C contains sample checklists that may be used or modified to meet specific requirements for a particular balloon system or flight operation.

Weather Brief

As previously discussed, a weather brief must be obtained prior to flight; as a procedural matter, the information from the briefing sets the parameters for the flight (launch site, flight track, and potential landing areas).

All electronic sites and briefings represent the starting point for how weather affects the flight. A balloon pilot should *not* make the mistake of gathering weather data, getting a forecast, deciding to fly, and then paying less attention to actual weather cues. Weather monitoring is an ongoing process that starts on the drive to the launch field and continues until the balloon is packed after landing. Continue gathering weather data for the flight until safely back on the ground.

Despite using the most current and complete electronic weather data available to consider before driving out for the flight, each pilot must never lose sight of the fact that hot air balloons are visual flight rules (VFR) aircraft. Many pilots lose all access to unlimited or even selected electronic weather information en route to or at the launch site. Unless the launch site is a reporting station for any weather service, a pilot is required to visually interpret and evaluate weather data specific to the launch site. Local surface conditions—particularly winds below 500'–800'—show the greatest variability of both speed and direction. These winds are perhaps of greatest importance in determining the conditions and risks for launch and landing. A number of simple observations enable the balloon pilot to gather this data at the launch site, as well as during the flight. *[Figure 6-2]*

Simple Wind Observations
Smoke begins drifting off of vertical at 2–3 mph
Most deciduous tree leaves begin rustling above 4–5 mph (except aspens, birch, and other broadleaf trees, which respond at lower wind speeds)
Twigs sway and flags extend around 7–8 mph
Dust and paper kick up off the ground above 12 mph
Wires and ropes whistle above 20 mph

Figure 6-2. *Simple wind observations.*

The balloon can sometimes provide valuable weather information provided the pilot knows what to look for. Fabrics on a laid-out envelope begin rippling around eight miles per hour (mph), while 10–12 mph sends small bubbles of air past a closed throat and up along the load tapes. Such information, perhaps unavailable or inaccurate elsewhere, can prove invaluable, especially during the initial phase of flight.

Each pilot must sharpen his or her ability to read and interpret local surface conditions for one primary reason: up to 1 hour may elapse between gathering weather data online and laying out at the launch site. Weather patterns can change in far less time than this, and local conditions can change almost instantly.

Some pilots choose to rely on a portable weather station or handheld wind meter to monitor local conditions at the launch site. Cost, size, portability, and features vary among brands, but each can provide useful information on wind speed/direction, maximum gusts, and other weather data. These can range from a small $100 handheld wind meter to a $500 weather station with 15" mast and dashboard data display. *[Figure 6-3]*

Figure 6-3. *Checking wind velocity with a handheld wind meter.*

Perhaps the cheapest yet most valuable launch site weather aid is the pibal or pilot balloon, a 10–14" helium-filled balloon released at the launch site for visual weather cues and forecast confirmation. Some pibal actions clearly mean cancel the flight: rolling along the ground, not clearing obstacles, climbing then diving, or dramatic speed/direction changes (which indicate wind shears). To read and confirm winds at different altitudes, release several pibals at 10–20 second intervals. Fill the first with 100 percent helium, the second with mostly helium and a breath of air from your mouth, and the third with helium and two breaths. This creates different rates of ascent which helps gain a better perspective of the low level winds. Every pilot benefits from having pibals in their vehicle and using them before launch. Appendix D lists various rates of climb for different sizes of pibals; using this knowledge, the pilot can determine the altitudes at which the winds change.

Every pilot and crewmember can build a mental database of visual cues that report weather data. More important than prelaunch weather is inflight weather, which is further removed in time from forecast conditions (electronic or briefings) but can change just as fast and dramatically. Wind speed or direction changes can signal developing rain or frontal movement. Radar returns require human interpretation and can miss developing conditions or provide false echoes off birds or buildings. Weather-savvy pilots and crew gather weather information from whatever source is available to them anywhere, at any time.

Performance Planning/Fuel Planning

The good balloon pilot checks the balloon's performance charts, considering that day's conditions, and comes to a reasonable conclusion regarding the limitations of the balloon for that particular flight. Prior to flight (and perhaps even prior to layout), fuel gauges should be checked to ensure there is sufficient fuel on board to conduct the planned flight.

The Chase Crew

Generally, there are two different areas of responsibility for a crew: inflation/launch and chase/recovery. Both are usually referred to as ground crew. Passengers often serve as inflation crew, become passengers for the flight, and crew again after the balloon has landed. Before the noise and activity make discussion difficult, the pilot should give the crew briefing and discuss any requirements before inflation.

Number of Crewmembers

The number of crewmembers is a matter of individual preference and depends upon the size of the balloon, purpose of the flight, terrain, and other factors. When using a small balloon for instruction, a crew of three people, including the instructor, student, and one crew, may be sufficient. For passenger flights in large balloons, a larger number of crewmembers are generally required. Having too many people in the chase vehicle may be a distraction for the driver. The recommended minimum crew for the vehicle is two: one person to drive and the other to navigate and serve as a balloon "spotter." This is perhaps the safest method.

The optimum size inflation crew for a sport balloon is four people—the pilot operating the burner, two people holding the mouth open, and one person on the crown line. Some pilots opt to have a fifth person serve as a fan operator; this is a matter of technique and personal preference. Many pilots prefer a larger number of crewmembers; however, it is important to be aware that too many crewmembers may often be working against each other due to lack of coordination. In windy or crowded situations, it is important to have a person holding the crown line. If inflation requires more crew than usual, due to the windy conditions, consideration should be given to canceling the flight. Although the balloon may get airborne, chances are that flying out of one's comfort zone and having to prepare for a very windy landing may impair concentration. The distraction may hinder safe, enjoyable flying.

Clothing

Crewmembers, as well as the pilot, should be clothed for safety and comfort. Cover or restrain long hair. Scarves, hanging jewelry, or loose eyeglasses can interfere with smooth setup, and can potentially be very dangerous, particularly near the inflation fan. Long sleeves and long trousers made of cotton instead of synthetic material are recommended. Try to wear clothes in layers since temperatures can change quite a bit from before sunrise to the recovery. Proper clothing protects participants from burns, poison oak/ivy, and other harmful plants.

All crewmembers should wear gloves, preferably smooth leather, loose fitting, and easy to remove. Wear comfortable and protective footwear. If it becomes necessary to walk or hike from a landing site inaccessible by the chase vehicle, proper clothing and footwear make the task easier and less hazardous.

Types of Flight

Knowing the function or purpose of a flight is important to the crew, so they know the goals of the operation; possible time aloft; probable direction(s) of flight; probable altitudes; communications, if any, to be used; and useful maps or charts. Balloon flights can be classified into several different types: paid passenger, instruction, race, rally, advertising/promotion, and fun.

Many commercial balloon pilots defray the cost of the sport by offering paid passenger rides. The crew should know that these passengers are paying for the privilege, and may have been promised a certain type and length of flight. Instructional flights require the crew follow the direction of the instructor, so the student may see and participate as much as possible. The crew should work closely with the instructor and student and not take over any portion of the operation, thus denying the student the opportunity to learn. For competitions, crew responsibilities may be different. The pilot may have only a single goal in mind and focus on that goal. The crew's job is always to help the pilot, but in the case of the competition flight, the crew should try to relieve the pilot of some of the routine tasks so he or she may concentrate on the goals. Regardless of the type of flight, the crew is there to support the pilot in conducting a safe and successful flight operation.

Direction of Flight

The first element of the flight the chase crew must know is the direction the balloon is going. It is important to understand that the balloon's direction is very difficult to detect from a moving vehicle. Many pilots recommend the chase crew drive the chase vehicle away from the launch site only far enough to get the vehicle out of the way of the balloon (and other balloons) and to be clear of any possible spectator crowds. As soon as the crew is sure they are clear of other traffic, they should park in a suitable place with a good view of the balloon, and determine the balloon's direction of flight. There is no point in rushing after the balloon until the direction it is going is known. The balloon changes direction shortly after launch if the winds aloft are different from the surface winds.

After a while, the crew should proceed to a point estimated to be in the balloon's path. In other words, get in front of the balloon so it flies over the chase vehicle. If the balloon is moving at five knots, the chase crew need drive only a short time to get in front of the balloon. The direction of flight is much easier to determine if the balloon is floating directly toward the vehicle rather than flying parallel to the vehicle's path. If a radio is not being used, as the balloon flies over the chase vehicle, the pilot and crew can communicate by voice or with hand signals. In this instance, the crew should be outside the vehicle with the engine turned off.

The Crew Briefing

Crew briefings vary from a few last minute instructions (to an experienced, regular crew), or a long, detailed discourse on how to layout, assemble, inflate, chase, recover, and pack a balloon. *[Figure 6-4]* A pilot can give crew briefings by telephone the night before, or in the chase vehicle on the way to the launch site, but most crew briefings are done at the launch site prior to the flight. It is important for the pilot to remember who is ultimately responsible for the entire operation and that the crew is the pilot's representative on the ground.

Figure 6-4. *Crew briefing.*

Whether this is the crewmembers' first time or one-hundredth time crewing, they should be briefed before each flight. Instructions contained in the briefing may be less detailed for an experienced crew. The following instructions should be given for each flight:

- Estimated length of flight and any information that aids the chase and recovery
- Anticipated direction of flight
- Position and duties during inflation
- Duties once the balloon has reached equilibrium

A typical flight briefing may be "I intend to make a 1 hour flight and I have about 2 hours of fuel on board. From my weather briefing and the pibal, I should travel in a southeasterly direction; but if I go west, I will land before getting to the freeway. I will probably do a lot of contour flying, but may go up to 2,000 feet to look around. Let's use channel six on the radio. There is a county road map on the front seat."

"Patricia, you will be the crew chief for today's flight, as well as the driver. The keys to the van are in the ignition, and there is a spare set in the console. I would like you and Bob to do the mouth this morning. Bob, you will be on the side away from the fan. Pat will show you what needs to be done, and I will double check you during the inflation. You will also be navigating, so we will have a look at the map together in a minute. Susan, you will be on the fan. Be sure to keep people away from the plane of the fan, and please do not move the fan while it is running. Leslie, you are on the crown line today. You have done it before, so I know that you know the procedures, but as a reminder, do not wrap any lines around your hand, arm, or body. I will check with you a couple of times during the inflation to make sure you are positioned properly."

"Any questions? Good. Sleeves down, gloves on, and let's go!"

The Crown Line

The duty of the crown line crewmember is to hold the end of the line, lean away from the envelope, and use body weight to stabilize the envelope. As the air is heated and the envelope starts to rise, the crewmember holding the crown line should allow the line to pull him or her towards the basket, putting only enough resistance on the line to keep the envelope from swaying or moving too fast. Release the tension slowly after the envelope is vertical. *[Figure 6-5]*

Figure 6-5. *Crew can often stabilize a rolling envelope by taking a single crown line 45° off of downwind and walking directly upwind on stand-up.*

Figure 6-6. *Double crown lines forming a 90° or wider angle offer the greatest inflation stability provided crew walk them directly upwind on stand-up.*

The crown line varies in length. Some pilots let the line hang straight down; some pilots connect the end of the line to the basket or burner frame. Other pilots keep the line only long enough to assist with a windy inflation, or deflation in a confined area. Usually, there are no knots in the crown line, but you might find a type of loop attached to it. Some pilots put knots in their line, or attach flags or other objects. These may snag in trees and cause problems. Lines tied to the basket form a huge loop that may snag a tree limb and should be secured with a light, breakaway tie.

To improve control, some pilots use a double crown line during very windy conditions. This technique, as shown in *Figure 6-6,* allows for better control of the envelope when winds may be gusting, but still within reasonable flight limits, as may be present in an afternoon launch. Many pilots launch in afternoon winds, knowing that winds generally decrease significantly as sunset approaches. One advantage of this technique is that the balloon can be kept stable in fairly gusty winds, even if held on the ground for an extended period of time. This may be helpful for the pilot who may be doing a static (non-flying) tether, or perhaps during an evening "balloon glow" when winds may remain gusty until well after sunset.

Launch Site

When selecting a launch site, factors to consider are obstacles in direction of flight (powerlines, buildings, towers, etc.), available landing sites, and the launch site surface.

Location and Obstacles

The launch site is selected based on surface winds and winds aloft, in conjunction with the desires of the balloon pilot. Consideration is given to what type of flight is being made and to the overall goal of the flight. Upon arriving at a particular launch site, most pilots release another pibal to verify the wind direction, and ensure that the forecasts or earlier readings have not changed significantly.

Ideally, the launch site should be a large, open grassy area. Obstacles in the flight path are a consideration. As a general rule, the balloon should be placed as far as possible upwind of any obstructions to flight (powerlines, trees, buildings, etc.), using a minimum of 100 feet of clearance for each knot of wind. *[Figure 6-7]*

Figure 6-7. *Examples of good (top) and poor (bottom) launch sites. In the top picture, the pilot would have adequate clearance above the power lines at the far end of the field, as well as more than adequate area to layout the envelope. In the bottom picture, the pilot would have a very short area to lay out. There are obvious obstacles, such as the trees and cell phone tower, which create hazards to a safe launch.*

Landing Sites

The best launch site is of little use if there are no appropriate landing sites downwind. Many pilots examine their flight path from the perspective of the landing area, particularly in an area of few landing fields, and "reverse plan" their launch site in order to make the appropriate choice of launch sites. Occassionally, this may require using a launch site that may be somewhat less than desirable. As long as the launch can be made safely, with appropriate obstacle clearance, this is perfectly acceptable.

Launch Site Surface

After determining the flight direction, the next condition that determines the details of the balloon layout is the actual launch site surface. Of course, all pilots want to lay out their balloon on clean, dry, short, green grass. Most pilots are not that fortunate unless they have their own launch site and never fly from different places. Wise pilots modify their techniques according to available conditions, or they have more than one layout procedure to adapt to various launch sites.

Whether flying from a regular launch site, a brand new location, or from an assigned square at a rally, check the ground for items that may damage or soil the balloon. Look for and remove nails, sharp rocks, twigs, branches, and other foreign objects. If there are patches of oil or other substances, cover them with pieces of carpet, floor mats from the chase vehicle, tarps, or the envelope bag. Some pilots cover the ground where they lay out their balloon with a large tarp every time they fly.

Unless flying at a known site, do not assume it is all right to drive the chase vehicle directly onto the launch area. There are some locations (a soft athletic field, for example) where it is necessary to carry the balloon onto the launch area. If using private property, it is necessary to get permission to use the property for flight activities. This may be the first experience with landowner relations, and the considerate pilot takes action to minimize his or her impact on the landowner's property. Landowner relations are an important part of ballooning, and are discussed in greater detail later in this chapter.

Balloon Placement and Wind Direction

Consider the wind direction before the balloon is unloaded from the chase vehicle. Take into account the surface wind at the time of cold inflation to avoid carrying a heavy balloon bag and basket around. A "Murphy's Law" type of rule is that the wind always changes during inflation. Local knowledge is invaluable. If other balloons are around, check with the most experienced local pilot.

A wind change at or shortly after sunrise is normal in many places. If the balloon is laid out prior to sunrise, a wind change is likely. If flying in a new area, it is beneficial to watch the local pilots, as they have knowledge of local conditions and idiosyncrasies. Some general trends are that air usually flows downhill or down valley first thing in the morning, and air usually flows from cold to warm in the morning. The air drainage may stop very shortly after the sun rises and starts heating the ground. The early morning wind may come from a different direction than the prevailing or predicted wind. Some local pilots may lay their balloons out in a direction that does not match the airflow at the time, but that corrects 15 to 30 minutes later when the sunrise change occurs and the inflation starts.

Removing the Balloon from the Vehicle

Once the specific location on the launch field has been determined, the balloon should be removed from the vehicle, or unloaded from its trailer, and prepared for flight. *[Figure 6-8]* Good procedures and habit patterns formed here minimize the amount of lifting of balloon components, and reduces the wear and tear on the crew and the pilot, as well as the equipment itself. Despite being a lighter-than-air aircraft, the packed balloon and basket are heavy; the average sport balloon system can weigh over 550 pounds.

Figure 6-8. *Unloading the basket and moving the vehicle forward minimizes equipment carrying.*

One method is for the pilot to establish the specific location and direction of layout, and direct that the basket be placed there. The transport vehicle maneuvers to allow the basket to be removed and placed at the pilot's discretion, and then pulls forward about 15 or 20 feet to allow for the envelope bag and inflator fan to be removed and placed on the ground. The vehicle may then be removed from the launch site, or moved to the back of the basket to serve as a tie-off point for the balloon's quick-release system. This method minimizes the amount of lifting and carrying of heavy components. When used in reverse, it allows for the system to be packed up at the end of the flight, and kept ready for a future flight.

Assembly

If the basket is disassembled for transport and handling, it must be assembled in accordance with the flight manual prior to layout. The pilot, using the checklist, should either perform these functions or carefully supervise them.

Basket

Once the basket is appropriately placed, it is necessary to attach the uprights and burner frame (as appropriate). Each brand of balloon is slightly different, and the balloon's flight manual should be checked for the specific procedures. Some balloon baskets must be placed on their side to install the uprights and burner, while some may be left standing upright. Either way, after the components are installed, they should be checked for installation, integrity, security, and general condition.

Burner and Fuel System

If not accomplished in the previous step, the burner(s) (or heater) should be mounted and connected. Retaining pins, as applicable, must be mounted, and fuel lines connected between the burner and the tanks. The practice of using an adjustable wrench on fuel line fittings is discouraged; while convenient, these wrenches usually cause excessive wear on the fittings over time, and may result in damage or early replacement. It is usually better to use the correct size open-end wrench. Fittings do not need to be heavily torqued into place; this is a common error with newer pilots. A good fitting usually seals with little more than finger-tight force. Anything more may indicate corrosion or a damaged flange seal in the fitting. *[Figure 6-9]*

Figure 6-9. *Burner installation on a Lindstrand.*

It is appropriate to check for leaks in the fuel system at this point; this should be a checklist item. The pilot opens the fuel tank valves and checks each connection, at both tank and burner, by using a "sniff test." The odorant in propane, ethyl mercaptan, has a very distinctive odor and is relatively easy to detect. The burner should also be examined to ensure that no fuel is leaking from fittings, pressure gauges, or interior plumbing. Once satisfied that there are no leaks, the fuel may be turned off and the pressure bled off. The basket should then be placed in a vertical position to prepare for the next function check.

Burner Check

Most manufacturers require burner checks as a necessary part of an equipment function check and are indicative of good habit patterns. Each system has a specific procedure, appropriate to that particular system, which is outlined in the flight manual for that model of balloon. In general, each burner, including the pilot light, is checked individually for correct operation to ensure that the burner is working properly. Each burner is checked using each fuel tank, and all backup systems should also be checked for correct function and operation.

As an example, assume a single burner balloon with three tanks connected into a common fuel line (tanks in parallel). The pilot would first light the pilot lights and check for correct operation. Then, one tank would be opened, and the blast valve activated. The pilot checks for correct operation, smooth blast valve movement, and for pressure in the fuel system, as excessive loss of pressure may be indicative of a fuel tank or line problem. This procedure should be repeated for each tank, closing off the first tank when moving to the second, and so on. After all tanks are checked, then the backup system (Fire2™, Whispervalve™, Liquid Fire™, Metering Valve, etc.) should also be tested. Once all systems are checked for functionality, then the tanks should be closed, pressure released from the fuel system by burning off the fuel, and the basket placed on its side in preparation for attaching the envelope. (NOTE: At no time should all of the tanks in a parallel-plumbed system be opened at once. In the event of a fuel leak or ruptured hose, there are three or more tanks feeding the leak. This is an extremely hazardous practice.)

Pilots should also note that there is a growing trend towards the use of inflation, or "pony" tanks, used to supplement on-board fuel for inflation. Use of an inflation tank allows the pilot to launch with completely full tanks, as the additional tank is either temporarily plumbed into the balloon's fuel system by the use of a quick-disconnect fitting, or hoses are switched around at the burner fitting. There is no manufacturer's written procedure available for the use of an inflation tank, and the procedure is strongly discouraged. Of perhaps greater concern is the mindset that the one or two gallon advantage this gives the pilot, is of greater concern than the safety aspect of the procedure. If a pilot consistently requires these one or two gallons of fuel to safely conduct a flight, then there are possibly other performance planning factors being neglected.

Most pilots conduct a burner check at the beginning of each flight. Others, if conducting multiple flights over a relatively short period of time (a weekend balloon rally, for example), may do a full burner check for the first flight only, and then only check for leaks on each remaining flight. This practice, while common, may violate the manufacturer's operations manual and invites problems; it is best avoided.

Instrument Checks

It is appropriate to check the instruments at this point. Virtually all instrument systems currently in use require battery power; it is wise to check the instruments, and ensure that the batteries provide sufficient power to operate them. There is nothing more frustrating than being ready to fly and finding that the instruments do not work. After the instruments are checked for operation, they may be mounted in the basket. Some pilots choose to set the altimeter at this time; the barometric pressure should be set according to the information gathered during the weather briefing. In the absence of a barometric pressure setting, the altimeter should be set to the appropriate field elevation. The only time that the altimeter may be set to zero is when the balloon is being flown from an ocean beach; 14 CFR part 91, section 91.121 provides clarification on this requirement. This is necessary to avoid airspace issues. *[Figure 6-10]*

Figure 6-10. *The altimeter should be set to barometric pressure, if available, or to the field elevation.*

At this point, the basket is on its side with the instruments mounted and ready for the envelope to be attached. The pilot and crew should open the envelope bag, remove the suspension cables and attachment hardware (carabiners, "A-blocks," or toggles), and attach the bottom of the envelope to the basket superstructure in accordance with the manufacturer's instructions. Care should be taken that cables are not twisted and are free of damage, kinks, loose wires, or missing sheathing (found on Kevlar® suspension cables). Once the envelope is attached, the hardware should be checked for security and the envelope laid out for inflation.

Layout

There are many variations of laying out a balloon and preparing it for inflation. The manufacturer of the balloon or the way the balloon is assembled sets some of these inflation styles. The launch site surface, the order in which the balloon is assembled, and how the balloon is removed from the chase vehicle have a bearing on the way the preflight layout and inspection proceeds. There is no one best way to lay out a balloon, as there is no one best way to inflate a balloon. The two most common ways to prepare the balloon for cold inflation are to spread it out or inflate from a long strip.

The pilot should be mindful of current conditions prior to removing the envelope from its bag. It is wise to take a moment and walk around the launch site to check the surface for any debris that may damage the balloon.

As a cautionary note, do not drag the envelope bag along the ground when pulling the envelope out. Many envelopes have holes and tears caused by being dragged over sharp objects while getting the envelope out of the bag. Lift the bag clear of the ground and carry it unless the launch surface is perfect with no sharp objects or dirty spots.

Spread Layout

The spread layout method for inflation is the most widely used method. By handling the envelope with the load tapes, the fabric is pulled away from the center until the envelope takes its normal shape while still flat on the ground. Exercise care when sliding the fabric across the ground to avoid causing damage.

All balloons have an even number of load tapes. By using the number on the load tape when spreading the envelope, the envelope can be arranged in a proper position for inflation. With one crewmember on each side of the envelope fabric, start at the mouth and travel the length of the appropriate load tape, pulling the fabric taut up to the equator. This allows the bottom to be laid out flat. Be careful to handle only the load tapes when positioning the envelope because pulling on the fabric could cause damage. Holding a load tape and flapping up and down gently helps avoid tears and other damage while spreading the envelope. This action traps a small bubble of air which floats, rather than drags fabric and keeps the balloon cleaner and easier to pull. *[Figure 6-11]*

Figure 6-11. *A spread layout.*

Check the deflation system at this time and properly position it in accordance with the balloon flight manual. While the envelope is filling with air, the crew can assist the process by lifting upward on the load tapes, allowing more cold air to pack the envelope. This method allows the envelope to deploy smoothly and easily, even with a small inflation fan.

Strip Layout

When inflating on pavement or a small or narrow launch site, many pilots prefer not to deploy the envelope on the ground. Instead, they prefer to pull the envelope straight out from the basket, making sure the top gore is on top its full length, and to inflate the envelope entirely with the fan. This may require a larger fan, depending on the size of the balloon and envelope material. *[Figure 6-12]*

Figure 6-12. *A strip layout.*

Once the balloon is stretched out, ensure the control lines (deflation, cooling, or rotating) are correctly attached to the basket according to the manufacturer's instructions. This method minimizes handling the fabric on a rough or dirty surface. It requires more diligence by the ground crew to ensure it deploys correctly.

Progressive Fill

Inflation out of the bag is a coordinated technique in which the envelope is cold inflated directly out of the bag. With the envelope suspension connected to the basket, it is pulled progressively out of the bag as the fan is running, inflating the envelope as the crew slowly, with pauses, carries the balloon bag away from the basket while the envelope fills. Another variation, the progressive fill, has the crew holding the envelope in their arms, gradually releasing more and more of the envelope, from mouth to top, until the very top of the envelope is released for filling. The progressive fill packs the lower portion of the envelope with cold air, and exposes less of the fabric to the ground. This method generally results in a faster inflation, but may create problems with attaching a parachute top or ensuring that lines are not tangled.

Inflator Fan Placement

The type of inflator fan needed for different layout techniques depends primarily on the amount of work required. The strip layout method requires a large, strong fan to force the envelope into shape, while the spread layout method requires less fan energy.

Fan placement and techniques are as varied as the pilots who fly. It may appear that every pilot has a slightly different technique for placement and use of the fan, but the general principle is the same. The fan should be placed to the side of the basket *[Figure 6-13]*, with the air flow directed into the mouth of the balloon to inflate the envelope, but far enough

Figure 6-13. *Placement of the inflation fan.*

back to allow the airflow to "seal" the throat. This allows the most efficient use of the fan.

A method pilots use to determine the best placement of the fan is to attach small pieces of fabric or surveying tape on the lower portion of the mouth, using safety pins in such a manner as to not create damage or holes. Then, the crew holds the mouth of the balloon open with the fan running; if all pieces of the tape are pointing into the envelope, then the fan is properly placed; however, if any of the tapes are fluttering back out the mouth of the envelope, this demonstrates that the airflow is going into the envelope and coming back out again. This does not allow for a good inflation, and the fan should be adjusted to correct this deficiency. This demonstration should be done without use of the burner, as the pilot may inadvertently cause damage or start a fire. *[Figure 6-14]*

Figure 6-14. *Determining correct inflation fan placement. Note that the strips of tape are generally pointing into the mouth. In this case, the pilot would do well to incline the fan slightly towards the right side of the mouth, in order to better "seal" the envelope opening, and get a more efficient inflation.*

It is generally better to pack the balloon full of cold air. This makes a tighter mouth, which avoids burning fabric, and the balloon may be less affected by a light wind when it is round and tight.

Safety Restraint/Quick-Release

The quick-release or safety restraint should be secured, with little or no slack in the line, to prevent possible damage or injury in the event of a break or premature release. Every major balloon manufacturer specifies the approved method for tying off. Balloon baskets, suspensions, load plates, and burner supports have been destroyed by improper tie-off in light winds. When tying down for inflation, reference the balloon's flight manual for the recommended procedures and techniques. The use of a safety restraint is a recommended procedure for enhanced safety.

Passenger Briefing

Prior to inflation, a pilot should give passengers their first briefing for behavior during the flight and landing. Inform passengers that during the landing they should stand in the rear of the basket where shown by the pilot (based on wind conditions), facing the direction of flight, with feet and knees together, knees slightly bent, and holding tightly to the sides of the basket. They are not to exit the basket until instructed by the pilot to do so. This procedure allows the passengers to see the landing sequence, and is generally accepted throughout the ballooning community. In larger balloons, which usually have compartmentalized baskets, some pilots encourage the passengers to face away from the direction of flight. Either procedure is acceptable, as long as it is briefed prior to flight, and reinforced prior to landing.

The balloon is ready for inflation when the preflight preparation is complete. Excess equipment is stowed in the basket and chase vehicle, and a radio check is completed.

Inflation

The inflation procedure takes the balloon envelope from a pile of fabric to an aircraft capable of lifting a load. The pilot's goal should be a smooth, controlled inflation that does not damage the environment, balloon, or harm the crew. At the end of the inflation, the balloon should be upright, close to equilibrium, and ready to fly.

Inflation Styles

There are many different styles of inflation. Some pilots use one or two large fans to inflate the balloon fast and get it tight. Some pilots prefer to fill the envelope slowly to give them time for preflight preparation. Some use many crewmembers;

some use only a few crewmembers. Balloon size, available crew, weather, location, and personal preference are factors that determine procedures and number of crewmembers.

The Inflation

After the balloon is correctly laid out, place the inflation fan to the side of the basket within arm's reach of the pilot, facing into the center of the envelope mouth, making sure the fan blades are not in line with the pilot, crew, or spectators. If the fan is well designed and maintained, it will not move around or require constant attention during operation. Exact fan placement depends on the type of fan, burner, and size of the envelope, as previously discussed. Pump air into the envelope and not under, over, or to the side of the mouth.

A crewmember should be placed at each side of the mouth of the balloon to lift the material and create an opening for air to enter the envelope. During cold inflation (i.e., with the fan only) hold the mouth open wide enough to admit the airflow from the inflator fan. It is best to inflate the balloon as full as possible. At a minimum, inflate the balloon to approximately 75 percent full of cold air.

As the envelope inflates, the pilot should check to see that control lines are correctly deployed and the deflation panel is correctly positioned. This can all be done through or in the vent or from the top; it is not necessary to walk on the fabric. During this phase of the inflation, the envelope should also be checked again for damage that would disqualify the balloon from flight.

Once the preflight inspection and cold inflation are complete, and the pilot is satisfied that the envelope contains enough ambient air to begin hot inflation, the two crewmembers at the mouth should hold it open as wide and as tall as possible to keep the fabric away from the burner flame. The crewmembers should face away from the burner. At this point, fan speed may be reduced to approximately one-half or two-thirds full throttle and the pilot may light the burner's pilot lights in preparation for inflation. Before activating the blast valve, the pilot should make eye contact with each crewmember at the mouth and make sure each is ready. Crew readiness is paramount. The crew at the mouth of the envelope must be aware the burner is about to be used.

Allow the fan to run at a reduced speed until the balloon mouth lifts off the ground and is no longer receiving air. If the fan is turned off too soon, envelope air comes back out of the mouth and the backwash distorts the flame at the beginning and end of each blast. Do not hurry to turn off the fan. Some pilots elect to shut off the fuel to the inflator fan, which accomplishes two things. This procedure normally allows a fan to continue running for about a minute, which should be long enough to inflate the balloon, and also allow fuel in the fan's carburetor to be used, eliminating the likelihood of gas fumes should the fan be stored in an enclosed space during transport.

The first burn or blast of the burner should be a short one to confirm the correct direction of the flame and to check the readiness of the mouth crew. If they are startled by the flame or noise and drop the fabric, the short burn prevents or minimizes damage. To reduce discomfort of the crew, it is best to inflate the balloon with a series of short burns and pauses, rather than one continuous blast. Inflate using standard burns, with short pauses of about 2 or 3 seconds between burns. The pauses give the fabric and skin a chance to cool and allow communication between the pilot and the crew, if necessary. Under some circumstances, contraction and inflation of the balloon mouth may be seen. Burns should be timed to match the expansion of the mouth. These mouth movements are called "breathing"; burns should be timed to match the open time to avoid damaging the fabric.

Some pilots prefer to inflate the balloon with one long blast of the burner. The advantage of this type of inflation is that the balloon inflates a few seconds faster and the mouth tends to stay fully open during the process. There are several disadvantages. Voice communication is nearly impossible due to the noise of the burner. Anyone or anything within a few feet of the burner may get burned. Also, some burners could be discolored or damaged by long burns.

The next step is to continue the burn-and-pause routine until the balloon is nearly ready to leave the ground. The crew should be standing by the basket ready to hold the balloon ("hands on" or "weight on"), in case of a miscalculation, allowing the balloon to start lifting off the ground before the pilot is ready. The use of the safety harness prevents unplanned departures.

Many pilots fail to achieve equilibrium or neutral buoyancy immediately after inflation. If equilibrium is not achieved, the balloon is much more susceptible to wind. For example, if the envelope is not full, a slight wind can cave in a side causing a spinnaker effect. If the balloon is standing, but not ready to fly, the pilot has only one option should the balloon start to move horizontally; the pilot must deflate. If the balloon is only 5 or 10 seconds of heat away from lifting off, the pilot has the choice of deflation or launch.

The inflation is the first action of ballooning that requires a pilot in command (PIC). The inflation should be safe and efficient. Now, the balloon is almost ready to launch.

Prelaunch Check

After the balloon is inflated and upright, the pilot should perform a pre-launch check. Ensure that loose equipment is properly stowed and secured. For balloons using multiple tanks, it may be appropriate to shut off the tank that was used for inflation, and open the tank that will be used first for flight. Some pilots make another quick check of the burner to ensure that there are no leaks or deficiencies present. The top cap should be activated to release the tabs (as necessary), and ensure that the mechanism is functioning properly. The altimeter, if not previously set, should be set to the proper barometric pressure or field elevation, and the temperature indicating system should be checked for a proper reading. If used, radios should be turned on and secured. The pilot should note the time of inflation, and quickly check the fuel level in each tank to ensure that there is sufficient fuel for the planned flight. This pre-launch check should be brief and verified through the use of a pre-launch checklist.

Launch

If carrying passengers, now is the time to invite them in the basket. Immediately compensate for the additional weight with sufficient heat to regain equilibrium. The passengers have already been briefed on the correct landing procedure. Brief them again on behavior in the basket. Advise passengers not to touch any control lines, take care of their possessions, stay within the confines of the basket, and, above all, to obey the PIC.

At least one crewmember should remain near the basket in case the pilot or passengers need assistance. This is a good time to give the crew a final briefing regarding the expected distance and length of the flight, radio channels, and other last minute instructions. If other balloons are launching from the same area, ask a crewmember to step back from the balloon to check that it is clear above.

Two or three standard burns in a row from equilibrium usually provide a slow departure from the ground. If there are no nearby, downwind obstacles to clear, a slow ascent rate is preferred to test wind direction and detect subtle wind changes. Climbing at a slow rate is the best way to avoid running into balloons above. There is an unwritten rule in ballooning (not regulatory) that the balloon below has the right of way (due to lack of visibility above). Although the balloon below has the right of way, the higher balloon needs time to climb out of the way, if necessary. Pilots must maintain awareness of other balloons operating near them, particularly in crowded or rally situations. *[Figure 6-15]*

Figure 6-15. *The balloon below has the right of way while ascending. The pilot of the higher balloon should, as a courtesy and in the interest of safety, yield the right of way to the ascending balloon, as that pilot is probably unable to see the balloon above.*

A fast ascent rate from launch is only to avoid ground obstacles or to pass quickly through an adverse wind, and only when it is clear above. Should circumstances require a fast ascent rate, the pilot should set up for the lift off by having the ground crew put their weight on the basket by hanging their arms over the side, not holding on. The pilot should then heat the balloon to a temperature beyond that needed for equilibrium; 20° above the neutral buoyancy point may be a good starting point. After getting a check of possible traffic above the balloon, the pilot instructs the crew, "weight off," and the crew responds by removing their weight from the basket. It is imperative that the crew clears the basket, and that no crewmembers are left hanging outside. The balloon then rises at a fairly rapid ascent rate. The pilot must be aware that the balloon is marginally under control at this point, and that too excessive a climb rate may result in a condition known as "floating the top." This is a scenario where the air pressure created by the climb may push the top cap of the balloon down, causing an out-of-control descent. Maximum climb rates are specified in the operating limitations section of the balloon's flight manual.

It is very easy to be distracted during launch and make an unintentional descent. Make sure all ground business is taken care of, such as instructing the chase crew and stowing all equipment correctly, before leaving the ground.

The pilot should be aware of the possibility of uncommanded lift (often referred to as "false lift") and the possibility of an unplanned descent caused by surface wind or an ascent from a sheltered launch site. Pay attention to people and obstacles, including the chase vehicle, fences, and particularly to powerlines. Realize where all powerlines are and visually locate them as soon as possible.

Uncommanded (False) Lift

One consideration that must be made at the outset of any balloon flight is the possibility of experiencing phenomena variously referred to as "false lift," "false heavy," or "uncommanded buoyancy." These terms all describe an onset of various factors and conditions, which, despite the differences in terminology, all relate to the result of air moving over or under the balloon. The most important thing to remember is a balloon encountering one or more of these factors is not under the full and complete control of the pilot, and is therefore a hazard. Pilots should be aware of these conditions, avoid them if possible, and be aware of procedures and practices to minimize their effect on the balloon's flight.

Three areas of focus warrant discussion:

- False lift
- False heavy (air flow under the lower portion of balloon, creating downward lift)
- Envelope distortion causing diminished capacity

It is important to understand the total physics involved. While the balloon is at neutral buoyancy on the ground, there are two lift forces at work. The first is from the heating of the air, creating buoyancy inside the envelope. The second is the flow of air over the top. The lift from these two elements combine to create the lift necessary to be at equilibrium. The addition of a small amount of heat, through a short burn, increases the total lift and allows the balloon to rise.

False Lift

During initial flight training, pilots are taught about the effects of air flowing over the top of the envelope. While the balloon is static on the ground, the shape of the top forces the flow of air to compress over the top creating a low pressure area. *[Figure 6-16]* This low pressure area creates lift in much the same way an airplane wing does. There are two components of lift: heated air inside the envelope and the lift created by the air passing over the top. As the balloon takes off and accelerates to the speed of the air mass, the flow of air over the top diminishes, thus any lift created by it is no longer available. If the balloon is at equilibrium at launch, and there is not some response by the pilot to add more heat, there will not be sufficient lift to stay in the air, as a portion of the total lift has diminished.

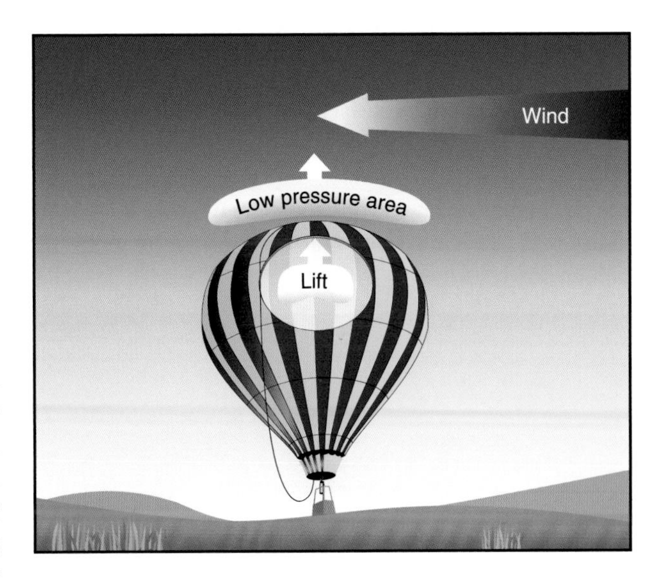

Figure 6-16. *False lift dynamics.*

Pilots are usually taught that the lift created by air flow over the top is to be considered "false lift," because it was not created by applying heat to the envelope. The lift is real; as long as the speed of the air flow and the balloon remain sufficiently different, the lift continues to be generated. As the balloon accelerates, the lift created by the air flow is lost and more heat must be added to maintain the same rate of ascent.

The same condition can exist when descending through a low level wind shear or jet. As the balloon penetrates the lower boundary of the wind shear, the top of the balloon is momentarily in a faster moving layer of air which increases the air flow lift component of the total lift generated. For a short period, there may be equilibrium with the two components of lift, heat and air flow. As the descent continues, the top of the balloon moves out of the faster moving air (the value of one of the lift components), and air flow is diminished. The rate of descent increases, unless the pilot takes action to increase the buoyancy portion of the total lift by making a burn. Some pilots, when experiencing this, believe it to be a "false heavy" situation, an incorrect perception.

False Heavy (Downward Lift)

False heavy is a condition which is the exact opposite of the false lift scenario described above, except that it is happening at another part of the balloon and the force generated has a downward component.

This phenomenon occurs when descending into a faster moving air mass or wind shear. The lower half of the balloon enters the shear and the surface of the balloon allows the air flow to generate lift. This lift is tangent to the surface of the balloon. Because it is below the equator, where the tangent line points in a downward direction, the lift has a downward component. This downward component of lift pulls the balloon down. *[Figure 6-17]*

Figure 6-17. *False heavy dynamics.*

This effect is more severe on a fully loaded balloon than a lightly loaded one, by reasons of skin tension. A lightly loaded balloon has more slack in the fabric on the lower portion of the balloon. On a heavily loaded balloon, the fabric below the equator has greater tension providing a surface where a low pressure can develop.

A scenario having much less impact is fast moving air across the mouth of the balloon. This creates a dynamic low pressure, similar to a venturi, which may cause the air to be pulled out of the envelope. As the air moves over the mouth of the balloon, it creates the dynamic low pressure, which pulls the static air inside the envelope out. Of all the possibilities discussed here, this point has the least impact on the lifting capability of the balloon, but is important when standing on the ground following the initial inflation to equilibrium.

Diminished Capacity

Another way in which a wind shear can increase the rate of descent is by diminishing the capacity of the envelope. For example, when descending, below is a low level wind shear with the air near the surface moving much faster or slower than the air mass in which the balloon is traveling. As the balloon enters the lower air mass, the side of the balloon is pushed in, decreasing the capacity of the envelope and pushing the air out the mouth. The larger the difference between the speed of the two air masses, the greater the effect. The lift created by buoyancy is decreased and the balloon starts to descend. If this happens at a low altitude, and the pilot has not responded in a timely manner, this may result in a hard landing.

If a balloon descends abruptly from a 30 mph wind into a 15 mph wind, it experiences an effective abrupt increase in wind across its surface from no wind to a 15 mph wind. This removes the boundary layer on its surface and greatly increases heat lost by conduction, while causing distortion in the form of a "dish." This condition is very dangerous to the low level flight of a fully loaded balloon. It should be noted that as the envelope lowers into the slower wind and begins to distort and slow-up, the effective wind speed over the top begins to increase. Air moving over the top of an envelope produces false lift. Combined with diminished capacity, this again presents a hazardous condition which may result in an extremely hard landing.

"Dishing" usually does not affect the flight path of a lightly loaded balloon as much as a more heavily loaded balloon because no internal lifting heat is dumped; it is only redistributed inside the distorted envelope. To understand what a "dish" can do to lift, the balloon pilot needs to understand how the "heat line" fluctuates under different loading conditions. A balloon "floats" in the air because the hot air inside it weighs less than the volume of air it displaces. Usually the bottom of the heated air is down close to its mouth. Notice that a normal two-second single burn on a heavily loaded balloon adds only a small percentage of heat compared to its total required hot air volume. The same burn in a lightly loaded balloon adds a much larger percentage of lifting heat compared to its total required hot air volume. This is a simple way to visualize responsiveness.

If a heavily loaded envelope experiences a major "dish," important lifting heat can be squeezed out. This condition is extremely serious if on a final approach to landing because there is not enough time and altitude to add enough heat. Remember, not only must the heat loss be replaced to make the balloon neutrally buoyant, but more must be added to stop any downward momentum.

Heat loss can change the slope of the approach and, pilot unaware, make it steeper. The stronger the shear, the greater the change in slope. Knowledge of this and the importance of adding heat quickly could prevent an excessively hard landing or an accident.

Some signs of shear to watch for are any movement of crown lines, handling lines, throat ropes, skirt, or even basket movement. It is important to realize that as the balloon lowers into the slower moving air, distorts, and starts slowing, the effect of heat loss can be masked by false lift. Even if the shear is a mild one, false lift momentarily exists. When the balloon slows enough and exits the faster moving air, the hidden false lift and dishing disappear and the balloon descends out of control.

Some pilots intentionally create a situation of diminished capacity when making a high performance descent. In low or stable wind conditions, this can be successfully executed. However, in the presence of strong wind shears, this technique can prove disastrous.

If a pilot is in a false lift scenario, the first action should be to continue to fly the balloon. If the situation is encountered during lift-off, or is believed to exist, the pilot should maintain a positive rate of climb until the false lift dissipates. If the false heavy scenario exists during the landing, the pilot must be prepared for an acceleration in the descent rate. An appropriate action would be to add heat to slow the descent, unless a steep approach to landing is desired.

In reality, there is nothing false about any of these situations. They are real and may create hazardous flight dynamics. In many cases, more than one of these elements is at work. It is important for the pilot to be aware of them, their effects, and consider what actions are necessary when they are encountered. The best prevention is anticipating these conditions and maintaining situational awareness. If the conditions are extreme, it may be said that the best and first consideration is staying on the ground.

Landowner Relations

An otherwise perfect flight can be marred without the use of the proper relationship skills needed to foster good landowner relations. Often neglected, these skills provide the balloon pilot the locations necessary to inflate, launch, and land. Without these properties, ballooning would be severely limited. Taking the time to explain one's actions to a landowner, or dealing with a farmer whose livestock have been spooked by an ill-timed contour flight, can create lasting impressions that have tremendous long-term negative impact on the continued evolution of the sport.

During the launch phase of a flight, building/landowner relations is an easy task. The pilot should select launch sites that avoid flight paths and landings around sensitive areas, such as livestock, expensive crops, nature preserves, etc. Once a launch site is selected, the pilot should make an effort to identify the property as public or private. Generally, school fields and local parks may be used without further inquiry unless there have been previous problems with balloons. In such a case it would be appropriate to check with the local authorities for the use of these facilities.

Private property, however, is another issue. A balloon pilot and crew should never assume the right to use a private location to launch or recover a balloon. To do so exhibits a degree of arrogance that has no place in ballooning and subjects everyone participating to trespass laws. The landowner usually lives on the property and has paid for that right. In the event that no one is immediately available, the pilot should either select another launch site or perhaps inquire of neighbors who may be able to inform you of the landowner's location. Finding the landowner and obtaining permission to use a particular field may be one of the most important tasks of the launch process for the crew chief, if one is assigned. Undoubtedly, the one time the pilot does not have appropriate authorization for use of a launch site and uses another's property, someone will be watching and problems later ensue.

The positive side of this is that most landowners welcome the balloon pilot and his crew, want to learn a little about balloons, and gladly allow the use of their property for the launch. Many see this as an opportunity and actively participate in the process. Others may grant permission, but stand back from the activity. Whichever type of landowner is encountered, they usually respond positively to a pilot and crew that respect rights and protects landowner interests.

Chapter Summary

It is frequently said that every pilot sets up equipment and prepares for flight in a different manner. The purpose of this chapter is not to emphasize those differences, but rather to illustrate the underlying similarities and procedures that every balloon pilot must follow to safety begin a flight.

Chapter 7

Inflight Maneuvers

Introduction

This chapter discusses various aspects of inflight maneuvers. It covers the standard burn, level flight, ascents and descents, maneuvering and lateral control, and contour flying. This chapter also includes information on chase crew management, landowner relations, and tethering.

The Standard Burn

To fly with precision, the balloon pilot needs to know how much heat is going into the envelope at any given time and how that heat affects the balloon's performance. Balloon pilots have few outside sources, instruments, or gauges to help them fly. When a balloon pilot uses the burner, there is no direct way of knowing exactly how much lift is increased. Because there are few mechanical aids to help balloon pilots fly, some methodology must be devised to standardize a pilot's input action, so the outcomes of those actions are predictable and the balloon is controllable. The standard burn is one way to gauge in advance the balloon's reaction to the use of the burner. *[Figure 7-1]*

The standard burn is an attempt to calibrate the amount of heat being used, and is defined as a burn of four seconds. If each burn can be made identical, the balloon pilot can think and plan, in terms of number of burns, rather than just using random, variable amounts of heat with an unknown effect. The standard burn is based on using the blast valve or trigger valve found on most balloon burners. Some brands use a valve that requires only a fraction of an inch of movement between closed and open, and some require moving the blast valve handle 90°. While the amount of motion required to change the valve from fully closed to fully open varies, the principle remains the same—make each burn identical to another.

The inexperienced pilot should begin with the premise that the average balloon requires one standard burn of four seconds every 25 to 40 seconds in order to maintain level flight. Experience determines the exact length of time between burn intervals, and how other variables such as weight and ambient air temperature affect those intervals. The primary goal is to determine the rhythm of burns necessary to maintain level flight. All other maneuvers, then, become a departure from this point. The burn begins with the brisk, complete opening of the blast valve and ends with the brisk, complete closing of the valve at the end of the burn. During their training, some pilots count "one, one thousand; two, one thousand; three, one thousand; four, one thousand" to develop the timing.

The standard burn does not mean a burn that is standard between pilots, but rather, it is an attempt for the individual pilot to make all burns exactly the same length. The goal is not only to make each burn exactly the same length, but also to make each burn exactly the same. The pilot must open and close the valve exactly the same way each time. Most balloon burners were designed to operate with the blast valve fully open for short periods of time. When the blast valve is only partially opened, two things happen:

1. The burner does not operate at full efficiency, and
2. The pilot is not sure how much heat is being generated.

Burner Ratings

During any discussions of burner output, talk usually turns to the issue of "burner ratings"; that is, the amount of heat actually produced by a specific burner. Invariably, the talk turns into a "my burner is better than yours" type of exchange. In reality, there is virtually no difference between the burners produced by the major manufacturers and their respective output.

A pilot needs to understand the concept of British Thermal Units, or BTU, as a measurement of output. A BTU is defined as the quantity of heat required to raise the temperature of one pound of water by one degree Fahrenheit. While not particularly relevant to heating air, this has come to be a generally accepted form of measure for burner output. Of the four major manufacturers, three rate their burners at 18 to 19 million BTU output; the fourth at 43.9 million (with the caveat that that is produced at a supply line pressure of 200 PSI, well above the pressures that most balloonists use). A pilot needs to be aware that BTU output in a balloon burner equates to the amount of propane that is burned in an hour; BTU output is usually computed on this basis. Referring to the Propane Primer in Chapter 2, propane has a nominal value of 91,600 BTU per gallon under ideal circumstances. A simple mathematical calculation indicates that the average burner (a 18.5 million BTU rating) consumes approximately 202 gallons of fuel per hour. (18.5 million BTU divided by 91,600 BTU per gallon of propane = 201.96 gallons of fuel)

However, balloon burners are not used for an hour at a time; instead, they are used for very short periods of time, as discussed on page 7-2. If the numbers above are recalculated, it can be said that the standard burn of four seconds, as described in this section, puts approximately 5,139 BTU of heat into the balloon's envelope. Remember, this is a theoretical statement of heat available, as no burner is 100 percent efficient.

There are numerous factors involved in burner design, such as volume backpressures, restrictions in the internal plumbing design, and more, all of which contribute to the final output rating. It is helpful to remember that the mathematics of burner design reveals the fact that output is a linear equation, when compared to pressure input; there is a direct correlation between the fuel line pressure and the burner's output. This is not the case with the noise output of the burner, as the noise output increases exponentially as the pressure is increased. Reduction of burner noise represents the next hurdle in burner design; perhaps someday it will be possible to enjoy an almost silent balloon flight.

—Mark West and Martin Harns, Aerostar International, Inc.

Figure 7-1. *Burner ratings and calculations.*

A partially opened valve is producing a fraction of the heat available, but there is no way of knowing what the fraction is. *[Figure 7-2]*

Figure 7-2. *Activated burner.*

Another advantage of briskly opening and closing the valve is to minimize the presence of a yellow, soft flame. During inflation, for instance, a strong, narrow, pointed flame that goes into the mouth opening without overheating the mouth fabric or crew is desirable. A partial-throttle flame is wide and short and subject to distortion by wind or the inflation fan. If less than a full burn is desired, shorten the time the valve is open and not the amount the valve is open. Due to burner design (and the inefficiency of a partially opened valve), four 1-second burns do not produce as much heat as one 4-second burn.

If the mechanical aspects of flying can be learned, the systematic cadence can be converted into a rhythm that is smooth and polished. With practice, the rhythm becomes second nature and pilots fly with precision, without thinking about it. Using the standard burn, pilots can better predict the effect of each burn, minimize the potential danger of a burn to the envelope, and have a better flame pattern. The standard burn is referred to when discussing specific maneuvers.

Straight and Level Flight

Level flight, or equilibrium, is probably the most important of all flight maneuvers, as it serves as a baseline from which all other maneuvers are derived. A good pilot maintains level flight with a series of standard burns.

Level flight is achieved when lift exactly matches weight and the balloon neither ascends nor descends, but remains at one altitude. For every altitude, there is an equilibrium temperature. If a pilot is flying at 500 feet mean sea level (MSL) and wants to climb to 1,000 feet MSL, the balloon temperature must be increased. This is not only to attain equilibrium at the new (higher) altitude, but some excess temperature must also be created to overcome inertia and get the balloon moving.

Theoretically, if a pilot were to hold a hot air balloon at a constant temperature, the balloon would float at a constant altitude. However, there is no practical way to hold the envelope air temperature constant. Each time the pilot burns, the balloon tends to climb. The air in the envelope is always cooling and the balloon tends to descend. If the subsequent burns are perfectly timed, the balloon flies in a series of very shallow sine waves. *[Figure 7-3, Line A]* Of course, any variable changes the balloon flight. A heavier basket load, higher ambient temperature, or sunny day all require more fuel (by shortening the interval between burns) to maintain level flight.

In discussions of level flight, the terms "flying light" and "flying heavy" are sometimes used and bear explanation. A balloon is said to be "flying light" when the sine wave being described in the air is predominately on the high side of the desired altitude *[Figure 7-3, Line B]*. Many new or inexperienced pilots tend to fly light, and use the vent line in order to return to the desired altitude; this may create a situation where the pilot gets into a constant overcontrolling exercise and is best avoided. "Flying heavy" can be described as a scenario where the sine wave is predominately on the low side of the desired altitude *[Figure 7-3, Line C]*; if the balloon is left alone, it tends to fall. Flying heavy can be hazardous when contour flying. A balloon pilot must use all visual cues available and exercise a "finesse" type of control when contour flying (described later in this chapter).

Through experimentation, standards can be established that may be used as a basis for all flights. With practice and using the second hand of a wristwatch, a new pilot can fly almost level. The exercise of learning the pattern of burns (each day and hour is different) is an interesting training exercise, but not a practical real-life technique. The ability to hold a hot air balloon at a given altitude for any length of time is a skill that

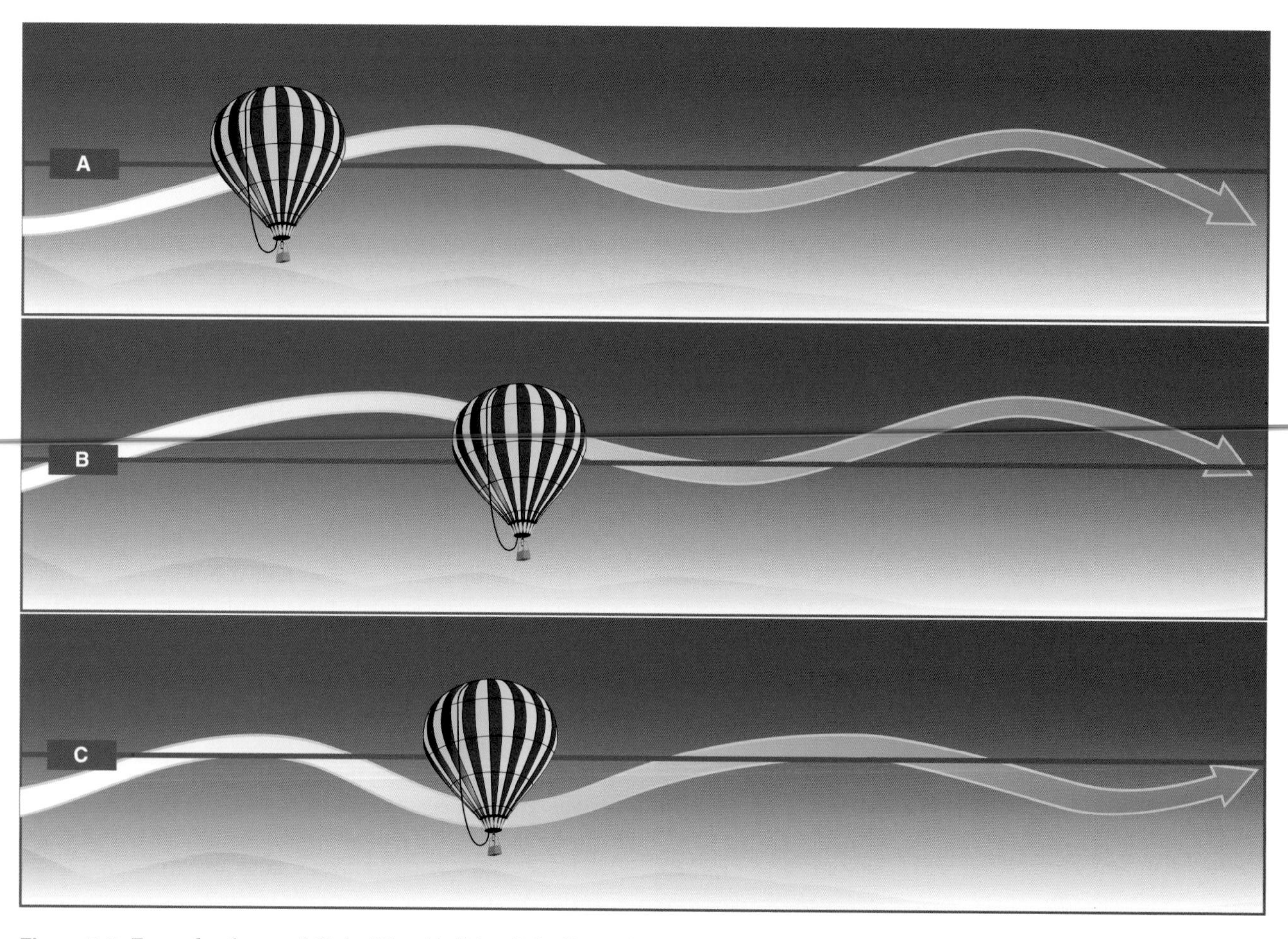

Figure 7-3. *Example of normal flight (Line A), flying light (Line B), and flying heavy (Line C).*

comes only with serious practice. Unfortunately, most pilots do not spend enough time practicing level flight. During the practical test, given a choice of altitudes, it is easier to fly level at a lower altitude, due to the ease of acquiring visual references. However, the pilot must exercise care not to violate minimum safe altitude requirements (explained later in this chapter), and must be constantly alert for obstacles such as power lines in the flightpath.

Ascents and Descents

The temperature of the air inside the envelope controls balloon altitude. A balloon that is neither ascending nor descending is in equilibrium. To cause the balloon to ascend, increase the temperature of the air inside the envelope. If the temperature is increased just a little, the balloon seeks an altitude only a little higher and/or climb at a very slow rate. If the temperature is increased significantly, the balloon seeks a much higher altitude and/or climbs faster. If the balloon is allowed to cool or hot air is vented, the balloon descends.

Even without the input action by the pilot, it must be remembered that the air inside the envelope is dynamic. The air mass is constantly moving within the confines of the envelope, attempting to seek a level of equalization. While it varies with each envelope, input action by the balloon pilot can take from 6 to 15 seconds to be realized as a reaction by the balloon. Planning the maneuver, anticipating the reaction time, inputting the proper burn, and observing the reaction must result in smooth and natural movement by the pilot.

Ascents

Using evenly spaced, identical standard burns to fly level, a pilot needs to make only two consecutive burns to have added excess heat to make the balloon climb. For example, if the pilot can maintain level flight with a standard burn every 35 seconds, and then makes two burns in succession instead of one, the balloon has an extra burn and climbs. How fast the balloon climbs depends on how much extra heat has been added. Under nominal conditions, if the standard burn has been made to hold the balloon at level flight and a second burn is within a few seconds (not waiting the 35 seconds), the average balloon starts a slow climb. Three consecutive burns results in a faster climb.

Once the desired climb rate is established, return to the level flight routine to hold the balloon at that rate. The higher the altitude, or the faster the rate of climb, the shorter the interval between burns. In an average size balloon (usually a 77, 000 or 90,000 cubic foot envelope) at 5,000 feet, the pilot may be required to make a standard burn every 15 to 20 seconds to keep the balloon climbing at 500 feet per minute (fpm). At sea level, the same rate may require burning only every 30 to 40 seconds. Burn rates cannot be predicted in advance, but practice provides a basis to begin with and experimentation finds the correct burn rate for a particular day's ambient temperature, altitude, envelope size, and balloon weight.

Another skill to develop in ascents is knowing when to stop burning so the balloon will slow and level at the chosen altitude. The transition from a climbing mode to level flight involves estimating the momentum and coasting up to the desired or assigned altitude. One methodology is illustrated in *Figure 7-4*. During this flight, the pilot has decided to leave his current altitude (A) and climb to another (C). While climbing, the pilot adjusts burn times to be at a 100 fpm rate of climb by the time he or she is at a point 100 feet below the desired altitude (B). Under most circumstances, the balloon "coasts" the additional 100 feet and rounds out, or resumes level flight, at the desired altitude. It may be necessary to use the vent for a short time to stop the ascent with caution not to vent excessively, thereby causing the balloon to go into a descent. The balloon pilot must remember that each manufacturer has a specific interval that the vent can be opened while in flight. Should that interval be exceeded, it can have a disastrous result. With practice and application, the pilot can learn this skill without the use of the vent, not only conserving fuel, but also flying a much more controlled flight.

Generally, to achieve a smooth transition to the new altitude, the rate of ascent should not exceed the distance to the new altitude. For example, if the balloon pilot is 500 feet below the desired new altitude, the rate of climb should not exceed 500 fpm. When the pilot is 300 feet below the desired altitude, the rate of climb should not exceed 300 fpm. An ascent of 200 to 300 fpm is slow enough to detect wind changes at different altitudes, which is helpful in maneuvering. Above 500 fpm, it is possible to fly through small, narrow wind bands or wind with very small direction changes without noticing. It is a good idea to launch and climb at a slow speed (100 to 200 fpm) to make an early decision regarding which direction to fly.

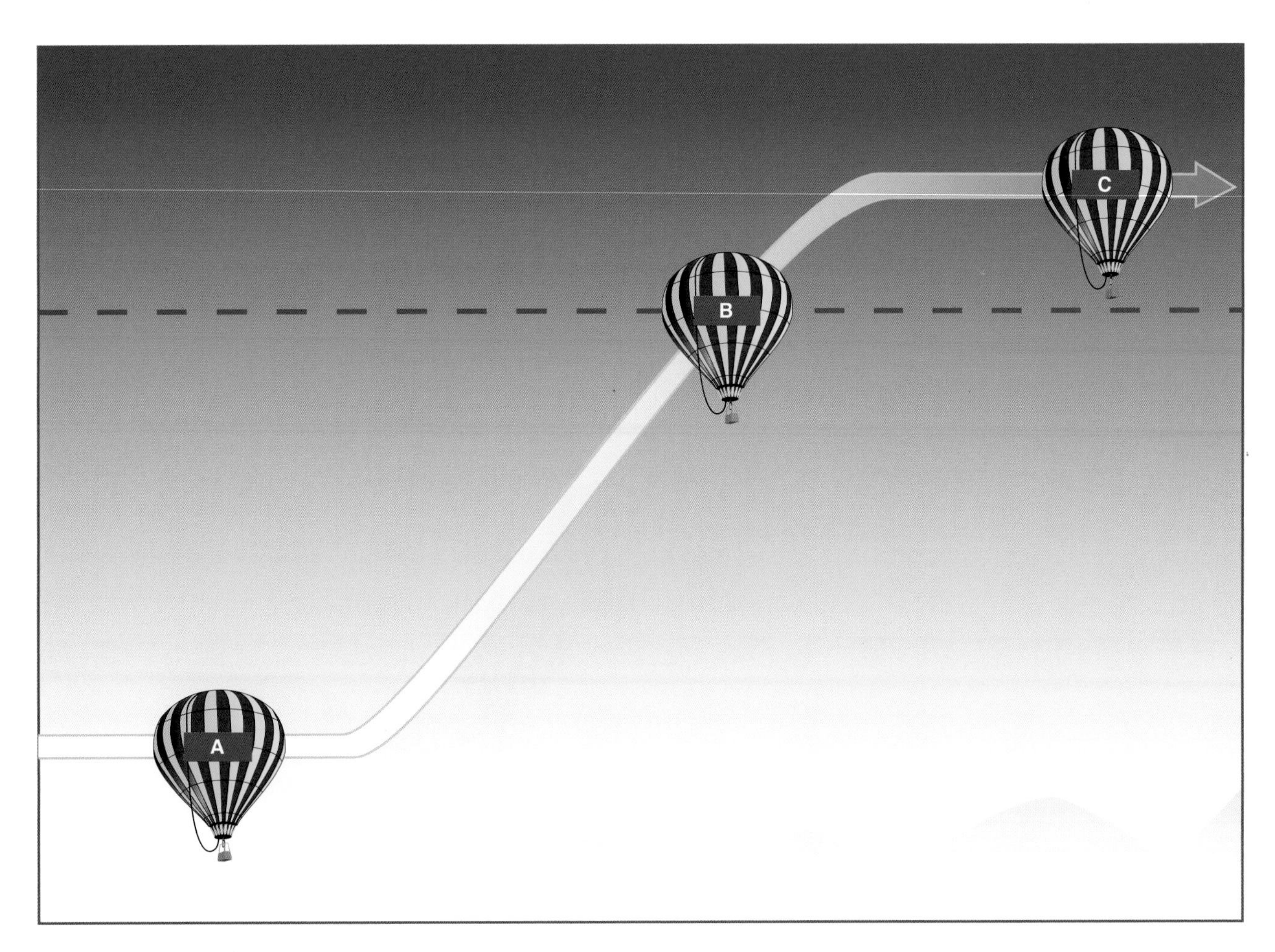

Figure 7-4. *Ascent schematic.*

Descents

The only direct control of the balloon the pilot has is vertical motion; the pilot can make the balloon go up by adding heat. Therefore, the pilot can make it come down by venting or not adding heat. For horizontal or lateral motion, a balloon pilot must rely on wind, which may or may not be going in the desired direction. A good pilot learns to control vertical motion precisely and variably for maximum lateral choice.

To start a descent from level flight, skip one burn, and then return to the level flight regimen to hold the descent at a constant rate. Alternatively, hot air balloons are equipped with a vent. When opened, the vent releases hot air from the envelope and draws cooler air in at the mouth, thus reducing the overall temperature and allowing the balloon to descend. A pilot should learn to calibrate the use of the vent, as well as the burners, and know how much air is being released in order to know what effect to expect. For predictability, time the vent openings and open the vent precisely. Parachute vent balloons usually have a manufacturer's limitation on how long the vent may be open. Use the vent sparingly; it should not be used instead of patience, unless a rapid change in altitude is needed.

A new pilot can learn the classic balloon flare by matching the vertical speed indicator (VSI) to the altimeter, i.e., descend 500 fpm from 500 feet above ground level (AGL), 400 fpm from 400 feet AGL. Below 200 feet, a pilot should not use instruments, but look below for obstacles, especially power lines. This maneuver is illustrated in *Figure 7-5*. Departing from the altitude indicated at "A," the pilot allows the balloon to start a descent. During the descent, the pilot maintains a reasonable rate of descent, slowing to a rate of 200 fpm at B. By the time the balloon reaches C, the pilot should have a rate of descent of 100 fpm, and by using one or two standard burns, should be able to level off ("round out") at the desired new altitude.

Rapid/Steep Descents

A rapid, or steep, descent in a balloon is a relative term. A 700 fpm descent started at 3,000 feet AGL is not necessarily rapid; however, if started at 300 feet AGL, it is rapid and may be critical. Rapid descents should be made with adequate ground clearance and distance from obstacles.

To execute a steep descent, the balloon pilot must be well aware of the balloon's response times and have sufficient altitude for the maneuver. In *Figure 7-6,* the pilot has initiated

Figure 7-5. *Descent schematic.*

a steep descent at Point A. For this discussion, assume that altitude is 1,500' AGL. The descent may be initiated either through the use of the vent or by holding two or three burns. By the time the balloon reaches the point indicated at B, it may be descending in excess of 500 fpm; the pilot should make a standard burn at this point, as the air passing over the envelope fabric accelerates cooling. A standard burn ensures that the balloon maintains a temperature sufficient to keep it under control throughout the maneuver. This burn should not slow the descent.

At a point halfway between the ground and the previous burn *[Figure 7-6, Point C]*, the pilot should make a long (twice standard length) burn; if there is no reaction from the balloon, then the pilot should do another burn. Immediately upon sensing a reaction from the balloon, the pilot should stop the burn and allow the balloon to descend to a proper pullout altitude. *[Figure 7-6, Point D]* If done properly, the deceleration burns stops the balloon's descent just above the desired altitude or the ground. This maneuver requires experience and practice; timing is critical.

During the initial stages of learning this maneuver, the pilot should set a "floor" (altitude lower limit) to practice with. As the pilot gains more skill, as well as confidence, that "floor" may be lowered until the pilot is able to execute the maneuver to ground level.

Maneuvering

The art of controlling the horizontal (lateral) direction of a free balloon is the highest demonstration of ballooning skill. The balloon is officially a nonsteerable aircraft. Despite the fact that balloons are nonsteerable, some pilots seem to be able to steer their balloons better than others. Being knowledgeable of the wind at various altitudes, both before launch and during flight, is the key factor for maneuvering.

Maneuvering, or steering, comes indirectly from varying one's time at different altitudes and different wind directions. The pilot must have knowledge of the winds at different levels, as previously discussed in Chapter 3, Preflight Planning, as well as being able to determine the balloon's direction in flight.

Winds Below

When in flight, winds below can be observed in many ways. Observe smoke, trees, dust, flags, and especially ponds and lakes to see what the wind is doing on the ground. To

Figure 7-6. *Steep descent schematic (not to scale).*

determine what is happening between the balloon and the ground, watch other balloons, if any.

Another means of checking winds below is to drop a very light object and watch it descend to the ground. However, exercise caution with this method. Title 14 of the Code of Federal Regulations (14 CFR) part 91 allows objects to be dropped from the air that will not harm anything below. 14 CFR section 91.15, Dropping Objects, states "No pilot in command of a civil aircraft may allow any object to be dropped from that aircraft in flight that creates a hazard to persons or property. However, this section does not prohibit the dropping of any object if reasonable precautions are taken to avoid injury or damage to persons or property."

Some items that may dropped without creating a hazard are small, air-filled toy balloons, small balls made of a single piece of tissue, or a small glob of shaving cream from an aerosol can. A facial tissue, about 8" x 8", rolled into a sphere about the size of a ping-pong ball works well. These balls fall at about 350 fpm, can be seen for several hundred feet, and are convenient to carry. Counting as the tissue ball descends, a pilot can estimate the heights of wind changes by comparing times to the ground with the altimeter reading. Experiment by dropping some of these objects, practice reading the indications, and plan accordingly.

Direction

There are several methods of accurately determining the balloon's direction at any given moment, particularly with the advent of low-cost hand-held global positioning system (GPS) units. However, the pilot should be familiar with "old-school" methods, as GPS units break or fall overboard, and batteries fail.

The pilot should first be positioned in the front of the basket, relative to the direction of flight. Then, the pilot should create a "sight picture," using part of the basket's superstructure, making an allowance for any possible spin of the basket, and then select a landmark along the route of flight. This process takes a few seconds if the landmark does not change position in relation to the "sight picture." Then, the line that may be drawn between the balloon's current position and the selected landmark is the ground track of the balloon. It is important that the pilot maintain a constant altitude while performing this evaluation; any change in altitude may put the balloon into a different wind flow, which requires the evaluation to be performed again.

It can be helpful to carry a magnetic sighting compass on the flight. Identifying the bearings to a couple of distant landmarks and comparing this to the ground track can provide useful information to the balloon pilot.

Examine *Figure 3-12* in Chapter 3. If the pilot can make the mental extrapolation between known wind directions at different altitudes and the desired direction of travel, maneuvering the balloon becomes a simple exercise in direct control, that is, vertical movement.

Contour Flying

Contour flying may be the most fun and most challenging, but, at the same time, may also be the most hazardous and misunderstood of all balloon flight maneuvers. A good definition of contour flying is flying safely at low altitude, while obeying all regulations, considering persons, animals, and property on the ground. Safe contour flying means never creating a hazard to persons in the basket or on the ground, or to any property, including the balloon. *[Figure 7-7]*

At first glance, the definition is subjective. One person's hazard may be another person's fun. For instance, a person who has never seen a balloon before may think a basket touching the surface of a lake is dangerous, while the pilot may think a splash-and-dash is fun.

Minimum Safe Altitude Requirements

Legal contour flying has a precise definition. While the FAA has not specifically defined contour, it has specified exactly what minimum safe altitudes are. 14 CFR part 91, section 91.119, refers to three different areas: anywhere, over congested areas, and over other than congested areas, including open water and sparsely populated areas.

More balloonists are issued FAA violations for low flying than for any other reason. Many pilots do not understand the minimum safe altitude regulation. Many balloonists believe the regulation was written for heavier-than-air aircraft and that it does not apply to balloons. That is a false belief; the regulation was written to protect persons and property on the ground and it applies to all aircraft, including balloons. Since this regulation is so important to balloonists, the following is the applicable portion of 14 CFR part 91, section 91.119—Minimum safe altitudes: General.

"Except when necessary for takeoff or landing, no person may operate an aircraft below the following altitudes:

(a) Anywhere. An altitude allowing, if a power unit fails, an emergency landing without undue hazard to persons or property on the surface.

(b) Over congested areas. Over any congested area of a city, town, or settlement, or over any open air assembly of persons, an altitude of 1,000 feet above the highest obstacle within a horizontal radius of 2,000 feet of the aircraft.

Figure 7-7. *Contour flying.*

> (c) *Over other than congested areas. An altitude of 500 feet above the surface, except over open water or sparsely populated areas. In those cases, the aircraft may not be operated closer than 500 feet to any person, vessel, vehicle, or structure."*

14 CFR part 91, section 91.119(a) requires a pilot to fly at an altitude that allows for a power unit failure and/or an emergency landing without undo hazards to persons or property. All aircraft should be operated so as to be safe, even in worst-case conditions. Every good pilot is always thinking "what if...," and should operate accordingly. This portion of the regulation can be applied in the following way. When climbing over an obstacle, a pilot can make the balloon just clear the obstacle, fly over it with room to spare, or give the obstacle sufficient clearance to account for a problem or miscalculation. An obstacle can be overflown while climbing, descending, or in level flight. Descending over an obstacle gives the greatest opportunity to misjudge clearance over an obstacle. In level flight, the danger is reduced. Hazards are minimized by climbing. Most instructors teach minimizing the hazard by climbing when approaching an obstacle, thus giving room to coast over the obstacle in case of a burner malfunction.

14 CFR part 91, section 91.119(b) concerns flying over congested areas, such as settlements, towns, cities, and gatherings of people. There is no standard definition of "congested area" or "open air assembly of persons" but case law has indicated that a subdivision or homes, constitute a congested area, as does a small rural town.

A balloon pilot must stay 1,000 feet above the highest obstacle within a 2,000-foot radius of the balloon. This is a straightforward regulation and easy to understand. Note that the highest obstacle is probably an antenna, tower, or some other tall object, not the rooftops. Two thousand feet is almost one-half mile. This portion of the regulation is often forgotten or ignored. *[Figure 7-8]*

A conscientious pilot includes livestock of any form—dairy cows, horses, poultry—in the 1,000-feet above rule. Domestic animals, while not specifically mentioned in the regulations, are considered to be property; and experienced pilots know that almost all poultry, exotic birds, swine, horses, and cows may be spooked by the overflight of a balloon. Livestock in large fields seem to be less bothered by balloons; however, it is always a good idea to stay at least 1,000 feet away from domestic animals. This is discussed in detail on page 7-16.

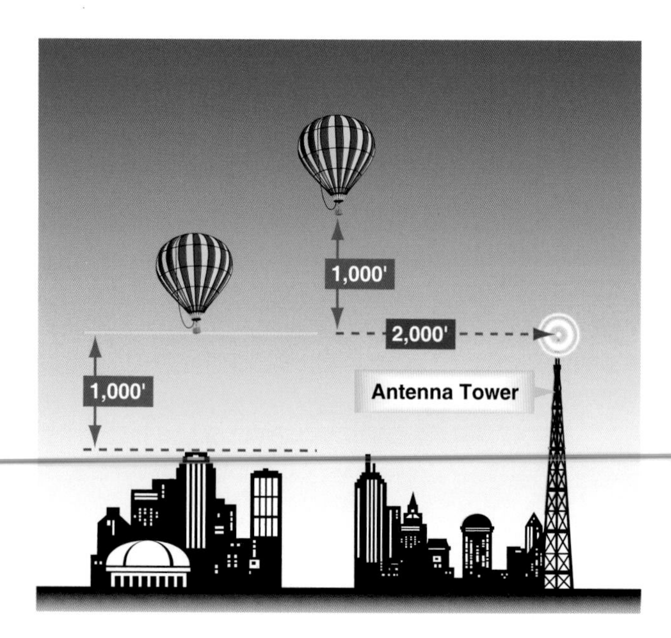

Figure 7-8 *Minimum safe altitudes over a congested area.*

Figure 7-9 *Minimum safe altitude over a sparsely populated area.*

14 CFR part 91, section 91.119(c) refers to two area types: sparsely populated and unpopulated. Here, the pilot must stay at least 500 feet away from persons, vehicles, vessels, and structures. "Away from" is the key to understanding this rule. The regulation specifies how high above the ground the pilot must be and also states the pilot may never operate closer than 500 feet. There exists a possibility for misunderstanding in interpreting the difference between congested and other than congested, as neither of these terms are defined by the FAA regulations. As an example, operating below 1,000 feet AGL within 2,000 feet of a congested area is in violation of 14 CFR part 91, section 91.119(b), even though the bordering area may be used only for agricultural purposes. Therefore, when flying over unpopulated land near a housing tract, the balloon must fly either above 1,000 feet AGL or be 2,000 feet away from the houses.

To stay 500 feet away from an isolated farmhouse, imagine a 1,000-foot diameter clear hemisphere centered over the building. *[Figure 7-9]* If the balloon is 400 feet away from the structure on the horizontal plane, the balloon pilot only need fly about 300 feet AGL to be 500 feet away from it. If the balloon passes directly over the building, then there must be a minimum of 500 feet above the rooftop, chimney, or television antenna to be legal.

In summary, regulations require:

1. Flying high enough to be safe if a problem occurs;
2. 1,000 feet above the highest obstacle within a 2,000-foot radius above a congested area; and
3. An altitude of 500 feet above the surface, except over open water or sparsely populated areas.

In those cases, the balloon may not be operated closer than 500 feet to any person, vessel, vehicle, or structure. This is an easy to understand regulation that requires compliance from all pilots.

Contour Flying Techniques

Aside from the legal aspects, contour flying is probably the most difficult flying to perform. The balloon pilot must see all obstacles on or near the balloon path, remember their location, and maintain situational and spatial awareness. Terrain or obstacle height must be estimated and allowed for, and the pilot must always be prepared for an unexpected situation. A relationship must be established between the balloon altitude and the terrain or obstacle height. All these mental calculations must occur in a few seconds, in a continuous cycle, as the pilot executes a complicated flight profile.

The balloon practical test standards (PTS) asks the applicant to demonstrate contour flying by using all flight controls properly to maintain the desired altitude based on the appropriate clearance over terrain and obstacles, consistent with safety. The pilot must consider the effects of wind gusts, wind shear, thermal activity and orographic conditions, and allow adequate clearance for livestock and other animals.

Since most contour flying is done in unpopulated areas, the balloon is rarely higher than 300 feet AGL and frequently much lower; therefore, the balloon's flight instruments are seldom observed. Because mechanical instruments have several seconds lag and electronic instruments are very sensitive, pilots must rely on their observation and judgment.

When flying at low altitude, the pilot must be vigilant for obstacles, especially powerlines and traffic, and not rely solely on instruments inside the basket. The pilot should always face the direction of travel, especially at low altitude. The pilot's feet and shoulders should be facing forward. The pilot should turn only his or her head from side to side (not the entire body) to gauge altitude and to detect or confirm climbs and descents. Facing the direction of flight cannot be overemphasized; there are many National Transportation Safety Board (NTSB) and FAA accident reports describing balloon contacts with ground obstacles because the pilot was looking in another direction.

Contour flying requires somewhat shorter burns than the standard burn. To fly at low altitudes requires half or quarter burns. One disadvantage in using small burns is the possibility of losing track of the heat being created. Precise altitude control requires special burner techniques. Another hazard of a series of too small burns is the accumulation of added heat before the effects of the last burn have been evaluated. The balloon actually responds to a burn 6 to 15 seconds after the burner is used.

One technique to determine if the balloon is ascending, flying level, or descending is to sight potential obstacles in the flightpath of the balloon, such as the power lines shown in *Figure 7-10*. As the balloon approaches the wires, the pilot should determine how the wires (or other obstacles) are moving in his or her field of vision relative to the background. If they are moving up in the pilot's field of vision (shown by the red arrow), or staying stationary, then the balloon is on a descent that may place the pilot and passengers at risk. Conversely, if the wires are moving down in the pilot's field of vision (as indicated by the green arrow), then the balloon is either in level flight or ascending, and able to clear the obstacle.

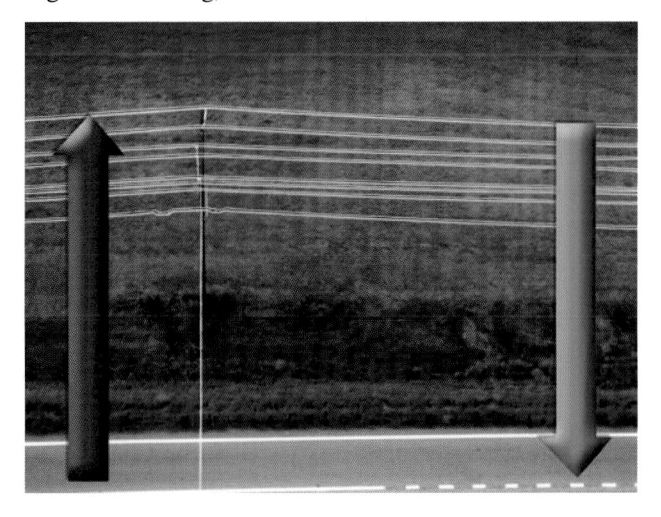

Figure 7-10 *The movement of the power lines either up or down in the pilot's field of vision can indicate whether or not the balloon has sufficient obstacle clearance.*

Some favorite sighting objects are a power pole as the near object and the line of a road, field, or orchard as the far object—lines may be observed moving up and down the poles. Water towers with checkerboard or striped markings are also good sighting objects. Vigilance is required for constant scanning of the terrain along the flightpath, and the pilot must be alert to avoid becoming fixated on sighting objects. Again, the pilot should look where he or she is going, not where he or she has been. When flying in proximity to other balloons, particularly at lower altitudes, it is easy to become fixated on the other balloon(s) and attempt to follow their lead. The balloon pilot should remember to fly his or her balloon and let the other pilots fly theirs.

The particular pleasures of contour flying can best be enjoyed in a balloon. It is wonderful to fly at low level over the trees, drop down behind an orchard, float across a pond just off the water, watch jackrabbits scatter—see sights up close. No other aircraft can perform low level contour flying as safely as in a balloon, and in no other aircraft is the flight as beautiful. Contour flying can be great fun, but remember that the balloon should always be flown at legal, safe, and considerate altitudes.

Contour Flying Cautions—Aborted Landings

The line between contour flying and unsafe, inconsiderate, and misunderstood practices can sometimes be very fine. Observers often misinterpret aborted landings on the ground as buzzing or rude flying. Sometimes landing sites seem to be elusive. A typical situation is the pilot descending to land at an appropriate site, but miscalculating the winds below and the balloon turns away from the open field toward a farmhouse. The farmer sees the balloon descend, turn towards the house, and, with noisy burners roaring, zoom back into the air and proceed. The pilot was not being rude or inconsiderate, just inexperienced. The pilot did not mean to swoop down to buzz the house; the wind had changed. If the pilot had watched something drop from the basket to gauge the winds below or been more observant, the pilot would have known the balloon would turn towards the house as it descended. A squirt of shaving foam from an aerosol can or a small piece of rolled up tissue could have alerted the pilot of the different wind at lower altitudes.

Two or three of these swoops over a sparsely populated area, and people on the ground may not only think the pilot is buzzing houses, some people may think the pilot is having a problem and is in trouble. That is when the well-meaning landowner calls the police to report a "balloon in trouble." Flying too close to a house (a friend's house, for example) to say hello, dragging the field, or giving people a thrill by flying too low over a gathering are examples of buzzing, which is illegal and can be hazardous.

Use of Instruments

14 CFR part 31 and the balloon manufacturers' equipment lists specify certain instruments to be in the balloon. However, most pilots find they use instruments less and less as they gain experience and familiarity with the balloon.

For instance, while the VSI and the altimeter can be used to execute a smooth descent and transition to level flight, the experienced pilot refers only occasionally to the instruments during maneuvers. This is especially so in maneuvers involving descents where more reliance is placed on sight pictures and visual references.

Some beginner pilots become fixated on the instruments and forget to scan outside for obstacles. If a pilot spends too much time looking at the flight instruments, the instructor may cover the instrument pack with a spare glove or a hand to try to break the formation of a bad habit. Instruments are required and useful, but should not distract the pilot from obstacle avoidance. Always practice see-and-avoid.

InFlight Emergencies

An emergency is a sudden, unexpected situation or occurrence that requires immediate action. In aviation, an emergency is a critical, possibly life-threatening or property-threatening occurrence that may require outside assistance. In an emergency, the pilot may violate any regulation as necessary to safely resolve the emergency, but must be prepared to justify his or her actions.

Because of its basic simplicity, there are few catastrophic failures or emergencies in a hot air balloon. Virtually all emergency situations involving balloons can be grouped into three categories:

- Loss/malfunction of vent or deflation line
- Loss/malfunction of pilot light
- Fuel leak

Loss/Malfunction of Vent or Deflation Line

This is usually thought of as a loss of the envelope valve control line. This could occur after a particularly windy inflation, during which the pilot has inadvertently burned through the control line and failed to notice it prior to launch. Another variation could be a control line stuck in a guide pulley, preventing the pilot from being able to control the parachute vent.

Regardless of the reason, loss of a control line must be handled. A pilot experiencing this problem must be prepared to land the balloon with a minimal amount of control, as he or she has use only of the burner to affect the descent. The pilot should maintain minimum altitudes and land in the largest area possible. If a high wind landing is expected, the pilot should anticipate rebound and dragging after touchdown; the pilot also needs to consider the prospect of deliberately landing the balloon in a less than desirable area in order to avoid potential power line contacts.

Loss/Malfunction of Pilot Light

Hot air balloon burner pilot lights are extremely reliable; however, they do fail at times. This is usually caused by a failure in the valve controlling the pilot light, or clogging of the orifice in the pilot regulator due to contamination in the fuel supplying the pilot light.

In a dual burner system, the failure of a pilot light is not a serious issue, as the balloon can still be controlled through the use of the second burner. The pilot may attempt to relight the pilot light by using both burners at the same time. To some extent, the pilot may be able to utilize this procedure to fly the balloon until able to land in a suitable location, should the extinguished pilot light not relight.

Pilot light failure in a single burner balloon is of more concern, as the pilot's options for continued flight are reduced. The pilot needs to take immediate action to relight the pilot light, either through the use of a piezoelectric igniter (if equipped), or by using a striker or other ignition source. Many balloon burners are not designed to be conveniently relit while standing in the basket; this should be practiced on a periodic basis by the pilot, at a minimum during flight reviews. Should the pilot light not relight immediately, many burners can still be utilized by "cracking" the main blast valve very slightly, and then lighting the fuel stream from the burner's jets. Other sources may include the backup system fitted in that particular balloon. The pilot may slightly open the valve controlling the backup system, and attempt to light that stream of propane. In any case, a landing as soon as practical is probably the best course of action

Fuel Leak

Fuel leaks in flight have the potential for catastrophic results and must be acted upon immediately. Because of the design of most balloon fuel systems, there are numerous potential "leak points" which may be the source for a fuel leak. Fuel lines and fuel line fittings, tank valves, and blast valves all have the potential for failure, and the balloon pilot must be aware of the various circumstances, and be prepared to deal with them.

Virtually all in-flight fuel leak emergencies in a balloon can be dealt with by shutting off the fuel source. Small fires from leaks around a fuel line fitting may be extinguished by shutting off the fuel source and then wrapping a gloved hand around the fitting and "snuffing" out the flame before it can spread further.

Because of the variety of systems, valve types, and differences in operations, a pilot should review the flight manual for his or her particular balloon system, and be familiar with and regularly practice the emergency procedures. The previously listed information is general in nature, and is not specific to any particular type of balloon, nor should it be taken as a specific procedure to be followed in the event of an in-flight emergency. In all cases, the information contained in the manufacturer's flight manual should be followed.

In all emergencies, it is imperative that the pilot maintain control of the balloon. Many minor problems can quickly become major problems if the pilot fails to continue to fly the balloon. Additionally, the use of a checklist for in-flight emergencies is not appropriate. First, the pilot must resolve the situation, and then refer to an appropriate checklist or the balloon's flight manual to verify the appropriate action.

Tethering/Mooring

Tethering a hot air balloon, despite its apparent simplicity, is perhaps the most demanding and stressful operation in ballooning, both in terms of equipment and the pilot. Balloons are designed to be free flown, not tied to the ground, and tethering incurs forces on a balloon that can, under certain circumstances, exceed design limits. The pilot conducting a tether is often called upon to conduct two or three hours of precise, "finesse" flying, which many times he or she is not prepared for. The crew must endure 2 to 3 hours of non-stop handling to manage this complex and often exhausting operation. Safe conditions fall within narrow ranges and a safe tether often demands more attention and management than flying in marginal conditions.

Every tether situation is unique enough to require tailoring the operation to specific needs and equipment. However, the idea of a "simple" tether should not lure the pilot into underestimating the very real demands and risks that come with it. A skilled and knowledgeable crew allows a pilot to take advantage of the many benefits tethering offers. Under suitable conditions, a well planned multi-hour commercial tether can last for hours, reach a media audience of millions, and offer several hundred people their first brief balloon ride. Regardless of the reasons for tethering a balloon, all forms of tethering require the same basic preparations and guidelines for safety.

The balloon pilot contemplating a tether operation should remember that the requirements of 14 CFR Part 91, General Operating and Flight Rules, apply to all operations conducted with a type-certified hot air balloon. There is a popular misconception that tether operations are conducted under the provisions of 14 CFR Part 101, Applicability; this is an incorrect assumption. Tether operations must be conducted by a certificated pilot, and may not, under any circumstances, be performed by an individual not in possession of an airman's certificate.

Laying out the balloon for a properly executed tether operation requires a little more preparation than a normal launch layout. Initial planning needs to take into consideration the winds, both current and forecast, for the period of time the tether is to be conducted. The tether site should be at least twice, and preferably three times, the size of a normal launch area for that particular balloon. It may be surrounded by trees, buildings, and other obstacles. Under certain circumstances, this may work to the pilot's advantage, as these may block low-level, lower velocity winds. Higher winds may create turbulence that significantly affects the balloon. *[Figure 7-11]*

There are two primary methods of tethering, the three-line method and the six-line method. In the three-line method, lateral lines are attached at the top of the basket, generally to the suspension of the balloon, to transfer stress loads from the balloon directly to the lateral lines, and not through the basket's superstructure. In the six-line method, three vertical "bridle" lines are connected from the top of the balloon to the top of the basket's superstructure and three additional lines

Figure 7-11 *Layout schematic for a balloon tether.*

(laterals) are attached to the vertical bridles. *[Figure 7-12]* The six-line system is preferable in windier conditions, as it allows the balloon to move within the framework created by the bridles and laterals, but keeps the balloon upright and confined to a relatively small area. The specific setup required is dependent on the intent of the tether and whether rides are being offered.

In almost all cases, the manufacturer's flight manual specifies procedures and techniques for tethering the balloon, and the prudent pilot is aware of these instructions and possible limitations prior to conducting a tether.

Tether safety often depends more on flight team skills and preparation than anything else. The following tethering safety tips are drawn from the book, *Hot Air Balloon Crewing Essentials,* by Gordon Schwontkowski, and is an excellent reference text for balloon pilots as well as crew.

- Whether you use a commercially available tether system or your own quality ropes matters less than how well and how thoroughly you prepare. Securely attach two or three tether lines minimum to load rings, envelope carabiners, or other brand specific point on your equipment.
- Triangulated wind protection is the goal. Attach tether lines to secure objects such as parked vehicles so that two lines always distribute wind loads. This protects you on all sides from turbulence and potential downdrafts from buildings.
- Focus! Constantly watch weather conditions and line tensions for changes or needed adjustments. Listen to the crew for feedback and instruction. Direct spectators and participants according to your needs and their safety rather than chatting with them.
- Note and adjust for weather changes. Weather places great unnatural stresses on tethered balloons. Be prepared to suspend or cancel a tether at any moment due to changing weather. Allow extra time for packing lines and other tether equipment when considering whether to shut down.
- Plan for any one line to fail or come untied with no notice. Create a backup plan you can implement immediately to maintain control and safety.
- Watch for spectators and children who want to hang onto tether lines and ride them off the ground. Keep all noncrew away from lines; suddenly tight or rising lines pose risks for all concerned. At NO TIME should any crew member ever leave the ground (riding or hanging on the basket, holding ropes, etc.).
- Devote one crew member exclusively to organizing passengers in a line far back from the balloon and tether lines. Select groups by number, weight, age, or an appropriate combination of factors.
- Adding weight is necessary on landing and during passenger switches, giving many opportunities for someone's toes or a foot to slip beneath a fully loaded basket. Keep everyone similarly clear of lines running to the basket or lower envelope.
- Inspect lines frequently to ensure they are attached securely at both ends, particularly under heavy loads or in stronger wind conditions.

Figure 7-12 *A six-line tether.*

Inflight Crew Management

The balloon chase crew does much more during the flight than merely drive and wait for the balloon to land. As an integral part of the flight team, the crew can often have a significant impact on flight safety. Rather than passively follow behind the balloon's perceived route, a knowledgeable chase crew can assist in the execution of the flight and play a large role in flight safety and management.

Immediately after the launch, the crew should promptly load equipment and prepare to leave the launch field. At a large fun fly of local pilots, or perhaps a competitive event, there may be a temptation for the crew to remain on the field and socialize; a crew on the field is of little or no use to a pilot who may be in difficulty. The crew chief, in conjunction with the pilot, should have a planned route off the launch site and a methodology for conducting the chase. Some pilots prefer for the chase crew to be ahead of the balloon, while other pilots have the chase crew follow the balloon throughout the entire flight. This is a matter of personal preference and may have an impact on the comfort level of the pilot. Either way is acceptable, as long as the pilot and crew remain in contact, and there is a plan in place for the eventual recovery of the balloon.

Chasing the balloon is, at times, grossly oversimplified. The crew should always bear in mind that they may have to plan on how to be in one of two places at any given moment: where the balloon currently is and where it appears to be heading in the next few minutes. A good rule during the first half of the planned flight is to leapfrog ahead of the balloon, let the balloon catch up, and then repeat the process. Even in familiar flying areas, the crew should frequently check maps to determine the flightpath, likely landing areas, and the best routes to get there. The crew must constantly be aware of changing conditions that may affect the flight; winds increasing, decreasing, or changing directions, or perhaps traffic situations that affect how the crew gets to the balloon after landing.

Crew Behavior

During the chase, the crew should remember to drive legally and politely. Driving at high rates of speed with no justification creates a sense of emergency where there is not one, and can draw unwanted attention to the chase crew. The chase crew is useless to the pilot if they have been stopped by the authorities.

While chasing, the crew should observe all "No Trespassing" and "Keep Out" signs, and stay on public, paved roads. Vehicular trespass is common and the laws are very restrictive regarding vehicles on private property. Pilots and their chase crews should adhere to local trespass laws.

If possible, the chase crew should try to keep the chase vehicle in sight of the balloon. The pilot needs to know the crew is nearby and not stuck in a ditch or off somewhere changing a flat tire. When the vehicle is stopped at the side of the road, park it so the entire vehicle is visible to the balloon pilot.

Pilot/Crew Communications

Radio communications between the balloon and the chase vehicle are fairly common. If radios are used, obey all Federal Communications Commission (FCC) regulations. Use call signs, proper language, and keep transmissions short. Many balloonists prefer not to use radios to communicate with their chase vehicle because it can be distracting to both pilot and chase vehicle driver. In any case, it is a good idea to agree on a common phone number before a flight in case the chase crew loses the balloon. *[Figure 7-13]*

Figure 7-13 *This crew chief has stopped safely off the road to talk to her pilot. Communications are important, but safety is paramount.*

Use of a Very High Frequency (VHF) Radio

There is confusion among pilots regarding frequencies that may be used from air-to-ground, balloon-to-chase crew, for instance. Many balloonists use 123.3 and 123.5 for air-to-ground (pilot-to-chase crew), as these frequencies are for glider schools and not many soaring planes are in the air at sunrise. Since all users of the airwaves must have an ID or call, ground crews may identify themselves by adding "chase" to the aircraft call sign. For example, the chase call for "Balloon 3584 Golf" would be "3584 Golf Chase," or perhaps simply "84 Golf Chase."

The air-to-air frequency is 122.75. Remember that everyone in the air is using this frequency; transmissions should be kept brief. A balloon pilot trying to contact a circling airplane would try 122.75 first.

Weather information is available on VHF radio. A balloon pilot could obtain nearby weather reports by tuning to the Automatic Terminal Information Service (ATIS). The appropriate frequency is listed on the cover of the sectional chart and in the airport information block printed on the chart near the appropriate airport.

Landowner Relations

Identification of Animal Populations

Balloonists must learn how to locate and identify animals on the ground. Even though it may be legal to fly at a certain low altitude, animals do not know the laws—nor do most of their owners. If a pilot causes dogs to bark, turkeys to panic, or horses to run, even while flying legally, it may provide legitimate cause for complaint.

Horses and ponies can be anywhere from the smallest paddock and roughest fields to the largest pasture. Boarding stables and breeding farms are easily identified by their painted wooden fences, stacked bales of hay and straw, and horse trailers. Assume all horses are valuable: race horses, breeders, and those privately boarded. While horses generally behave the same way when frightened, each one's alarm level differs. Anything from the glow of the back-up burner to the balloon's shadow can spook them. In a pasture, they may buck, neigh, or even charge; in extreme cases, they might try to jump the fence and may injure themselves. As horses are accustomed to hearing human voices, talking to them may help calm them down. *[Figure 7-14]*

Figure 7-14 *Flight in areas of horse activity requires caution and consideration.*

Cattle need more space than horses, about one acre per cow. Dairy cows tend to stay near the barn; corn near the pasture and a muddy yard usually means milk production. If they are out in more remote pastures, they are probably beef producers. When startled, cattle usually bunch together to face a threat but can just as easily panic and run. A stampede can break down a fence or locked gate. Once out, cattle can be herded by driving them from behind where they need to go by blocking sideways means of escape (with people or vehicles). Brood cows can be unpredictable, especially during breeding, and are capable of damaging a truck. As with horses, the sound of human voices may help calm cattle. *[Figure 7-15]*

Figure 7-15 *While cattle in a field may appear undisturbed, they can spook and stampede at the least provocation.*

Pigs are perhaps the most difficult livestock for farmers to raise. Easily disturbed, it is best to avoid them at all costs. When frightened, they bunch together and bolt, and are virtually unstoppable, almost impossible to catch, and potentially dangerous (especially boars). Pigs run until they are exhausted; rounding them up can take days. Pork prices fluctuate dramatically, and monetary claims against a balloonist who upsets pigs can be quite high.

Poultry (chickens and turkeys) are more often raised in confinement buildings than free range. A long windowless building with fans on the ends and atop the roof usually indicates a poultry production facility. With the lights and fans off, tens of thousands of birds can be amazingly quiet. Any disruption of this tightly controlled environment, however, can severely affect production. A sudden blast of a balloon's burner can start an all-out panic; birds trying to flock in a confined space trample and crush each other, leading to mass-scale suffocation. Ducks and geese raised outside average about 200 birds per pen. When panicked, they also tend to flock which can lead to injury and death.

Livestock will appear along a balloon's flightpath, as well as landing sites. Follow these livestock tips to minimize disruptions during these encounters:

- Always watch for signs of livestock: barns, sheds, silos, fences, muck heaps, tractors, trailers, stacked bales of straw and hay, and mudded areas all point to some form of livestock nearby.

- Stay away from barns and building clusters – livestock are most likely inside or nearby.
- Livestock are most active in the morning and early evening. When hot, they seek shade near buildings or under trees. When cold, they gather together to conserve heat and seek shelter from wind anywhere.
- Livestock breaking out of their field or enclosures as the balloon flies by demands immediate attention. A cow in the road can total a car; the driver will not do well either. The chase crew should either find the farmer or take steps to contain or protect the livestock without endangering themselves.
- The sound of a human voice can often calm agitated or frightened livestock—talk to them. Use calm and easy tones.
- Horses with riders warrant particular caution. A highly visible wave from the basket to the rider acknowledges their presence and says that all possible care is being taken.

Chapter Summary

As with the previous chapter, this chapter explains techniques and procedures that have come to be generally accepted throughout most of the ballooning community. They are by no means an exhaustive explanation of "how to fly." Some of the procedures may seem unduly complex, but with practice and experience, these maneuvers come naturally.

A good pilot observes other pilots and incorporates techniques that are safe and acceptable. New pilots sometimes fly with different instructors, in order to gain a perspective on "how the other guy does it"; other, more experienced pilots fly with another commercial pilot during a flight review. Almost all flight methodology, as long as it is safely conducted, and with a plan and purpose behind it, is acceptable for use as long as the performance requirements of the PTS can be met.

Chapter 8

Landing and Recovery

Introduction

No other aircraft has as many different types of landings as a balloon. Rarely are two landings in a balloon the same, and each has its own unique characteristics. The prime consideration in any balloon landing is the safety of the pilot and passengers. While accessibility, ease of recovery, and possible damage to the balloon are certainly considerations in any balloon landing, these do not override the simple fact that preventing injury to occupants must be the primary goal of any balloon pilot.

Accident statistics indicate that the landing sequence is the portion of flight in which the most injuries occur.

The Approach

An axiom of ballooning is "the best altitude for landing is the lowest altitude." Anyone can land from one foot above the ground; it takes skill to land from 100 feet. The pilot's ultimate goal in making the approach is to set himself up to make a smooth, gentle landing in the best possible location without causing damage to the balloon or injury to the passengers.

Conventional wisdom indicates three types of approaches to landing in ballooning—stair step, straight line, and steep (previously discussed in Chapter 7, Inflight Maneuvers). In reality, the stair step and straight line approaches are considered the same type of approach, a controlled approach, while the steep approach is viewed as an accelerated approach. Any approach is considered to be a variation of one of these two or perhaps a combination.

Of more importance, and a point frequently missed by newer or inexperienced pilots, is that the approach is being performed in two different planes or dimensions. Pilots are accustomed to thinking of the approach in vertical terms, but frequently discount the fact that the approach is also being performed on the horizontal plane.

Refer to *Figure 8-1.* This is the approach path to the selected landing site for this pilot. Most pilots presume that this will happen after making a determination of lower level winds, evaluating obstacle clearance, noting the presence of powerlines, and making the decision to commit to the landing. After that, the primary focus becomes making the descent without further consideration of the lateral or horizontal movement.

In *Figure 8-2,* the pilot has determined the winds below, and has set up for the same approach ensuring obstacle clearance, powerline considerations, and making the commitment to land. The difference here is that the pilot has a higher level

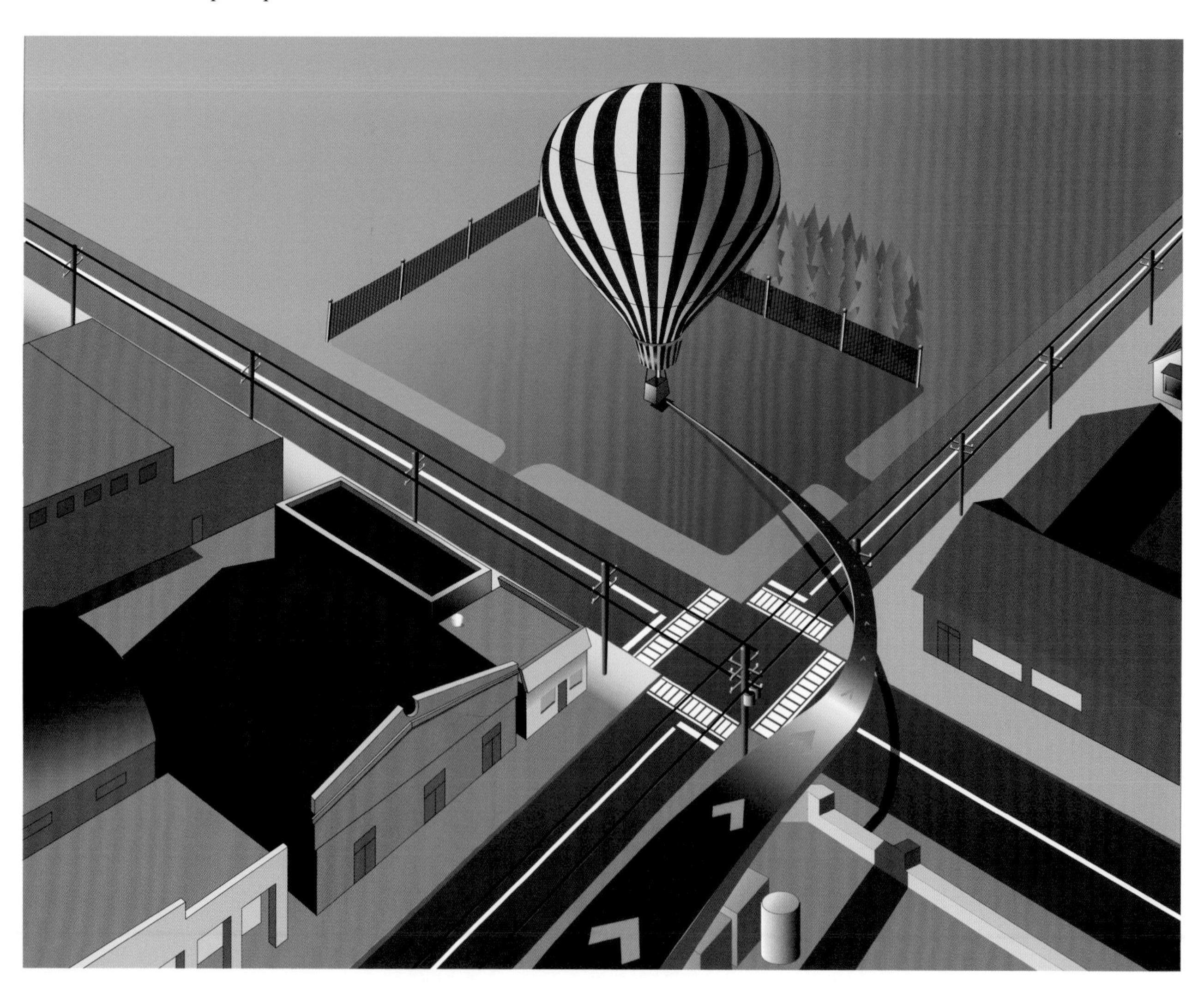

Figure 8-1. *Horizontal approach path as normally visualized and performed.*

Figure 8-2. *The horizontal plane's "reality" approach to landing.*

of situational awareness and understands the horizontal component, and therefore begins the approach with many more options available. In this example, if the pilot is tending more towards the road, he or she can "extend" the approach to remain in more favorable winds, and then continue the descent when the turn is more advantageous to the landing.

The vertical component of the landing profile is represented in *Figure 8-3.* This may be referred to as a "natural" descent profile. For example, the pilot may see a large field in the distance ahead and elect to initiate a shallow or low profile approach. The pilot would let the balloon cool naturally until it reached the point depicted as Point A in *Figure 8-3.* A normal descent rate at that point would project the balloon out on a line tangent with the descent curve at that moment. From that point, the pilot must maintain that line much like the control necessary for contour flying.

If, however, the winds are somewhat lighter or the landing site is tight, conducting the approach from Point B, *Figure 8-3* might be more appropriate. Again, the pilot should wait until the tangent of the descent curve aligns with the targeted

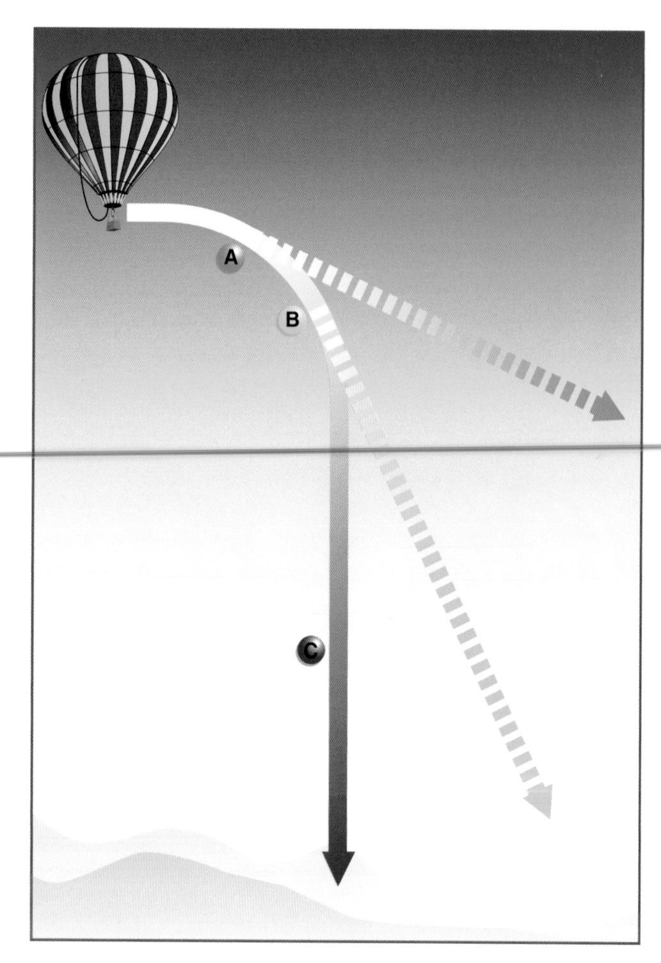

Figure 8-3. *Vertical profile for landing.*

landing site. At that point, the pilot should make a half burn and the balloon will follow the desired path.

As a final illustration, steep descents, as previously discussed in Chapter 7, are usually initiated and controlled from the point or area depicted as C in *Figure 8-3.* The pilot must take both planes of movement, the horizontal as well as the vertical, into consideration when making the approach.

Again, the balloon pilot must visualize the landing by imagining the path through the air and across the ground. Constant scanning in all directions while on the desired approach is imperative. Target fixation can cause a pilot to miss potentially dangerous objects or situations. Look for obstacles, especially powerlines, near the imagined track. Note surface wind velocity and direction by looking for smoke, dust, flags, moving trees, and anything else that indicates wind direction. Do not be influenced too much by a wind indicator at a distance from the proposed site if there is a good indicator closer.

As an example, imagine flying level at 700 feet above ground level (AGL) and it is time to land. Checking the fuel, there is 30 percent remaining in each of the two 20-gallon tanks. The balloon's track across the ground is toward the southeast, but a farmer's tractor is observed making a column of dust that is traveling nearly due east. The dust cloud rises from the ground at about a 45° angle. From this information, it is believed that the balloon turns left as it descends. This means the pilot is looking to the left of the line the balloon now travels. By dropping small tissue balls, the pilot determines that the wind changes about halfway to the ground and continues to turn left about 45°. As an initial plan, visualize the descent being no faster than about 500 feet per minute (fpm) initially and slowing to about 300 fpm about 400 feet AGL where the turn to the left can be expected. Because the balloon loses some lift from the cooling effect of the wind direction change, the rate of descent should be closely monitored during the turn.

With this imagined descent in mind, the pilot searches for an appropriate landing site. The next fallow field to the left of the present track is blocked by tall powerlines. The landing site is rejected as unsuitable. The next field that seems appropriate is an unfenced grain stubble field bordered by dirt roads with a 30-foot high power line turning along the west side to the left and parallel to the balloon's present track. Under the powerlines is a paved road with a row crop of sugar beets to the right and directly in the balloon's ground track.

The pilot selects a landing site at the intersection of the two dirt roads at the southeast corner of the field. The planned path would be across the field diagonally giving the greatest distance from the powerlines. Extending the final approach line back over the powerlines and into the sugar beet field, the pilot selects a point where the balloon is likely to begin its surface wind turn. Then, reverse planning from that point, a point should be determined in order to begin the initial descent.

Now, the balloon pilot performs the descent, turn, and landing that is visualized. If all goes as planned, the balloon cools and accelerates to about 500 fpm. The pilot applies some heat to arrest the descent while approaching the imagined turning point over the beets, levels off (or actually climbs a bit) while crossing over the power lines (about 100 feet above), and allows the balloon to cool again while setting up another descent across the stubble field. Due to the seven mile per hour (mph) estimated wind, the pilot allows the basket to touch down about 150 feet away from the dirt road intersection to lose some momentum as the basket bounces and skids over to the road, just as planned.

Imagine, how the landing might have occurred not knowing the surface wind was different from that of the flight path. Perhaps a less considerate pilot would plan on landing in the beets, believing the crop would not be hurt and the farmer would not care. Unfortunately, the balloon turns unexpectedly toward the power lines, causing the pilot to make several burns of undetermined amount, setting up a climb that prevents the pilot from landing on the dirt road. The road is directly in line with the balloon's track but disappearing behind the balloon. Another good landing site becomes unusable because of the lack of planning.

The prudent pilot makes a nice landing on a dirt road with 25 percent fuel remaining, while the unprepared pilot dives, turns, climbs, and is now back at 600 feet in the air looking for another landing site. For the next attempt, the pilot has less fuel and is under more pressure, all because of not noticing the power lines to the left and not checking the wind on the ground before making the descent.

Step-Down Approach

The step-down approach method involves varying descent rates. This procedure is used to determine lower level wind velocities and directions so that options may be considered until beginning the final descent phase to landing. There are other methods to evaluate lower level wind conditions, such as dropping strips of paper, small balloons, etc. While the descent path can be varied and sometimes quite shallow, it is important to avoid long, level flight segments below minimum safe altitudes without intending to land. Level flight at low altitudes could lead an observer to believe that the pilot has discontinued the approach and established level flight at less than a minimum safe altitude. *[Figure 8-4]*

Low Approach

The second type of approach is a low or shallow approach. If there are no obstacles between the balloon and the proposed landing site, a low or shallow approach allows a pilot to check the wind closer to the surface. The closer the balloon is to the surface, the easier it is to land. Low approaches are suitable in open areas with a wide field of view and few obstacles.

Obstacles and Approach Angles

Most good landing areas are not very large and generally have obstacles somewhere close by. The classic approach requires the balloon pilot to fly the balloon at a descent angle that clears the highest obstacle in its path to the intended landing site. However, if the pilot allows the balloon to descend below that angle, a rapid adjustment of that angle is necessary to avoid the top of the obstacle. In many cases, the attempt to miss an obstacle by a close margin may result in a period of overcontrolling or excessive burns and vents.

Figure 8-4. *The step-down approach.*

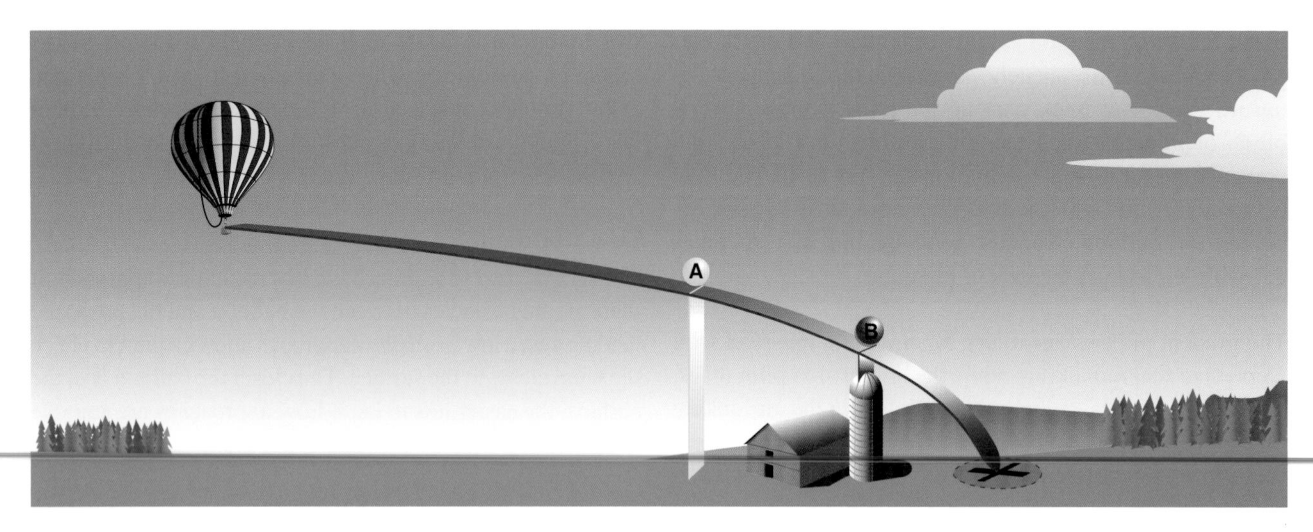

Figure 8-5. *In this approach, the pilot has elected to clear the obstacle at a minimal distance. Any overcontrol after Point A is likely to send the balloon past the intended landing area, and a very close clearance will be necessary over the obstacle at Point B in order to preserve the desired approach angle.*

The inexperienced pilot usually misses the intended landing site and is forced to find another or has an undesirably hard landing, if the pilot manages to make the landing site. See *Figure 8-5* for an illustration of this effect.

This type of approach is helpful by keeping the intended landing area in focus so that the descent results in the balloon touching down precisely where intended. Benefits to this methodology are straight line control and immediate visual feedback from the landing site. The disadvantage of using this specific methodology is that, again, any deviation from the depicted flight path may result in overcontrol or, at best, a "close shave" over the obstacle at Point B to preserve the approach.

Figure 8-6 illustrates the same vertical profile but with a few subtle and important differences. The initial approach is identical, but the initial target is different. Rather than focusing on the landing site past the barn requiring a clearance of two or three feet, this pilot has elected to "aim" for a point of safe obstacle clearance of the barn (about ten feet above the silo), knowing that a short vent on the other side "sharpens" the descent. It is clear that the balloon drops into the landing site somewhat closer to the obstacle, the conscientious pilot having already evaluated that possibility prior to committing to the landing. This methodology minimizes the possibility of overcontrolling and dramatically increases the chances of a successful landing.

Figure 8-6. *This approach is virtually the same as the one in Figure 8-5, but with subtle differences. In this case, the pilot has aimed for a location above the obstacle at Point B with appropriate obstacle clearance. Minor deviations after Point A on the approach path, if compensated for, still allow the pilot to make the intended landing area.*

To summarize, if there is an obstacle between the balloon and the landing site, the following are the three safe choices.

1. Give the obstacle appropriate clearance and drop in from altitude.
2. Reject the landing and look for another landing site.
3. Fly a low approach to the obstacle, fly over the obstacle allowing plenty of room, and then make the landing.

The first choice is the most difficult, requiring landing from a high approach and then a fast descent at low altitude. The second choice is the most conservative, but may not be available if the pilot is approaching the last landing site. The third choice is preferable. Flying toward the site at low altitude provides an opportunity to check the surface winds. By clearing the obstacle while ascending—always the safest option—the pilot ends up with a short, but not too high, approach.

Some Basic Rules of Landing

When a landing site is being considered, a balloon pilot should first think about the suitability of the site. "Is it safe, is it legal, and is it polite?"

Plan the landing early enough so that fuel quantity is not a distraction. Always plan on landing with enough fuel so that even if the first approach to a landing site is unsuccessful, there is enough fuel to make a couple more approaches.

The best landing site is one that is bigger than the balloon needs and has alternatives. If the balloon has three prospective sites in front of it, the pilot should aim for the one in the middle in case the surface wind estimate was off. If the balloon has multiple prospective landing sites in a row along its path, the pilot should take the first one and save the others for a miscalculation. Unless there is a 180° turn available, all the landing sites behind are lost. Before beginning the approach, the pilot should plan to fly a reasonable descent path to the landing site using the step-down approach, the low (shallow) approach, or a combination of the two.

Congested Areas

Making an approach in a congested area, and subsequently discovering the site to be impossible or inappropriate, is another example of a situation in which a pilot could be falsely accused of low flying. There are some inconsiderate pilots who fly too low in congested areas without reason, but they are rare. Every pilot has an occassional aborted landing situation. According to Title 14 of the Code of Federal Regulations (14 CFR) part 91, section 91.119, a balloon may fly closer to the ground than the minimum altitude, if necessary for landing. For example, during the approach the balloon turns away from the obviously preferred landing site, but there is another possible site only one-half mile in the proper direction. The pilot has two choices: (1) go back up to a legal altitude and try again, or (2) stay low in the wind that the balloon pilot is sure will carry the balloon to a good site and try to make the second landing site.

In making the first choice, the pilot could be accused of intentionally flying too low. However, with the second choice, he or she flies lower for a longer period of time, which might appear to be a blatant violation of the minimum altitude regulation. This is not an argument in favor of either technique. Many pilots prefer the second choice on the "once the pilot goes down to land, he or she had better land" theory. *[Figure 8-7]*

Figure 8-7. *A good landing is something to watch.*

Practice Approaches

Approaching from a relatively high altitude with a high descent rate down to a soft landing is a very good maneuver to practice periodically, regardless of the experience level of the pilot. This is used when landing over an obstacle and maneuvering to the selected field is possible only from a higher than desired altitude. This maneuver can also be used when, due to inattention or distractions, the pilot is at a relatively high altitude approaching the last appropriate landing site. Practicing this type of approach should take place in an uninhabited area and the obstacle should be a simulated obstacle.

A drop-in landing or steep approach to landing is another good maneuver to practice. Being able to perform this maneuver can get the balloon into that perfect field that is just beyond the trees or just the other side of the orchard. A pilot that is able to drop quickly, but softly, into the fallow field between crops is more neighborly than making a low approach over the orchard. Being able to avoid frightening cattle or other animals during an approach is a valuable skill.

Having the skill to predict the balloon's track during the landing approach, touching down on the intended landing target, and stopping the balloon basket in the preferred place can be very satisfying. It requires a sharp eye trained to spot the indicators of wind direction on the ground. Dropping bits of tissue, observing other balloons, smoke, steam, dust, and tree movement are all ways to predict the balloon track on its way to the landing site. During the approach, one of the pilot's most important observations is watching for power lines.

In ballooning, approaches can be practiced more often than landings. A good approach usually earns a good landing.

Thermal Flight

The experience of being caught in a thermal can certainly occur at any altitude, but is more common during the landing phase of balloon flight. This is primarily due to the balloon's proximity to the ground and being exposed to the conditions that cause thermals. *[Figure 8-8]*

There are indicators of thermal conditions that may be visible to the pilot, such as "dust devils" or small tornados visible from the ground. An uncommanded ascent, usually at a significant climb rate, may be experienced, or perhaps significant changes in direction. In the extreme, a balloon caught in a tight (compact) thermal may even create small horizontal circles in the sky.

There are two rules to thermal flight. First, the pilot must continue to fly the balloon. Many pilots are distracted by the thermal flight experience and may forget that the balloon is cooling at a faster than normal rate. The pilot should ensure that the temperature of the balloon's envelope is kept at or above that required to maintain normal flight, and under no circumstance should the pilot allow the balloon to cool. The temperature noted at equilibrum should be a minimum temperature to maintain. Additionally, the pilot should not

Figure 8-8. *Anatomy of a thermal. A thermal is created by the uneven heating of the surface of the Earth by solar radiation (insolation). As the sun heats the ground, the ground in turn warms the air above it. Warmer air is less dense, and therefore rises; as it rises, it cools due to expansion. This heating/cooling pattern sets up a cycle, whereby a downward flow is created outside the thermal column one the air has cooled to a temperature equal to that of the surrounding air. This can create a hazardous situation for a balloon pilot; the best action is to maintain buoyancy, and wait for the column to dissipate.*

vent in an effort to descend and land; if the balloon falls out of the thermal, the pilot may not have sufficient altitude to recover control.

The second rule of thermal flight is the axiom, "altitude is your friend." Thermals are usually short-lived phenomena and dissipate after a brief period of time. If the pilot keeps the balloon at flight temperature, the thermal usually "spits" the balloon out at a higher altitude. When this happens, the balloon can again be controlled by the pilot and a safe landing may be executed.

Landings

Landing Considerations

When selecting a landing site, three considerations in order of importance are: safety of passengers, as well as persons and property on the ground; landowner relations; and ease of recovery.

Some questions the pilot should ask when evaluating a landing site are:

- Is it a safe place for my passengers and the balloon?
- Would my landing create a hazard for any person or property on the ground?
- Will my presence create any problems (noise, startling animals, etc.) for the landowner?

High-Wind Landing

When a high-wind landing is likely, the pilot should explain to the passengers that a high-wind landing follows. It is better to alert them than allow them to be too casual. Passengers should be briefed again on the correct posture and procedures for a high-wind landing, to include wearing gloves and helmets, if available or required. Some pilots with dual burner systems elect to leave one burner operational through the landing process, in the event that heat must be applied to climb out of a bad approach. This decision should be made with a view towards the intensity of the impact; a pilot must also be aware that, instinctively, he or she may grab the burner handle during a rough landing. This could result in burner activation, damaging the balloon fabric.

The pilot should fly at the lowest safe altitude to a large field and check that the deflation line is clear and ready. Obstacles should be avoided and the pilot should ideally make an approach to the near end of the field. When committed to the landing, brief passengers again, turn off fuel valves, drain fuel lines, and turn off pilot lights. Depending on the landing speed and surface, open the deflation vent at the appropriate time to control ground travel. The passengers should be closely monitored to ensure they are properly positioned in the basket and holding on tightly.

Deflate the envelope and monitor it until all the air is exhausted. Be alert for fire, check the passengers, and prepare for recovery.

When faced with a high wind landing, the balloon pilot must remember that the distance covered during the balloon's reaction time is markedly increased. This situation is somewhat analogous to the driver's training maxim of "do not overdrive your headlights." For example, a balloon traveling at 5 mph covers a distance of approximately 73 feet in the 10 seconds it takes for the balloon to respond to a burner input—a distance equal to a semi-truck and trailer on the road. However, at a speed of 15 mph, the balloon covers a distance of 220 feet, or a little more than two-thirds of a football field. A pilot who is not situationally aware and fails to recognize hazards and obstacles at an increased distance may be placed in a dangerous situation with rapidly dwindling options.

Water Landings

Landing in water requires modification to the passenger briefing. In a ditching, make sure the passengers get clear of the basket in case it inverts due to fuel tank placement. Advise the passengers to keep a strong grip on the basket because it becomes a flotation device. Predicting the final disposition of the basket in water is difficult. If the envelope is deflated, the pilot and passengers can expect the envelope to sink, as it is heavier than water. Fuel tanks float because they are lighter than water, even when full. If the envelope retains air, and depending on fuel-tank configuration, the balloon basket may come to rest, at least for a while, on its side.

If the water landing is done on a river, the balloon may be dragged down river in the current. Generally, staying with the basket is the best course of action unless the balloon has landed close to the bank and those aboard are strong swimmers.

Passenger Briefings and Management

Prior to landing, the pilot should explain correct posture and procedure to the passengers. Many balloon landings are gentle, stand-up landings. However, the pilot should always prepare passengers for the possibility of a firm impact. The prelanding briefing should instruct passengers to do the following:

- Stand in the appropriate area of the basket.
- Face the direction of travel.
- Place feet and knees together, with knees bent.
- "Hold on tight" in two places.
- Stay in the basket.

Stand in the Appropriate Area of the Basket

Passengers and the pilot normally position themselves toward the rear of the basket. *[Figure 8-9]* This accomplishes three things.

1. The leading edge of the basket is lifted as the floor tilts from the occupant weight shift, so the basket is less likely to dig into the ground and tip over prematurely.
2. With the occupants in the rear of the basket and the floor tilted, the basket is more likely to slide along the ground and lose some speed before tipping.

Figure 8-9. *Landing with passengers.*

3. The passengers are less likely to fall out of the front of the basket.

In a high-wind landing, passengers should stand in the front of the basket because if the basket makes surface contact and tips over, passengers fall a shorter distance within the basket and are not pitched forward. They are more likely to remain in the basket, minimizing the risk of injury.

Face the Direction of Travel

Feet, hips, and shoulders should be perpendicular to the direction of flight. Impact with the ground while facing the side puts a sideways strain on the knees and hips, which do not naturally bend that way. Facing the opposite direction is more appropriate under certain conditions. Recall the discussion on passenger briefings on page 6-12.

Place Feet and Knees Together, with Knees Bent

To some people this may not seem to be a natural ready position; however, it is very appropriate in ballooning. The feet and knees together stance allows maximum flexibility. With the knees bent, one can use the legs as springs or shock absorbers in all four directions. With the feet apart, sideways flexibility is limited and knees do not bend to the side. With legs apart fore and aft, one foot in front of the other, there is the possibility of "doing the splits" and a likelihood of locking the front knee. Avoid using the word brace as in "brace yourselves," as that gives the impression that knees should be locked or muscles tensed. Legs should be flexible and springy at landing impact.

"Hold On Tight" in Two Places

This is probably the least followed of the landing instructions. Up to this point, the typical balloon flight has been relatively gentle, and most passengers are not mentally prepared for the shock that can occur when a 7,000 pound balloon contacts the ground. Passengers should be reminded to hold on tight. The pilot should advise the passengers of correct places to hold, whether they are factory-built passenger handles or places in the balloon's basket the pilot considers appropriate. The pilot should obey his or her own directions and also hold on firmly.

Stay in the Basket

Some passengers, believing the flight is over as soon as the basket makes contact with the ground, start to get out. Even a small amount of wind may cause the basket to bounce and slide after initial touchdown. If a 200-pound passenger decides to exit the basket at this point, the balloon immediately begins to ascend. All passengers should stay in the basket until individually told by the pilot to exit.

Monitoring of passengers is important because, after the balloon first touches down, passengers may forget everything they have been told. A typical response is for the passenger to place one foot in front of the other and lock the knee. This is a very bad position as the locked knee is unstable and subject to damage. Pilots should observe their passengers and order "feet together," "front (back) of the basket," "knees bent," "hold on tight," and "do not get out until I tell you." The pilot should be a good example to passengers by assuming the correct landing position. Otherwise, passengers may think, "If the pilot does not do it, why should we."

It is very important that the passenger briefing be reinforced more than once. Some balloon ride companies send an agreement to their passengers in advance which includes the landing instructions. Passengers are asked to sign a statement that they have reviewed, read, and understand the landing procedure. Many pilots give passengers a briefing and landing stance demonstration on the ground before the flight. This briefing should be given again as soon as the pilot has decided to land.

The pilot is very busy during the landing—watching the passenger's actions and reactions, closing fuel valves, draining fuel lines, cooling the burners, and deflating the envelope. The better the passengers understand the importance of the landing procedure, the better the pilot performs these duties and makes a safe landing.

Recovery

Landowner Relations

The greatest threat to the continued growth of ballooning is poor landowner relations. The emphasis should be to create and maintain good relations with those who own and work the land that balloonists fly over and land on. Balloon pilots must never forget that, at most landing sites, they are unexpected visitors. Balloonist may think the uninvited visit of a balloon is a great gift to individuals on the ground. The balloon pilot may not know the difference between a valuable farm crop and uncultivated land. A balloonist may not notice grazing cattle. A chase vehicle driver may drive fast down a dirt road, raising unnecessary dust. A crewmember may trample a valuable crop.

There are several situations where a balloon pilot or crew may anger the public. When people are angered, they demand action from the local police, county sheriff, Federal Aviation Administration (FAA), or an attorney. The FAA does not initiate an investigation without a complaint. The balloonist is the one who initiates a landowner relations problem.

A balloonist can ease the effect he or she has on people on the ground. First, the pilot should develop skills to allow the widest possible selection of landing sites. The pilot should ensure that the crew is trained to respect the land, obey traffic laws, and be polite to everyone they come in contact with. The crew should always get permission for the balloon to land and the chase vehicle to enter private property.

Each pilot should learn the trespass laws of his or her home area. In some states, it is very difficult for the balloon pilot and passengers to trespass, but very easy for the chase vehicle and crew to trespass. If the balloon lands on the wrong side of a locked gate or fence, the first thing the chase crew should do is try to find the landowner or resident to get permission to enter. If no one can be found, it may be necessary to carry the balloon and lift it over the fence.

Do not cut or knock down fences, as it is probably considered trespassing. In some places, even the possession of fence-cutting and fence-repairing tools may be interpreted as possession of burglary tools, creating liabilities. Pilot and crew should have a clear understanding of what is acceptable and what is not acceptable.

Sometimes local law enforcement officials arrive at the scene of a balloon landing site. Someone may have called them or they may have seen the balloon in the air. They may just be interested in watching the balloon (which is usually the case), or they may think a violation of the law has occurred. If law enforcement officials approach the balloon, the pilot should always be polite. Most emergency responders probably do not know much about federal law regarding aviation. However, even if they are wrong and the pilot knows they are wrong, there is no point in antagonizing them with a belligerent attitude. It is better to listen than to end up in court.

There is a wealth of information on landowner relations issues available from the Balloon Federation of America (BFA), as well as the British Balloon and Airship Club (BBAC). The BBAC web site, www.bbac.org, has an excellent video that was produced solely to deal with the issues of landowner relations in the United Kingdom. Despite being tailored for that country, it is worth viewing by the pilot and crew in the United States.

Packing the Balloon

After the crew has obtained the necessary landowner clearance to enter the landing site, the process of packing the balloon begins. Despite being the end of the flight, many pilots, as well as crew, consider this the point when the real work begins.

Deflating the balloon is the reverse of the inflation process. The crewmember designated to handle the crown line should secure the free end of the crown line and move downwind. The pilot should turn off the burner's pilot light and main fuel valves on the fuel tanks, and vent the fuel system. Once the fuel system has been shut off, and there is no risk of fire, the vent line can be activated to start the deflation. The crown line crewperson pulls on the crown line to assist in the deflation process and to help lay the balloon out in the desired direction. The pilot and crew should remember that it is virtually impossible to lay a balloon down against the wind. Other crewmembers may be stationed on the sides to keep fabric from draping over trees, bushes, and assist the balloon in coming down. Once the envelope is on its side, the crown line crewmember may move to the top of the balloon, and maintain slight tension on the load ring in order to continue the deflation. This crewmember should be reminded to leave gloves on, as the load ring is hot and may take some time to cool.

At this point, the balloon is ready to be "walked out," or "squeezed," meaning that the remaining air in the balloon is removed in preparation for repacking the envelope in the appropriate bag. The most common method is for the pilot or a crewmember to gather the envelope together at the throat, and, keeping their arms around the fabric, walk towards the top of the envelope squeezing the air out as they go. There is ample opportunity during this process to injure one's back. The person walking the envelope out should take care to not put excessive strain on their lower back during this process. There are some mechanical devices available to help in this process, and some pilots elect to use one of these, rather than put an individual's well being at risk. *[Figure 8-10]*

Figure 8-10. *Walking the balloon out with a "Squeeze-EZ."*

After the envelope is walked out and the crown line is secured, either in a separate bag or by folding it in half down the length of the envelope (which prevents tangling), the envelope is ready to be packed in its bag. The envelope bag, regardless of manufacturer, has a "flap" of fabric across the top. Some pilots prefer to pack the envelope so the flap is towards the balloon, while others prefer that the flap be away from the balloon. At first glance, this appears to be a random choice but there is sound reasoning behind it. If the balloon is usually launched from grassy, smooth fields, then it would be normal to have the flap away from the balloon. If, however, the balloon is laid out for inflation in areas with rocks, stubble, and other objects which may potentially damage the envelope, it is an accepted practice to pack the balloon with the envelope flap facing towards the balloon. Then, when the balloon is laid out on the next flight, it passes over the flap before contact with the ground and minimizes the risk of damage. *[Figure 8-11]*

Figure 8-11. *Packing the balloon. This pilot has elected to position the envelope bag near the basket and is bringing the balloon to the bag, rather than picking up the bag each time to move it towards the basket. Some pilots prefer this method, as it means less lifting for the crew.*

Crewmembers should be stationed on opposite sides of the bag at a 90° angle to the balloon. The pilot or designated crewmember lifts a section of the balloon and, with the crewmembers lifting the bag and bringing it to the envelope, place it in the envelope bag. This process continues until the envelope is almost completely packed and only the suspension cables are out. The suspension system is removed from the basket superstructure and secured, and placed in the envelope bag. The bag is closed and loaded on the chase vehicle.

Actions at this point are performed in the reverse order of the layout procedure. Generally, the envelope is secured onto the chase vehicle, the basket and burners are disassembled and secured on the vehicle, and a final check of the landing site is made to ensure that no loose items are left at the landing site. A final check with the landowner is usually appropriate. After a quick "thank you," the flight is officially "one for the books."

Refueling

Most pilots choose to refuel as soon after a flight as possible. This is probably appropriate for most circumstances, as this is one less issue to resolve when the time comes for the next flight. *[Figure 8-12]*

Figure 8-12. *Refueling. Note that this pilot is exercising safety precautions with sleeves down and gloves on.*

Each balloon has its own refueling procedures, which may be found in the appropriate flight manual for that balloon. Refueling involves connecting a supply line to the balloon's fuel lines, opening the refueling tank's main valve, opening a fixed liquid level gauge ("spit" valve) on the respective fuel tank in the balloon basket, and then opening the main supply valve on that tank. When the fixed liquid level gauge begins to "spit" liquid propane, the tank is full. Shut off the main tank valve first, then close the "spit valve," then close the main supply line, and then bleed the lines. This is a generalized description of the process; under all circumstances, the pilot should follow the procedure as outlined in the balloon's flight manual.

Safety during refueling is of paramount importance. While specific refueling procedures may vary, safety procedures do not. Propane vapor is a highly flammable and, under certain circumstances, explosive gas. There are many instances of accidents during refueling that have resulted in property damage, personal injury, and even death. The following safety rules serve the balloon pilot well to remember:

- No smoking around the balloon while refueling. This is an absolute.
- Never conduct refueling procedures from inside the basket.
- Disable strikers in the basket. Turn off cell phones and pagers. Synthetic clothing may also provide a source of ignition (static electricity) under certain circumstances.
- The chase vehicle should be shut off. Do not leave the engine running during the refuel process. In larger chase vehicles, such as RV conversions, water heater pilot lights must be shut off.
- Persons conducting the refueling should wear gloves at all times, preferably loose ones that can be removed quickly.
- Never refuel inside a closed trailer or inside a van. The vapor can quickly build up to a potentially combustible level. The basket should be moved to open air, as propane vapor is heavier than air.

For further safety recommendations, a pilot should consult with the propane supplier or see the appropriate section of the *Hot Air Balloon Crewing Essentials* publication previously cited in this handbook.

Logging of Flight Time

At some time subsequent to the flight, it is necessary for the pilot to make entries regarding the flight in their personal logbook, as well as the aircraft's logbook. It is an accepted practice in aviation that flight time is logged in tenths of a hour, as opposed to using hours and minutes. A tenth of an hour is represented by a 6-minute increment; remaining minutes are rounded up. This practice is used for both individuals and aircraft.

Pilot's Log

A pilot's flight time is required to be logged under the provisions of 14 CFR section 61.51. That section states, in part, that …"(1) Each person must document and record the following time…(a) training and aeronautical experience used to meet the requirements for a certificate, rating, or flight review of this part. (2) The aeronautical experience required for meeting the recent flight experience requirements of this part." (in this case, referring to currency requirements as stipulated under section 61.57) This is a relatively simple requirement and usually does not present issues for the pilot.

There are certain definitions relating to solo time, pilot in command (PIC) time, and instruction time, which are explained within the regulation. Pilots are advised that since they alone are responsible for maintaining the log books, knowledge of these definitions is necessary. Reviewing section 61.51 is appropriate. Should questions arise, a pilot should consult with the local Flight Standards District Office (FSDO) for interpretation.

Aircraft Log Books

14 CFR section 91.417 addresses the requirement for aircraft maintenance records (commonly referred to as a log book), and what the record must contain. All time recorded on the aircraft must be reflected in the maintenance record, and it is the responsibility of the owner or operator of the aircraft to ensure that the records are correctly maintained.

Flight time in a balloon is defined as that time beginning when the balloon is first made buoyant and lasts until the balloon is deflated. Balloons do not have a hour meter, as an airplane or other aircraft does, and the pilot must consider the fact that burner activation on inflation would equate to an engine start on an airplane. *[Figure 8-13]*

Figure 8-13. *Aircraft log books or maintenance records are for aircraft time and not for passenger names and flight experiences.*

There are some pilots who do not, for whatever reason, record aircraft time conducted while on tether. This is an incorrect practice. A balloon on tether is in flight, or more precisely, in a flight configuration, and conducting operations under the provisions of part 91. All inflated time of a certificated balloon should be noted in the maintenance records to ensure that inspection times are met.

Crew Responsibilities

The general theory regarding expectations of crew help at landing is if the pilot cannot land the balloon without help, he or she should not be flying. It is not always possible for the chase crew to be at the landing site, so plan to land without assistance.

Most balloon pilots have strong opinions regarding whether or not the crew must be present for a landing. Those who do not see a need for crew to be present have some solid reasons.

- It is not always possible for crew to be there. Traffic, wrong turns, and pilot decisions to land early or fly on can thwart their best efforts.
- With no formal training, enthusiastic crew might interfere with the landing or do the wrong thing and create a problem where none would have existed.
- Crew racing to every potential landing site may drive inconsiderately and even recklessly, endangering everyone on the road and creating a poor impression for ballooning.

Pilots who favor crew being on hand have a number of equally convincing arguments.

- In any branch of aviation, takeoff and landing are the most critical maneuvers. In ballooning, landing is number one. The vast majority of ballooning accidents and injuries occur on landing. A mistake during inflation or launch usually means calling a repair station; burned throat fabric is the leading repair nationwide. A mistake during landing often means calling the insurance agent. One insurance survey found accidents involving bodily injury on hard landings occur twice as often as damage to equipment or even power line strikes. This trend clearly identifies a need to improve safety at the end of every flight.
- The leading factor in accidents is wind. Highly variable surface winds often speed up, slow down, stop, turn, and even go backward. Crew onsite can radio the pilot regarding surface wind speeds and directions prior to landing, which factor into landing at a particular site, if at all.
- Landing approaches put the balloon at its closest to power lines, trees, buildings, and the ground. Potential risks increase, and the closest first responders able to handle the balloon and trouble are most likely to be the crew.
- When flying low or into the sun, a pilot may not see hidden power lines, antennas, or other obstacles which crew can easily spot while assessing the landing site.
- Crew can confirm the quality of a landing site. From a distance, a pilot cannot see what may rule the site out: livestock gathered under a tree, standing water,

a newly planted crop, another balloon deflating in the field, etc. The crew may find a site the pilot might not have seen from the air.

- Many pilots ask crew to get landowner permission before committing to a site or actually landing. This is often a crew's top priority. Finding the landowner can take a few minutes; one crewmember can do this while another walks out to assist the pilot. If it is a "no," time and fuel might have been better spent searching for the next suitable option.
- Landings create a workload challenge for the pilot. Consider all the tasks in the seconds before and after landing: flying the balloon, making an approach, briefing passengers, watching for obstacles, bracing for landing, deciding to move closer to a road or driveway, shutting down tanks, bleeding fuel lines, radioing crew, directing passenger unloading, deflating the envelope, keeping the envelope off obstacles, dealing with a hostile landowner or dog, and more. A single oversight, distraction, or poor decision can compromise safety. Extra hands may be required to protect everyone onboard and on the ground
- Crew serve as a pilot's redundancy and work to minimize flight risks of any nature. The FAA and every balloon manufacturer know the safety value of redundancy and the greatest risks (accidents and injuries) come at landing. There is no better time for crew to promote safety than in the flight's last minutes.

While crew-assisted landings may make some pilots uncomfortable, the reality is they happen all the time. This crew role can perhaps have more impact on flight safety than any regulation, equipment, or technique, provided the pilot and crew decide this before the flight.

Sometimes the best landing assistance may be none at all. Every pilot is trained to land without help from their crew. The balloon pilot may have chosen a landing site—fully visible from the air, with no obstacles, and perfect as is—early in the flight. Conditions are often smooth enough for safe landings. An unassisted landing often gives the pilot a better feel for the balloon's equilibrium. Letting the ground absorb the balloon's energy by touching and/or dragging might be the safest option of all. The landowner might greet the pilot in person even before the crew arrives, and passengers may have equipment disassembled and packed before the crew arrives. Many of these conditions frequently occur. A thoroughly prepared crew is one that is trained and ready for the many occasions that they are needed.

Chapter Summary

Among the many adages in aviation is "takeoffs are optional, landings are mandatory." For every flight, there must be a landing, and the balloon pilot should be prepared to execute a landing with skill and precision at any time during the flight.

In the early days of ballooning, it was common to see pilots "stick" the balloon (execute a hard landing) on a routine basis as burner systems were generally inadequate to slow the balloon during a rapid descent. With newer systems, the balloon can be better controlled and a smooth, graceful landing is a routine result, provided the pilot has planned ahead and understands the mechanics and concepts of the landing process. The information contained here is by no means all inclusive, but illustrates certain methodologies to good landing techniques.

Chapter 9
Aeromedical Factors

Introduction

As a pilot, it is important to stay aware of the psychological and physical standards required for the type of flying performed. This chapter provides information on medical certification and on aeromedical factors related to flying activities.

Most pilots must have a valid medical certificate to exercise the privileges of their airman certificates. Balloon pilots are not required to hold a medical certificate.

Some operations conducted outside the United States may require the balloon pilot to have a medical certificate. For example, United States certificated balloon pilots participating in events in Canada may be required to carry a second class medical certificate in order to carry passengers. In this case, it would be wise to check with the appropriate authorities to determine the requirements for specific operations. Title 14 of the Code of Federal Regulations (14 CFR) part 61 covers medical certification of pilots. Aviation medical examiners (AMEs) may be found by using a local telephone book, or contacting the local Flight Service District Office (FSDO).

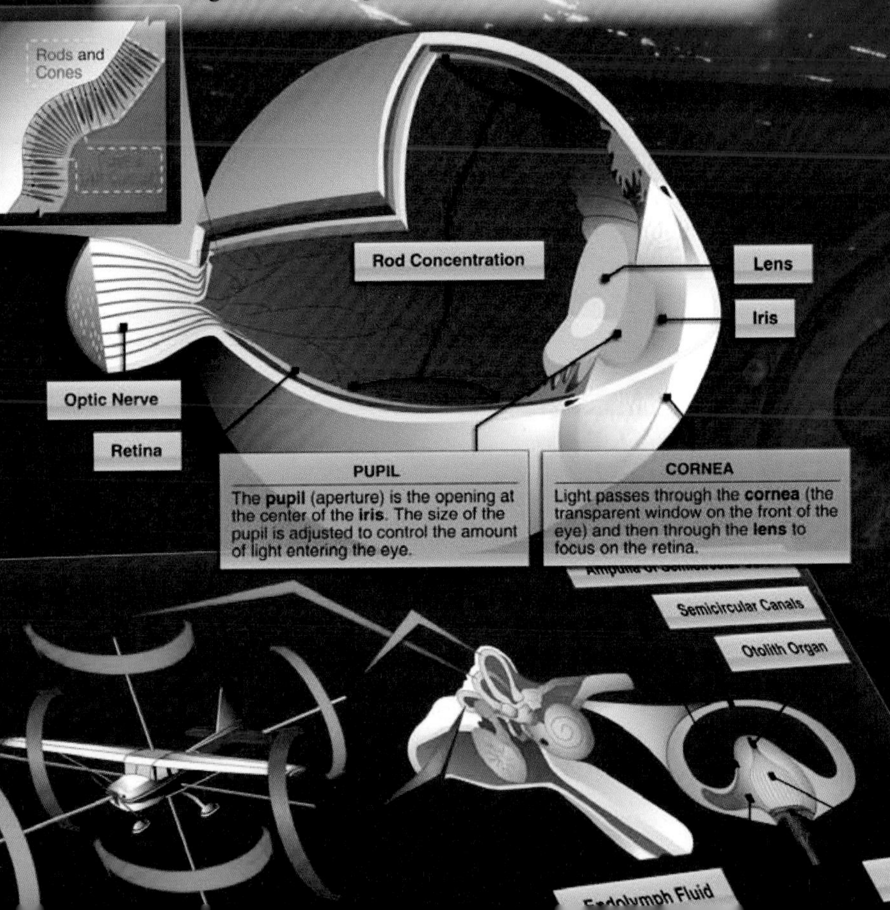

Environmental and Health Factors Affecting Pilot Performance

A number of health factors and physiological effects can be linked to flying. Some are minor, while others are important enough to require special consideration to ensure safety of flight. In some cases, physiological factors can lead to in-flight emergencies. Some important medical factors that a pilot should be aware of include hypoxia, hyperventilation, middle ear and sinus problems, motion sickness, stress and fatigue, dehydration, and heatstroke. Other subjects include the effects of alcohol and drugs, anxiety, and excess nitrogen in the blood after scuba diving.

Hypoxia

Hypoxia means "reduced oxygen" or "not enough oxygen." Although any tissue will die if deprived of oxygen long enough, the main concern is usually with getting enough oxygen to the brain, since it is particularly vulnerable to oxygen deprivation. Any reduction in mental function while flying can result in life-threatening errors. Hypoxia can be caused by several factors, including an insufficient supply of oxygen, inadequate transportation of oxygen, or the inability of the body tissues to use oxygen. Four forms of hypoxia based on their causes are: hypoxic hypoxia, hypemic hypoxia, stagnant hypoxia, and histotoxic hypoxia.

Hypoxic Hypoxia

Hypoxic hypoxia is a result of insufficient oxygen available to the body as a whole. A blocked airway and drowning are obvious examples of how the lungs can be deprived of oxygen, but the reduction in partial pressure of oxygen at high altitude is an appropriate example for pilots. Although the percentage of oxygen in the atmosphere is constant, its partial pressure decreases proportionately as atmospheric pressure decreases. As the aircraft ascends during flight, the percentage of each gas in the atmosphere remains the same; however, there are fewer molecules available at the pressure required for them to pass between the membranes in the respiratory system. This decrease in number of oxygen molecules at sufficient pressure can lead to hypoxic hypoxia.

Hypemic Hypoxia

Hypemic hypoxia occurs when the blood is not able to take up and transport a sufficient amount of oxygen to the cells in the body. Hypemic means "not enough blood." This type of hypoxia is a result of oxygen deficiency in the blood, rather than a lack of inhaled oxygen, and can be caused by a variety of factors. It may be because there is not enough blood volume (due to severe bleeding), or may result from certain blood diseases, such as anemia. More often it is because hemoglobin, the actual blood molecule that transports oxygen, is chemically unable to bind oxygen molecules. The most common cause of hypemic hypoxia is carbon monoxide poisoning. Hypemic hypoxia can also be caused by the loss of blood from a blood donation. While the volume of blood is restored quickly following a donation, normal levels of hemoglobin can require several weeks to return. Although the effects of the blood loss are slight at ground level, there are risks when flying during this time.

Stagnant Hypoxia

Stagnant means "not flowing." Stagnant hypoxia results when the oxygen-rich blood in the lungs is not moving, for one reason or another, to the tissues that need it. One form of stagnant hypoxia is an arm or leg "going to sleep" because the blood flow has accidentally been shut off. This kind of hypoxia can also result from shock, the heart failing to pump blood effectively, or a constricted artery. Cold temperatures can also reduce circulation and decrease the blood supplied to extremities.

Histotoxic Hypoxia

The inability of the cells to effectively use oxygen is defined as histotoxic hypoxia. "Histo" refers to tissues or cells, and "toxic" means poison. In this case, plenty of oxygen is being transported to the cells that need it, but they are unable to make use of it. This impairment of cellular respiration can be caused by alcohol and other drugs, such as narcotics and poisons. Research has shown that drinking one ounce of alcohol can equate to an additional 2,000 feet of physiological altitude. There are other issues concerning the use of alcohol in relation to flying in general; those will be discussed later in this chapter.

Symptoms of Hypoxia

High altitude flying can place a pilot in danger of becoming hypoxic. Oxygen starvation causes the brain and other vital organs to become impaired. One noteworthy attribute of the onset of hypoxia is that the first symptoms are euphoria and a carefree feeling. With increased oxygen starvation, the extremities become less responsive and flying becomes less coordinated. The symptoms of hypoxia vary with the individual, but common symptoms include:

- Cyanosis (blue fingernails and lips),
- Headache,
- Decreased reaction time,
- Impaired judgment,
- Euphoria,
- Visual impairment,
- Drowsiness,
- Lightheaded or dizzy sensation,

- Tingling in fingers and toes, and
- Numbness.

As hypoxia worsens, the field of vision begins to narrow, and instrument interpretation can become difficult. Even with all these symptoms, the effects of hypoxia can cause a pilot to have a false sense of security and be deceived into believing that everything is normal. The treatment for hypoxia includes descending to lower altitudes and/or using supplemental oxygen. Supplemental oxygen is required for certain operations above 12,500 feet mean sea level (MSL). (See 14 CFR part 91, section 91.211.)

All pilots are susceptible to the effects of oxygen starvation, regardless of physical endurance or acclimatization. When flying at high altitudes, it is paramount that oxygen be used to avoid the effects of hypoxia. The term "time of useful consciousness" describes the maximum time the pilot has to make rational, life-saving decisions and carry them out at a given altitude without supplemental oxygen. As altitude increases above 10,000 feet, the symptoms of hypoxia increase in severity, and the time of useful consciousness rapidly decreases. Many pilots have established an altitude lower than the required 12,500 MSL as their personal "do not exceed without oxygen" limit. All pilots are well advised to personalize their performance at altitude.

Since symptoms of hypoxia can be different for each individual, the ability to recognize hypoxia can be greatly improved by experiencing and witnessing the effects of it during an altitude chamber "flight." The Federal Aviation Administration (FAA) provides this opportunity through aviation physiology training, which is conducted at the FAA Civil Aerospace Medical Institute (CAMI) and at many military facilities across the United States. To attend an aviation physiology course at CAMI, call (405) 954-4837 or write:

Mike Monroney Aeronautical Center
Airman Education Program
CAMI (AAM-400)
P.O. Box 25082
Oklahoma City, OK 73125

Hyperventilation

Hyperventilation occurs when an individual is experiencing emotional stress, fright, or pain, and the breathing rate and depth increase, although the carbon dioxide level in the blood is already at a reduced level. The result is an excessive loss of carbon dioxide from the body, which can lead to unconsciousness due to the respiratory system's overriding mechanism to regain control of breathing.

Pilots encountering an unexpected stressful situation may subconsciously increase their breathing rate. If flying at higher altitudes, either with or without oxygen, a pilot may have a tendency to breathe more rapidly than normal, which often leads to hyperventilation.

Since many of the symptoms of hyperventilation are similar to those of hypoxia, it is important to correctly diagnose and treat the proper condition. If using supplemental oxygen, check the equipment and flow rate to ensure the symptoms are not hypoxia related.

Common symptoms of hyperventilation include:

- Headache;
- Decreased reaction time;
- Impaired judgment;
- Euphoria;
- Visual impairment;
- Drowsiness;
- Lightheaded or dizzy sensation;
- Tingling in fingers and toes;
- Numbness;
- Pale, clammy appearance; and
- Muscle spasms.

Hyperventilation may produce a pale, clammy appearance and muscle spasms compared to the cyanosis and limp muscles associated with hypoxia. The treatment for hyperventilation involves restoring the proper carbon dioxide level in the body. Breathing normally is both the best prevention and the best cure for hyperventilation. In addition to slowing the breathing rate, breathing into a paper bag or talking aloud helps to overcome hyperventilation. Recovery is usually rapid once the breathing rate is returned to normal.

Middle Ear and Sinus Problems

Ascents and descents can sometimes cause ear or sinus pain and a temporary reduction in the ability to hear. The physiological explanation for this discomfort is a difference between the pressure of the air outside the body and that of the air inside the middle ear and nasal sinuses.

The middle ear is a small cavity located in the bone of the skull. It is closed off from the external ear canal by the eardrum. Normally, pressure differences between the middle ear and the outside world are equalized by a tube leading from inside each ear to the back of the throat on each side, called the eustachian tube. These tubes are usually closed, but open during chewing, yawning, or swallowing to equalize pressure. Even a slight difference between external pressure and middle ear pressure can cause discomfort.

In a similar way, air pressure in the sinuses equalizes with the ambient or outside pressure through small openings that connect the sinuses to the nasal passages. An upper respiratory infection, such as a cold or sinusitis, or a nasal allergic condition can produce enough congestion around an opening to slow equalization. As the difference in pressure between the sinus and the ambient atmosphere increases, congestion may plug the opening. This "sinus block" occurs most frequently during descent. Slow descent rates can reduce the associated pain. A sinus block can occur in the frontal sinuses, located above each eyebrow, or in the maxillary sinuses, located in each upper cheek. It will usually produce excruciating pain over the sinus area. A maxillary sinus block can also make the upper teeth ache. Bloody mucus may discharge from the nasal passages.

Sinus block can be avoided by not flying with an upper respiratory infection or nasal allergic condition. Adequate protection is usually not provided by decongestant sprays or drops to reduce congestion around the sinus openings. Oral decongestants have side effects that can impair pilot performance. If a sinus block does not clear shortly after landing, a physician should be consulted. *[Figure 9-1]*

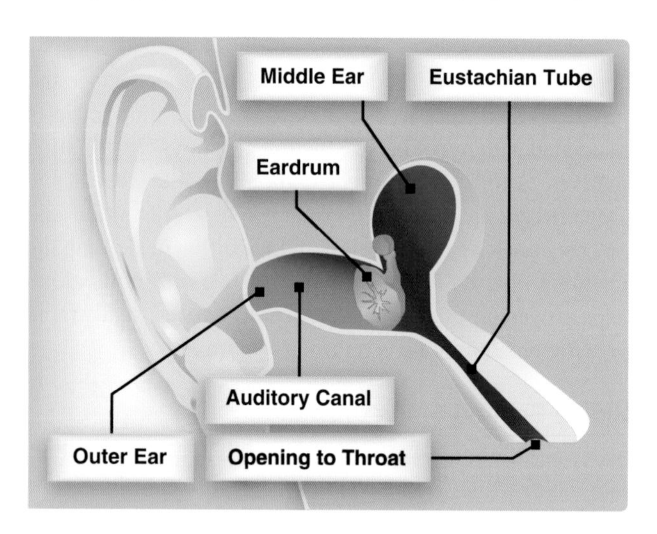

Figure 9-1. *The eustachian tube allows air pressure to equalize in the middle ear.*

During a climb, middle ear air pressure may exceed the pressure of the air in the external ear canal, causing the eardrum to bulge outward. Pilots become aware of this pressure change when they experience alternate sensations of "fullness" and "clearing." During descent, the reverse happens. While the pressure of the air in the external ear canal increases, the middle ear cavity, which equalized with the lower pressure at altitude, is at lower pressure than the external ear canal. This results in the higher outside pressure, causing the eardrum to bulge inward.

This condition can be more difficult to relieve due to the fact that the partial vacuum tends to constrict the walls of the eustachian tube. To remedy this often painful condition, which also causes a temporary reduction in hearing sensitivity, pinch the nostrils shut, close the mouth and lips, and blow slowly and gently in the mouth and nose. This is commonly referred to as the Valsalva procedure.

The Valsalva procedure forces air through the eustachian tube into the middle ear. It may not be possible to equalize the pressure in the ears if a pilot has a cold, an ear infection, or sore throat. A flight in this condition can be extremely painful, as well as damaging to the eardrums. If a pilot experiences minor congestion, nose drops or nasal sprays may reduce the risk of a painful ear blockage.

Spatial Disorientation and Illusions

Balloon pilots rarely experience issues with spatial disorientation while in flight, as virtually all balloon operations are conducted under visual flight rules (VFR) conditions. Knowledge of these conditions, however, is important in the event of unusual circumstances or situations, such as inadvertently being caught in fog, or perhaps in areas of low visibility. The balloon pilot should have an awareness of these issues, so that appropriate actions may be taken as necessary.

Spatial disorientation specifically refers to the lack of orientation with regard to the position, attitude, or movement of an aircraft in space. The body uses three integrated systems working together to ascertain orientation and movement in space. The eye is by far the largest source of information. Kinesthesia refers to the sensation of position, movement, and tension perceived through the nerves, muscles, and tendons. The vestibular system is a very sensitive motion-sensing system located in the inner ears. It reports head position, orientation, and movement in three-dimensional space.

All this information comes together in the brain and, most of the time, the three streams of information agree, giving a clear idea of where and how the body is moving. Flying can sometimes cause these systems to supply conflicting information to the brain, which can lead to disorientation. During flight in visual meteorological conditions (VMC), the eyes are the major orientation source and usually prevail over false sensations from other sensory systems. When these visual cues are removed, as they are in instrument meteorological conditions (IMC), false sensations can cause a pilot to become quickly disoriented.

The vestibular system in the inner ear allows the pilot to sense movement and determine orientation in the surrounding environment. In both left and right inner ears, three semicircular canals are positioned at approximate right angles to each other. *[Figure 9-2]* Each canal is filled with fluid and has a section full of fine hairs. Acceleration of the inner ear in any direction causes the tiny hairs to deflect, which in turn stimulate nerve impulses, sending messages to the brain. The vestibular nerve transmits the impulses from the utricle, saccule, and semicircular canals to the brain to interpret motion.

The postural system sends signals from the skin, joints, and muscles to the brain that are interpreted in relation to the Earth's gravitational pull. These signals determine posture. Inputs from each movement update the body's position to the brain on a constant basis. "Seat of the pants" flying is largely dependent upon these signals. Used in conjunction with visual and vestibular clues, these sensations can be fairly reliable. However, the body cannot distinguish between acceleration forces due to gravity and those resulting from maneuvering the aircraft, which can lead to sensory illusions and false impressions of an aircraft's orientation and movement.

Again, under normal flight conditions, most balloon pilots and passengers do not experience spatial disorientation (or vertigo) while flying.

Motion Sickness

Motion sickness, or airsickness, is caused by the brain receiving conflicting messages about the state of the body. A pilot may experience motion sickness during initial flights, but it generally goes away within the first few lessons. Anxiety and stress, which may be experienced at the beginning of flight training, can contribute to motion sickness. Symptoms of motion sickness include general discomfort, nausea, dizziness, paleness, sweating, and vomiting. Generally, under normal flight conditions, motion sickness is not an issue for balloon pilots or passengers.

Stress

Stress is defined as the body's response to physical and psychological demands placed upon it. The body's reaction to stress includes releasing chemical hormones (such as adrenaline) into the blood, and increasing metabolism to provide more energy to the muscles. Blood sugar, heart rate, respiration, blood pressure, and perspiration all increase. The term "stressor" is used to describe an element that causes an individual to experience stress. Examples of stressors include physical stress (noise or vibration), physiological stress (fatigue), and psychological stress (difficult work or personal situations).

Stress falls into two broad categories: acute (short term) and chronic (long term). Acute stress involves an immediate threat that is perceived as danger. This is the type of stress that triggers a "fight or flight" response in an individual, whether the threat is real or imagined. Normally, a healthy person can cope with acute stress and prevent stress overload. However, ongoing acute stress can develop into chronic stress.

Chronic stress can be defined as a level of stress that presents an intolerable burden, exceeds the ability of an individual to cope, and causes individual performance to fall sharply. Unrelenting psychological pressures, such as financial worries, difficult relationships, or work problems can produce a cumulative level of stress that exceeds a person's ability to cope with the situation. When stress reaches these levels,

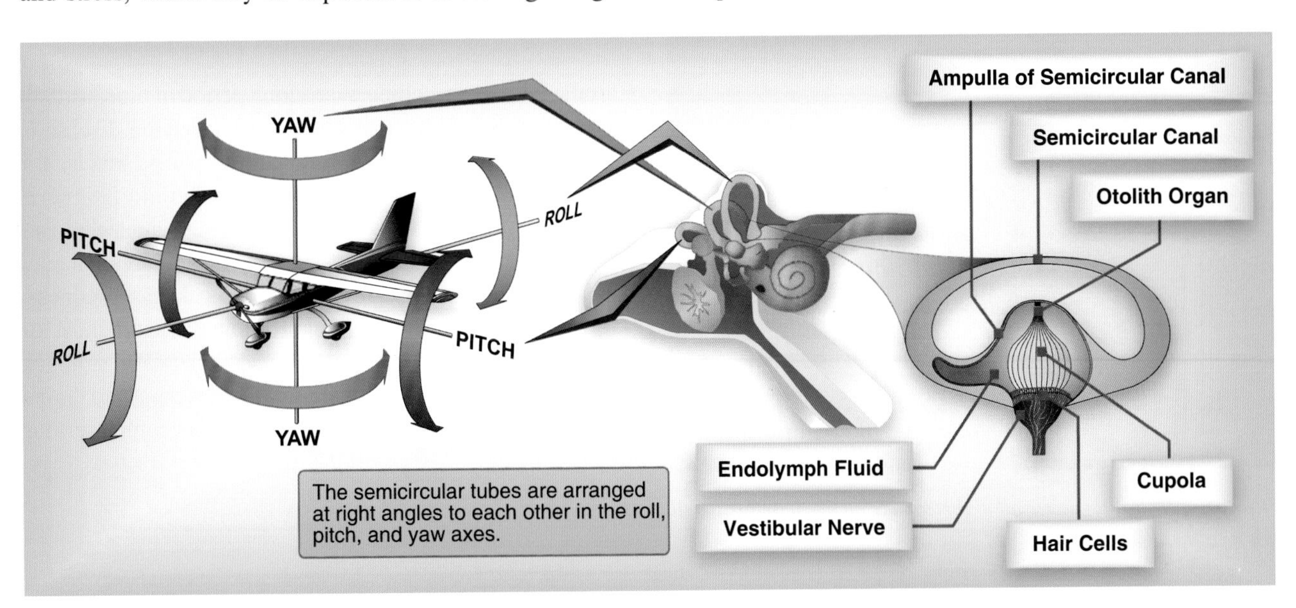

Figure 9-2. *The semicircular canals lie in three planes, and sense motions of roll, pitch, and yaw.*

performance falls off rapidly. Pilots experiencing this level of stress are not safe and should not exercise their airman privileges. Pilots who suspect they are suffering from chronic stress should consult a physician.

Fatigue

Fatigue is frequently associated with pilot error. Some of the effects of fatigue include degradation of attention and concentration, impaired coordination, and decreased ability to communicate. These factors can seriously influence the ability to make effective decisions. Physical fatigue can result from sleep loss, exercise, or physical work. Factors such as stress and prolonged performance of cognitive work can result in mental fatigue.

Like stress, fatigue also falls into two broad categories: acute and chronic. Acute fatigue is short term and is a normal occurrence in everyday living. It is the kind of tiredness people feel after a period of strenuous effort, excitement, or lack of sleep. Rest after exertion and 8 hours of sound sleep ordinarily cures this condition.

A special type of acute fatigue is skill fatigue. This fatigue can be readily seen in the balloon pilot who, for example, has driven most or all of the night in order to attend an event, or perhaps was up for most of the previous night. *[Figure 9-3]*

Figure 9-3. *This pilot drove all night to make an event, but still will not be able to fly.*

This type of fatigue has two main effects on performance:

- Timing disruption—appearing to perform a task as usual, but the timing of each component is slightly off. This makes the pattern of the operation less smooth, because the pilot performs each component as though it were separate, instead of as a part of an integrated activity.
- Disruption of the perceptual field—concentrating attention upon movements or objects in the center of vision and neglecting those in the periphery. This may be accompanied by loss of accuracy and smoothness in control movements.

Acute fatigue has many causes, but the following are among the most important to the pilot:

- Mild hypoxia (oxygen deficiency),
- Physical stress,
- Psychological stress, and
- Depletion of physical energy resulting from psychological stress.

Sustained psychological stress accelerates the glandular secretions that prepare the body for quick reactions during an emergency. These secretions make the circulatory and respiratory systems work harder, and the liver releases energy to provide the extra fuel needed for brain and muscle work. When this reserve energy supply is depleted, the body lapses into generalized and severe fatigue.

Acute fatigue can be prevented by proper diet and adequate rest and sleep. A well-balanced diet prevents the body from consuming its own tissues as an energy source. Adequate rest maintains the body's store of vital energy.

Chronic fatigue, extending over a long period of time, usually has psychological roots, although an underlying disease is sometimes responsible. Continuous high stress levels, for example, can produce chronic fatigue. Chronic fatigue is not relieved by proper diet and adequate rest and sleep, and usually requires treatment by a physician. An individual may experience this condition in the form of weakness, tiredness, palpitations of the heart, breathlessness, headaches, or irritability. Sometimes chronic fatigue even creates stomach or intestinal problems and generalized aches and pains throughout the body. When the condition becomes serious enough, it can lead to emotional illness.

If suffering from acute fatigue, stay on the ground. If fatigue occurs in the basket, no amount of training or experience can overcome the detrimental effects. Getting adequate rest and nutrition is the only way to prevent fatigue from occurring. Avoid flying without a full night's rest, after working excessive hours, or after an especially exhausting or stressful day. Pilots who suspect they are suffering from chronic fatigue should consult a physician.

Dehydration and Heatstroke

Dehydration is the term given to a critical loss of water from the body. The first noticeable effect of dehydration is fatigue, which in turn makes top physical and mental performance difficult, if not impossible. As a pilot, flying for long periods in hot summer temperatures or at high altitudes increases the susceptibility of dehydration since the dry air at altitude tends to increase the rate of water loss from the body. If this fluid is not replaced, fatigue progresses to dizziness, weakness, nausea, tingling of hands and feet, abdominal cramps, and extreme thirst. *[Figure 9-4]*

Figure 9-4. *Hydration is important both before and after participating in outdoor activities. While for obvious reasons during hot weather, an individual can dehydrate during cold weather, too.*

Heatstroke is a condition caused by inability of the body to control its temperature. Onset of this condition may be recognized by the symptoms of dehydration, but it has also been recognized only by complete collapse.

To prevent these symptoms, it is recommended that a pilot carry an ample supply of water to drink at frequent intervals on any long flight, whether thirsty or not.

Alcohol

Alcohol impairs the efficiency of the human body. Studies have proven that drinking and performance deterioration are closely linked. Pilots must make hundreds of decisions, some of them time critical, during the course of a flight. The safe outcome of any flight depends on the ability to make the correct decisions and take the appropriate actions during routine occurrences, as well as abnormal situations. The influence of alcohol drastically reduces the chances of completing a flight without incident. Even in small amounts, alcohol can impair judgment, decrease sense of responsibility, affect coordination, constrict visual field, diminish memory, reduce reasoning power, and lower attention span. As little as one ounce of alcohol can decrease the speed and strength of muscular reflexes, lessen the efficiency of eye movements while reading, and increase the frequency at which errors are committed. Impairments in vision and hearing can occur after consuming only one alcoholic drink.

The alcohol consumed in beer and mixed drinks is ethyl alcohol, a powerful central nervous system depressant. It acts on the body much like a general anesthetic. The "dose" is generally much lower and more slowly consumed in the case of alcohol, but the basic effects on the body are similar. Alcohol is easily and quickly absorbed by the digestive tract. The bloodstream absorbs about 80 to 90 percent of the alcohol in a drink within 30 minutes on an empty stomach. The body requires about 3 hours to rid itself of all the alcohol contained in one mixed drink or one beer.

When experiencing a hangover, a pilot is still under the influence of alcohol. Although a pilot may think that he or she is functioning normally, motor and mental response impairment is still present. Considerable amounts of alcohol can remain in the body for over 16 hours, so pilots should be cautious about flying too soon after drinking.

Altitude multiplies the effects of alcohol on the brain. When combined with altitude, the alcohol from two drinks may have the same effect as three or four drinks. Alcohol interferes with the brain's ability to utilize oxygen, producing a form of histotoxic hypoxia. The effects are rapid because alcohol passes quickly into the bloodstream. In addition, the brain is a highly vascular organ that is immediately sensitive to changes in the blood's composition. For a pilot, the lower oxygen availability at altitude and the lower capability of the brain to use available oxygen add up to a deadly combination.

Intoxication is determined by the amount of alcohol in the bloodstream. This is usually measured as a percentage by weight in the blood. 14 CFR part 91, section 91.17 requires that blood alcohol level be less than 0.04 percent and that 8 hours pass between drinking alcohol and piloting an aircraft. A pilot with a blood alcohol level of 0.04 percent or greater after 8 hours cannot fly until the blood alcohol falls below that amount. Even though blood alcohol may be well below 0.04 percent, a pilot cannot fly sooner than 8 hours after drinking alcohol. Although the regulations are quite specific, it is a good idea to be more conservative than the regulations.

Drugs

Pilot performance can be seriously degraded by both prescription and over-the-counter medications, as well as by the medical conditions for which they are taken. Many medications, such as tranquilizers, sedatives, strong pain relievers, and cough suppressants have primary effects that may impair judgment, memory, alertness, coordination, vision, and the ability to make calculations. Others, such as antihistamines, blood pressure drugs, muscle relaxants, and agents to control diarrhea and motion sickness have side effects that may impair the same critical functions. Any medication that depresses the nervous system, such as sedatives, tranquilizers, or antihistamines can make a pilot more susceptible to hypoxia.

Painkillers can be grouped into two broad categories: analgesics and anesthetics. Analgesics are drugs that reduce pain, while anesthetics are drugs that deaden pain or cause loss of consciousness.

Over-the-counter analgesics, such as acetylsalicylic acid (aspirin), acetaminophen (e.g., Tylenol), and ibuprofen (e.g., Advil) have few side effects when taken in the correct dosage. Although some people are allergic to certain analgesics or may suffer from stomach irritation, flying usually is not restricted when taking these drugs. However, flying is almost always precluded while using prescription analgesics, such as drugs containing propoxyphene (e.g., Darvon), oxycodone (e.g., Percodan), meperidine (e.g., Demerol), and codeine since these drugs may cause side effects such as mental confusion, dizziness, headaches, nausea, and vision problems. Anesthetic drugs are commonly used for dental and surgical procedures. Most local anesthetics used for minor dental and outpatient procedures wear off within a relatively short period of time. The anesthetic itself may not limit flying so much as the actual procedure and subsequent pain.

Stimulants are drugs that excite the central nervous system and produce an increase in alertness and activity. Amphetamines, caffeine, and nicotine are all forms of stimulants. Common uses of these drugs include appetite suppression, fatigue reduction, and mood elevation. Some of these drugs may cause a stimulant reaction, even though this reaction is not their primary function. In some cases, stimulants can produce anxiety and mood swings, both of which are dangerous when flying.

Depressants are drugs that reduce the body's functioning in many areas. These drugs lower blood pressure, reduce mental processing, and slow motor and reaction responses. There are several types of drugs that can cause a depressing effect on the body, including tranquilizers, motion sickness medication, and some types of stomach medication, decongestants, and antihistamines. The most common depressant is alcohol.

Some drugs, which can be classified as neither stimulants nor depressants, have adverse effects on flying. For example, some forms of antibiotics can produce dangerous side effects, such as balance disorders, hearing loss, nausea, and vomiting. While many antibiotics are safe for use while flying, the infection requiring the antibiotic may prohibit flying. In addition, unless specifically prescribed by a physician, do not take more than one drug at a time, and never mix drugs with alcohol because the effects are often unpredictable.

The dangers of illegal drugs also are well documented. Certain illegal drugs can have hallucinatory effects that occur days or weeks after the drug is taken. Obviously, these drugs have no place in the aviation community.

14 CFR part 65 prohibits pilots from performing crewmember duties while using any medication that affects the body in any way contrary to safety. The safest rule is not to fly as a crewmember while taking any medication, unless approved to do so by the FAA. If there is any doubt regarding the effects of any medication, consult an AME before flying.

Scuba Diving

Scuba diving subjects the body to increased pressure, which allows more nitrogen to dissolve in body tissues and fluids. The reduction of atmospheric pressure that accompanies flying can produce physical problems for scuba divers. Reducing the pressure too quickly allows small bubbles of nitrogen to form inside the body as the gas comes out of solution. These bubbles can cause a painful and potentially incapacitating condition called "the bends." (An example is dissolved gas forming bubbles as pressure decreases by slowly opening a transparent bottle of carbonated beverage.) Scuba training emphasizes how to prevent the bends when rising to the surface, but increased nitrogen concentrations can remain in tissue fluids for several hours after a diver leaves the water. The bends can be experienced from as low as 8,000 feet MSL, with increasing severity as altitude increases. As noted in the Aeronautical Information Manual (AIM), the minimum recommended time between scuba diving on nondecompression stop dives and flying is 12 hours, while the minimum time recommended between decompression stop diving and flying is 24 hours. *[Figure 9-5]*

Figure 9-5. *The reduction of atmospheric pressure that accompanies flying can produce physical problems for scuba divers.*

Vision in Flight

Of all the senses, vision is the most important for safe flight. Most of the things perceived while flying are visual or heavily supplemented by vision. As remarkable and vital as it is, vision is subject to some limitations, such as illusions and blind spots. The more a pilot understands about the eyes and how they function, the easier it is to use vision effectively and compensate for potential problems.

The eye functions much like a camera. Its structure includes an aperture, a lens, a mechanism for focusing, and a surface for registering images. Light enters through the cornea at the front of the eyeball, travels through the lens, and falls on the retina. The retina contains light sensitive cells that convert light energy into electrical impulses that travel through nerves to the brain. The brain interprets the electrical signals to form images. There are two kinds of light-sensitive cells in the eyes: rods and cones. *[Figure 9-6]*

The cones are responsible for all color vision, from appreciating a glorious sunset to discerning the subtle shades in a fine painting. Cones are present throughout the retina, but are concentrated toward the center of the field of vision at the back of the retina. There is a small pit called the fovea where almost all the light sensing cells are cones. This is the area where most "looking" occurs (the center of the visual field where detail, color sensitivity, and resolution are highest).

While the cones and their associated nerves are well suited to detecting fine detail and color in high light levels, the rods are better able to detect movement and provide vision in dim light. The rods are unable to discern color but are very sensitive at low light levels. However, a large amount of light overwhelms the rods, and they take a long time to "reset" and adapt to the dark again. There are so many cones in the fovea that the very center of the visual field has virtually no rods at all. Therefore, the middle of the visual field is not very sensitive in low light. Farther from the fovea, the rods are more numerous and provide the major portion of night vision.

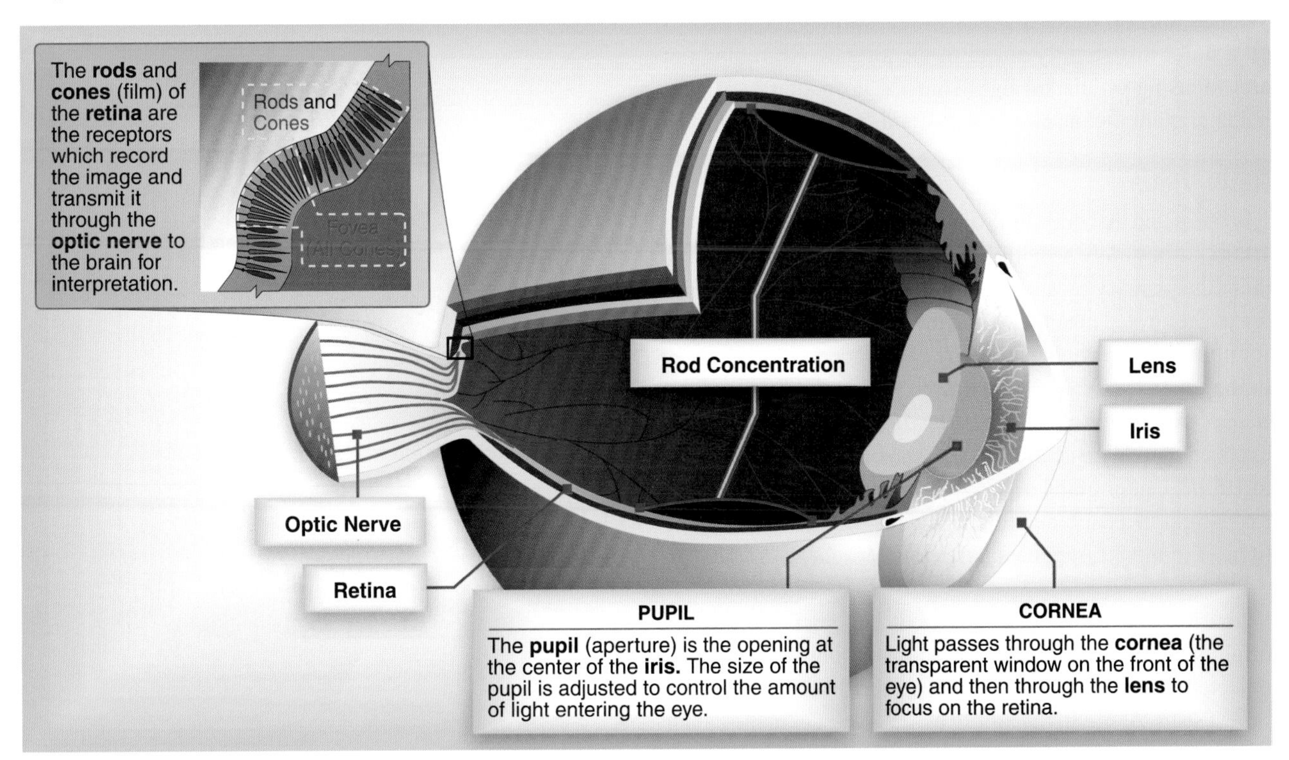

Figure 9-6. *The human eye.*

Figure 9-7. *The eye's blind spot.*

The area where the optic nerve enters the eyeball has no rods or cones, leaving a blind spot in the field of vision. Normally, each eye compensates for the other's blind spot. *Figure 9-7* provides a dramatic example of the eye's blind spot. Cover the right eye and hold this page at arm's length. Focus the left eye on the X in the right side of the windshield, and notice what happens to the balloon while slowly bringing the page closer to the eye.

Empty-Field Myopia

Another problem associated with flying at night or in reduced visibility is empty-field myopia, or induced nearsightedness. With nothing on which to focus, the eyes automatically focus on a point just slightly ahead of the aircraft. Searching out and focusing on distant light sources, no matter how dim, helps prevent the onset of empty-field myopia.

Night Vision

It is estimated that once fully adapted to darkness, the rods are 10,000 times more sensitive to light than the cones, making them the primary receptors for night vision. Since the cones are concentrated near the fovea, the rods are also responsible for much of the peripheral vision. The concentration of cones in the fovea can make a night blind spot in the center of the field of vision. To see an object clearly at night, the pilot must expose the rods to the image. This can be done by looking 5° to 10° off center of the object to be seen. This can be tried in a dim light in a darkened room. When looking directly at the light, it dims or disappears altogether. *[Figure 9-8]*

When looking slightly off center, it becomes clearer and brighter. Refer to *Figure 9-8.* When looking directly at an object, the image is focused mainly on the fovea, where detail is best seen. At night, the ability to see an object in the center of the visual field is reduced as the cones lose much of their sensitivity and the rods become more sensitive. Looking off center can help compensate for this night blind spot. Along with the loss of acuity (sharpness) and color at night, depth perception and judgment of size may be lost.

Balloon pilots, while not normally conducting flight operations at night, can experience similar issues when flying in low light conditions, particularly if there is haze or reduced visibility. In those instances where the balloon is operated at night, such as during a night "glow" or tether, night vision can be immediately destroyed by the light from the burner. It may take several minutes for the pilot to recover his or her vision, time in which complete awareness of his or her

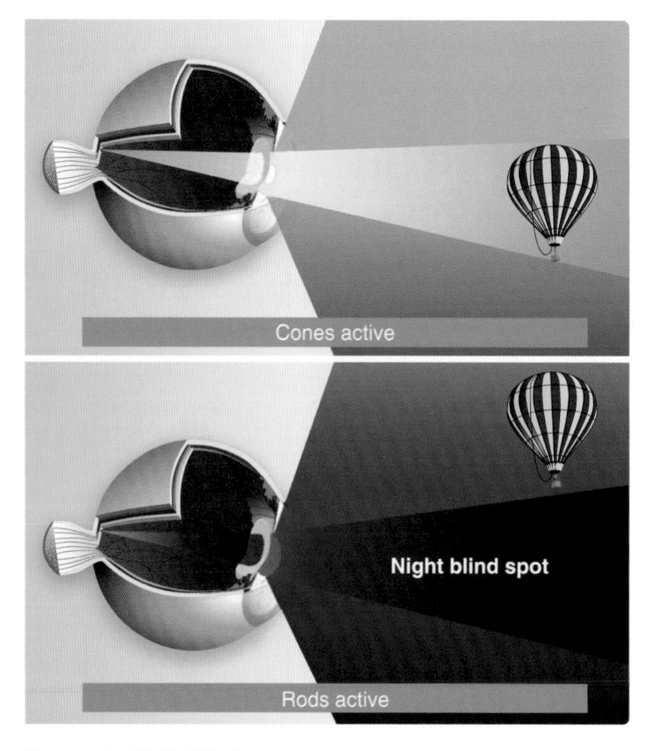

Figure 9-8. *Night blind spot.*

surroundings is lost. If those surroundings include people, a potentially dangerous situation can ensue. Closing one eye during a burn and not looking at the burner flame will minimize this momentary blindness.

Diet and general physical health have an impact on how well a pilot can see in the dark. Deficiencies in vitamins A and C have been shown to reduce night visual acuity. Other factors, such as carbon monoxide poisoning, smoking, alcohol, certain drugs, and a lack of oxygen also can greatly decrease night vision.

Chapter Summary

Balloon pilots and glider pilots are unique in that they "self-certify" they are physically fit to conduct flight duties. This is an individual responsibility and must not be abused. The ability to "self-certify" becomes particularly problematic after the balloon pilot has had a major medical issue arise, such as a heart attack, angina, major surgery, and other items in this category. While they may be perfectly capable of piloting a balloon after triple bypass surgery, for example, it may not be the recommended course of action.

The best recommendation is to be aware of the provisions of 14 CFR part 61. If a medical issue may be medically disqualifying for other pilots, the balloon pilot would be well advised to consult with an AME, and obtain recommendations on how best to proceed.

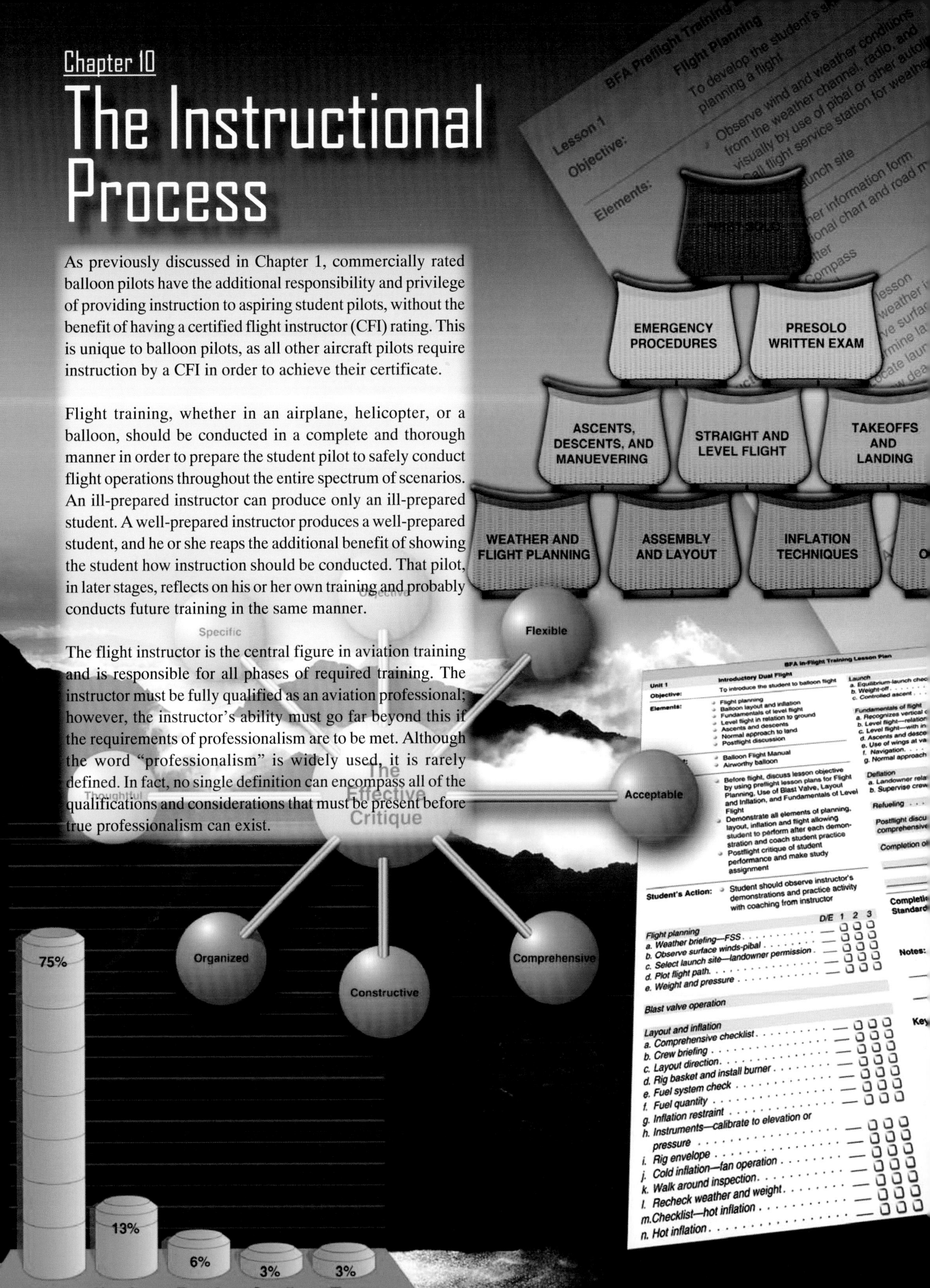

Chapter 10

The Instructional Process

As previously discussed in Chapter 1, commercially rated balloon pilots have the additional responsibility and privilege of providing instruction to aspiring student pilots, without the benefit of having a certified flight instructor (CFI) rating. This is unique to balloon pilots, as all other aircraft pilots require instruction by a CFI in order to achieve their certificate.

Flight training, whether in an airplane, helicopter, or a balloon, should be conducted in a complete and thorough manner in order to prepare the student pilot to safely conduct flight operations throughout the entire spectrum of scenarios. An ill-prepared instructor can produce only an ill-prepared student. A well-prepared instructor produces a well-prepared student, and he or she reaps the additional benefit of showing the student how instruction should be conducted. That pilot, in later stages, reflects on his or her own training and probably conducts future training in the same manner.

The flight instructor is the central figure in aviation training and is responsible for all phases of required training. The instructor must be fully qualified as an aviation professional; however, the instructor's ability must go far beyond this if the requirements of professionalism are to be met. Although the word "professionalism" is widely used, it is rarely defined. In fact, no single definition can encompass all of the qualifications and considerations that must be present before true professionalism can exist.

When discussing professionalism, most conversations omit any reference to the accountability of the instructor. An individual who takes on the responsibility of providing flight instruction must realize that they have a significant influence on the habits and actions of the student as a pilot. All of the student's impressions and perceptions towards the flight experience, balloon operations, and, *most importantly, safety* is drawn from the instructor's methods. It becomes imperative that the instructor understand that he or she, by default, actively and passively contributes to the future actions of the student and should make every effort to provide the most thorough training experience possible. It can be said that many students become "clones" of their instructor.

Though not all inclusive, the following list gives some major considerations and qualifications that should be included in the definition of professionalism.

- Professionalism exists only when a service is performed for someone or for the common good.
- Professionalism is achieved only after extended training and preparation.
- True performance as a professional is based on study and research.
- Professionalism requires the ability to make good judgment decisions. Professionals cannot limit their actions and decisions to standard patterns and practices.
- Professionalism demands a code of ethics. Professionals must be true to themselves and to those they serve. Anything less than a sincere performance is quickly detected by the student and immediately destroys instructor effectiveness.
- Professionalism requires that the individual instructor be able to conduct a self-assessment. A true professional must be able to critique his or her own performance with objectivity.

Flight instructors should carefully consider this list. Failing to meet these qualities may result in poor performance by both instructor and student. Preparation and performance as an instructor with these qualities in mind commands recognition as a professional in aviation instruction. Professionalism includes an instructor's public image.

A more complete discussion of the information contained in this chapter may be found in FAA-H-8083-9, Aviation Instructor's Handbook. There have been some minor changes in this chapter to address the specific aspects of balloon flight training. For expanded discussions of the areas of operation covered in the Commercial Pilot Practical Test Standards (PTS), (FAA-S-8081-18), the Aviation Instructor's Handbook is required reading. Commercial pilots who wish to excel at the instruction process may consider taking the Fundamentals of Instruction knowledge test administered through the FAA's knowledge testing program. While not a mandatory requirement, it is believed that the study necessary to successfully complete this exam assists the pilot/instructor in acquiring the knowledge necessary to plan and perform proper aviation training.

For the remainder of this chapter, it should be understood that the term "flight instructor" is meant to define those individuals who hold a commercial pilot certificate with a free balloon category rating. This chapter also makes the presupposition that virtually all balloon flight training is performed "one on one," and not as part of a group environment.

Flight Instructor Characteristics and Responsibilities

Students look to flight instructors as authorities in their respective areas. It is important that flight instructors not only know how to teach, but they also need to project a knowledgeable and professional image. In addition, flight instructors are on the front lines of efforts to improve the safety record of the industry. This section addresses the scope of responsibilities for flight instructors and enumerates methods they can use to enhance their professional image and conduct.

Instructor Responsibilities

The job of a flight instructor, or any instructor, is to teach. The learning process can be made easier by helping students learn, providing adequate instruction, demanding adequate standards of performance, and emphasizing the positive. *[Figure 10-1]*

Figure 10-1. *The four main responsibilities for flight instructors.*

Helping Students Learn

Learning should be an enjoyable experience. By making each lesson a pleasurable experience for the student, the instructor can maintain a high level of student motivation. This does not mean the instructor must make things easy for the student or sacrifice standards of performance to please the student. The student experiences satisfaction from doing a good job or from successfully meeting the challenge of a difficult task.

Learning should be interesting. Knowing the objectives, in clear and concise terms, of each period of instruction gives meaning and interest to the student, as well as the instructor. Not knowing the objective of the lesson often leads to confusion, disinterest, and uneasiness on the part of the student.

Learning to fly should be a habit-building period during which the student devotes his or her attention, memory, and judgment to the development of correct habits. Any objective other than to learn the right way is likely to make students impatient. The instructor should keep the students focused on good habits both by example and by a logical presentation of learning tasks.

Providing Adequate Instruction

The flight instructor should attempt to carefully and correctly analyze the student's personality, thinking, and ability. No two students are alike, and no one method of instruction can be equally effective for each student. The instructor must talk with a student at some length to learn about the student's background interests, temperament, and way of thinking. The instructor's methods also may change as the student advances through successive stages of training.

An instructor who has not correctly analyzed a student may soon find that the instruction is not producing the desired results. For example, this could mean that the instructor does not realize that a student is actually a quick thinker, but is hesitant to act. Such a student may fail to act at the proper time due to lack of self-confidence, even though the situation is correctly understood. In this case, instruction would obviously be directed toward developing student self-confidence, rather than drill on flight fundamentals. In another case, too much criticism may completely subdue a timid person, whereas brisk instruction may force a more diligent application to the learning task. A student may require instructional methods that combine tact, keen perception, and delicate handling. If such a student receives too much help and encouragement, a feeling of incompetence may develop.

Standards of Performance

Flight instructors must continuously evaluate their own effectiveness and the standard of learning and performance achieved by their students. The desire to maintain pleasant personal relationships with the students must not cause the acceptance of a slow rate of learning or substandard flight performance. It is a fallacy to believe that accepting lower standards to please a student produces a genuine improvement in the student-instructor relationship. An earnest student does not resent reasonable standards that are fairly and consistently applied.

Emphasizing the Positive

Flight instructors have a tremendous influence on their students' perception of aviation. The way instructors conduct themselves, the attitudes they display, and the manner in which they develop their instruction all contribute to the formation of either positive or negative impressions by their student. The success of a flight instructor depends, in large measure, on the ability to present instruction so that a student develops a positive image of aviation.

Most new instructors tend to adopt those teaching methods used by their own instructors. These methods may or may not have been good. The fact that one has learned under one system of instruction does not mean that this is necessarily the best way it can be done, regardless of the respect one retains for the ability of their original instructor. Some students learn in spite of their instruction, rather than because of it. Emphasize the positive because positive instruction results in positive learning.

Flight Instructor Responsibilities

All flight instructors shoulder an enormous responsibility because their students ultimately fly an aircraft. Flight instructors have some additional responsibilities including the responsibility of evaluating student pilots and making a determination of when they are ready to solo. Other flight instructor responsibilities are based on Title 14 of the Code of Federal Regulations (14 CFR) part 61 and advisory circulars (ACs). *[Figure 10-2]*

Evaluation of Student Piloting Ability

Evaluation is one of the most important elements of instruction. In flight instruction, the instructor initially determines that the student understands the procedure or maneuver. Then the instructor demonstrates the maneuver, allows the student to practice the maneuver under direction, and finally evaluates student accomplishment by observing their performance.

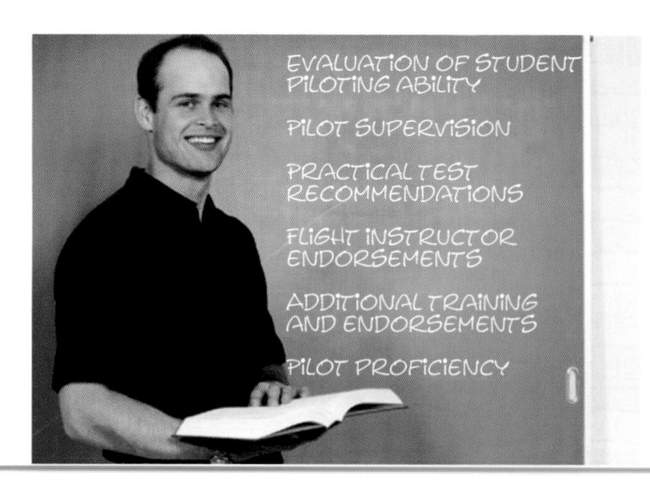

Figure 10-2. *The flight instructor has many additional responsibilities.*

Evaluation of demonstrated ability during flight instruction must be based upon established standards of performance, suitably modified to apply to the student's experience and stage of development as a pilot. The evaluation must consider the student's mastery of all the elements involved in the maneuver, rather than merely the overall performance.

Correction of student errors should not include the practice of taking the controls away from students immediately when a mistake is made. Safety permitting, it is frequently better to let students progress part of the way into the mistake and find their own way out. It is difficult for students to learn to do a maneuver properly if they seldom have the opportunity to correct an error. On the other hand, students may perform a procedure or maneuver correctly and not fully understand the principles and objectives involved. When the instructor suspects this, students should be required to vary the performance of the maneuver slightly, combine it with other operations, or apply the same elements to the performance of other maneuvers. Students who do not understand the principles involved probably are not able to do this successfully.

Student Pilot Supervision

Flight instructors have the responsibility to provide guidance and restraint with respect to the solo operations of their students. This is by far the most important flight instructor responsibility because the instructor is the only person in a position to make the determination that a student is ready for solo operations. Before endorsing a student for solo flight, the instructor should require the student to demonstrate consistent ability to perform all of the fundamental maneuvers.

Practical Test Recommendations

Provisions are made on the airman certificate or rating application form for the written recommendation of the flight instructor who has prepared the applicant for the practical test involved. Signing this recommendation imposes a serious responsibility on the flight instructor. A flight instructor who makes a practical test recommendation for an applicant seeking a certificate or rating should require the applicant to thoroughly demonstrate the knowledge and skill level required for that certificate or rating. This demonstration should in no instance be less than the complete procedure prescribed in the applicable PTS.

A practical test recommendation based on anything less risks the presentation of an applicant who may be unprepared for some part of the actual practical test. In such an event, the flight instructor is logically held accountable for a deficient instructional performance. This risk is especially great in signing recommendations for applicants who have not been trained by the instructor involved. For balloons, 14 CFR part 61 requires a minimum of two training flights of one hour each with an authorized instructor within 60 days preceding the date of the test for a private or commercial certificate. The instructor signing the endorsement is required to have conducted the training in the applicable areas of operation stated in the regulations and the PTS, and certifies that the person is prepared for the required practical test. In most cases, the conscientious instructor has little doubt concerning the applicant's readiness for the practical test.

FAA inspectors and designated pilot examiners rely on flight instructor recommendations as evidence of qualification for certification, and proof that a review has been given of the subject areas found to be deficient on the appropriate knowledge test. Recommendations also provide assurance that the applicant has had a thorough briefing on the practical test standards and the associated knowledge areas, maneuvers, and procedures. If the flight instructor has trained and prepared the applicant competently, the applicant should have no problem passing the practical test.

The recommended format for log book endorsements may be found in the current version of AC 61-65; an extract of those relevant to balloon instruction may be found in Appendix E.

Self-Improvement

Professional flight instructors must never become complacent or satisfied with their own qualifications and abilities. They should be constantly alert for ways to improve their qualifications, effectiveness, and the services they provide to students. Flight instructors are considered authorities on aeronautical matters and are the experts to whom many pilots refer questions concerning regulations, requirements, and new operating techniques.

Sincerity

The professional instructor should be straightforward and honest. Attempting to hide some inadequacy behind a smokescreen of unrelated instruction makes it impossible for the instructor to command the respect and full attention of a student. Teaching an aviation student is based upon acceptance of the instructor as a competent, qualified teacher and an expert pilot. Any facade of instructor pretentiousness, whether it is real or mistakenly assumed by the student, immediately causes the student to lose confidence in the instructor and learning is adversely affected.

Acceptance of the Student

With regard to students, the instructor must accept them as they are, including all their faults and problems. The student is a person who wants to learn, and the instructor is a person who is available to help in the learning process. Beginning with this understanding, the professional relationship of the instructor with the student should be based on a mutual acknowledgement that the student and the instructor are important to each other, and that both are working toward the same objective.

Personal Appearance and Habits

Personal appearance has an important effect on the professional image of the instructor. Today's aviation customers expect their instructors to be neat, clean, and appropriately dressed. Since the instructor is engaged in a learning situation, the attire worn should be appropriate to a professional status. *[Figure 10-3]*

Personal habits have a significant effect on the professional image. The exercise of common courtesy is perhaps the most important of these.

Demeanor

The attitude and behavior of the instructor can contribute much to a professional image. The professional image requires development of a calm, thoughtful, and disciplined, but not somber, demeanor.

The instructor should also present an attitude of enthusiasm, with respect to the training conducted, whether ground or flight. The enthusiasm expressed by the instructor is normally reflected in the student's response to the training, and makes it a much more enjoyable experience for all parties concerned.

Safety Practices and Accident Prevention

The safety practices emphasized by instructors have a long lasting effect on students. Generally, students consider their instructor to be a model of perfection whose habits they

Figure 10-3. *The flight instructor should always present a professional appearance.*

attempt to imitate, whether consciously or unconsciously. The instructor's advocacy and description of safety practices mean little to a student if the instructor does not demonstrate them consistently. For this reason, instructors must meticulously observe the safety practices being taught to students. A good example is the use of a checklist before takeoff. If a student pilot sees the flight instructor layout, inflate, and take off in a balloon without referring to a checklist, no amount of instruction in the use of a checklist convinces that student to faithfully use one when solo flight operations begin.

To maintain a professional image, a flight instructor must carefully observe all regulations and recognized safety practices during flight operations. An instructor who is observed to fly with apparent disregard for loading limitations or weather minimums creates an image of irresponsibility that many hours of scrupulous flight instruction can never correct. Habitual observance of regulations, safety precautions, and the precepts of courtesy enhances the instructor's image of professionalism. Moreover, such habits make the instructor more effective by encouraging students to develop similar habits.

Proper Language

In aviation instruction, as in other professional activities, the use of profanity and obscene language leads to distrust or, at best, to a lack of complete confidence in the instructor. To many people, such language is actually objectionable to the point of being painful. The professional instructor must speak normally, without inhibitions, and develop the ability to speak positively and descriptively without excesses of language.

Also consider that the beginning aviation student is being introduced to new concepts and experiences and encountering new terms and phrases that are often confusing. At the beginning of the student's training, and before each lesson during early instruction, the instructor should carefully define the terms and phrases to be used during the lesson. The instructor should then be careful to limit instruction to those terms and phrases, unless the exact meaning and intent of any new expression are explained immediately. In all cases, terminology should be explained to the student before it is used during instruction.

The Learning Process

To learn is to acquire knowledge or skill. Learning also may involve a change in attitude or behavior. Pilots need to acquire the higher levels of knowledge and skill, including the ability to exercise judgment and solve problems. The challenge for the flight instructor is to understand how people learn, and more importantly, to be able to apply that knowledge to the learning environment. *[Figure 10-4]*

Figure 10-4. *Effective learning shares several common characteristics.*

Definition of Learning

The ability to learn is one of the most outstanding human characteristics. Learning occurs continuously throughout a person's lifetime. To define learning, it is necessary to analyze what happens to the individual. Thus, learning can be defined as a change in behavior as a result of experience. This can be physical and overt, or it may involve complex intellectual or attitudinal changes which affect behavior in more subtle ways. In spite of numerous theories and contrasting views, psychologists generally agree on many common characteristics of learning.

Characteristics of Learning

Flight instructors need a good understanding of the general characteristics of learning in order to apply them in a learning situation. If learning is a change in behavior as a result of experience, then instruction must include a careful and systematic creation of those experiences that promote learning. This process can be quite complex because, among other things, an individual's background strongly influences the way that person learns. To be effective, the learning situation should contain the following four points:

1. Learning is purposeful. Most people have definite ideas about what they want to do and achieve. Their goals are sometimes short term, involving a matter of days or weeks, while others may have goals carefully planned for a career or a lifetime. Each student, then, has specific intentions and goals. Learning, then, becomes a means to those goals.
2. Learning is a result of experience. A student can learn only from personal experience; therefore learning and knowledge cannot exist apart from a person. Even when observing or performing the same procedures, two people can react differently; they learn different

things from it, according to the manner in which the event affects their individual needs.

3. Learning is multifaceted. The learning process includes verbal elements, emotional elements, and problem-solving elements all taking place at once.
4. Learning is an active process. The individual who wishes to be proficient at a particular task or skill should understand that they never stop learning.

Principles of Learning

Over the years, educational psychologists have identified several principles which seem generally applicable to the learning process. They provide additional insight into what makes people learn most effectively.

- Readiness—individuals learn best when they are ready to learn. However, they do not learn well if they see no reason for learning. Getting students ready to learn is usually the instructor's responsibility. If students have a strong purpose, a clear objective, and a definite reason for learning something, they make more progress than if they lack motivation.
- Exercise—the principle of exercise states that those things most often repeated are best remembered. It is the basis of drill and practice. The human memory is fallible. The mind can rarely retain, evaluate, and apply new concepts or practices after a single exposure. Students learn by applying what they have been told and shown. Every time practice occurs, learning continues. The instructor must provide opportunities for students to practice and, at the same time, make sure that this process is directed toward a goal.
- Effect—based on the emotional reaction of the student. It states that learning is strengthened when accompanied by a pleasant or satisfying feeling, and that learning is weakened when associated with an unpleasant feeling. Experiences that produce feelings of defeat, frustration, anger, confusion, or futility are unpleasant for the student. If, for example, an instructor attempts to teach precision maneuvering during the first flight, the student is likely to feel inferior and be frustrated.
- Primacy—the state of being first, often creates a strong, almost unshakable, impression. For the instructor, this means that what is taught must be right the first time. For the student, it means that learning must be right. "Unteaching" is often more difficult than teaching. Every student should be started right. The first experience should be positive, functional, and lay the foundation for all that is to follow.
- Intensity—a vivid, dramatic, or exciting learning experience teaches more than a routine or boring experience. A student is likely to gain greater understanding of steep approaches or short-field, high wind landings by performing them rather than merely reading about them. The principle of intensity implies that a student learns more from the real thing than from a substitute.
- Recency—things most recently learned are best remembered. Conversely, the further a student is removed in time from a new factor in understanding, the more difficult it is to remember. Instructors recognize the principle of recency when they carefully plan a summary for a ground school lesson, a flight period, or a postflight critique.

How People Learn

Initially, all learning comes from perceptions which are directed to the brain by one or more of the five senses: sight, hearing, touch, smell, and taste. Psychologists have also found that learning occurs most rapidly when information is received through more than one of the senses. *[Figure 10-5]*

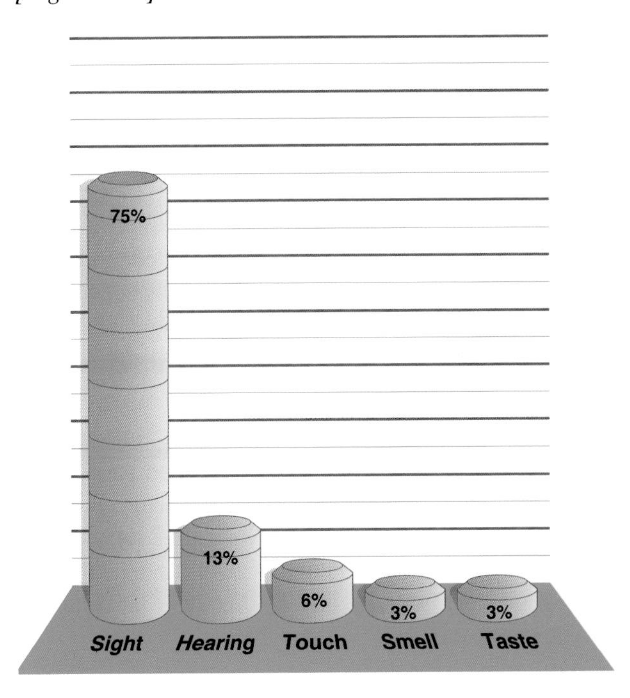

Figure 10-5. *Most learning occurs through sight, but the combination of sight and hearing accounts for about 99 percent of all perception.*

Perceptions

Perceiving involves more than the reception of stimuli from the five senses. Perceptions result when a person gives meaning to sensations. People base their actions on the way they believe things to be.

Real meaning comes only from within a person, even though the perceptions which evoke these meanings result from external stimuli. The meanings which are derived from perceptions are influenced not only by the individual's experience, but also by many other factors. Knowledge of the factors which affect the perceptual process is very important to the flight instructor because perceptions are the basis of all learning.

Factions Which Affect Perception

There are several factors that affect an individual's ability to perceive. Some are internal to each person and some are external.

- Physical organism—provides individuals with the perceptual sensors for perceiving the world around them.
- Basic need—a person's basic need is to maintain and enhance the organized self. A person's most pressing need is to preserve and perpetuate the self. To that end, an instructor must remember that anything asked of the student that may be interpreted by the student as endangering the self is resisted or denied.
- Goals and values—perceptions depend on one's goals and values. The precise kinds of commitments and philosophical outlooks which the student holds are important for the instructor to know, since this knowledge assists in predicting how the student interprets experiences and instructions.
- Self-concept—a powerful determinant in learning. If a student's experiences tend to support a favorable self-image, the student tends to remain receptive to subsequent experiences. A negative self-concept inhibits the perceptual processes which tend to keep the student from perceiving, effectively blocking the learning process.
- Time and opportunity—it takes time and opportunity to perceive. Learning some things depends on other perceptions which have preceded these learnings, and on the availability of time to sense and relate these new things to the earlier perceptions. Thus, sequence and time are necessary.
- Element of threat—does not promote effective learning. When confronted with a perceived threat, students tent to limit their attention to the threatening object or condition.

Insight

Insight involves the grouping of perceptions into meaningful groups of understanding. Creating insight is one of the instructor's major responsibilities. To ensure that this does occur, it is essential to keep each student constantly receptive to new experiences and to help the student realize the way each piece relates to all other pieces of the total pattern of the task to be learned.

As perceptions increase in number and are assembled by the student into larger blocks of learning, they develop insight. As a result, learning becomes more meaningful and more permanent. Forgetting is less of a problem when there are more anchor points for tying insights together. It is a major responsibility of the instructor to organize demonstrations and explanations, and to direct practice, so that the student has better opportunities to understand the interrelationship of the many kinds of experiences that have been perceived. Pointing out the relationships as they occur, providing a secure and nonthreatening environment in which to learn, and helping the student acquire and maintain a favorable self-concept are key steps in fostering the development of insight.

Motivation

Motivation is probably the dominant force governing the student's progress and ability to learn. Motivation may be positive or negative, tangible or intangible, subtle and difficult to identify, or it may be obvious.

Positive motivation is provided by the promise or achievement of rewards. These rewards may be personal or social; they may involve financial gain, satisfaction of the self-concept, or public recognition. Motivation which can be used to advantage by the instructor includes the desire for personal gain, the desire for personal comfort or security, the desire for group approval, and the achievement of a favorable self-image.

Negative motivation may engender fear, and be perceived by the student as a threat. While negative motivation may be useful in certain situations, characteristically it is not as effective in promoting efficient learning as positive motivation.

Positive motivation is essential to true learning. Negative motivation in the form of reproofs or threats should be avoided with all but the most overconfident and impulsive students. Slumps in learning are often due to declining motivation. Motivation does not remain at a uniformly high level. It may be affected by outside influences, such as physical or mental disturbances or inadequate instruction. The instructor should strive to maintain motivation at the highest possible level. In addition, the instructor should be alert to detect and counter any lapses in motivation.

Levels of Learning

Levels of learning may be classified in any number of ways. Four basic levels have traditionally been included in flight instructor training:

- Rote—the ability to repeat something back which was learned but not understood. An example of this may be the student who reads and can repeat back the applicable provision of 14 CFR section 91.119, Minimum Safe Altitudes, but has no concept of how this may affect their flight.
- Understanding—to comprehend or grasp the nature or meaning of something. Once a student has received proper instruction on performing a steep descent, and has some experience controlling the balloon in straight and level flight, he can consolidate those old and new perceptions into an insight on how to make a steep approach. At this point, the student has developed an understanding of the procedure for the steep approach.
- Application—the act of putting something to use that has been learned and understood. When the student understands the procedure for entering and performing a steep approach to the ground, has had the maneuver demonstrated, and has practiced the approach until consistency has been achieved, the student has developed the skill to apply what has been learned. This is a major level of learning, and one at which the instructor is too often willing to stop.
- Correlation—associating what has been learned, understood, and applied with previous or subsequent learning. The correlation level of learning, which should be the objective of aviation instruction, is that level at which the student becomes able to associate an element which has been learned with other segments or blocks or learning.

Most training is conducted in such a manner that the student never progresses past the rote and understanding levels. This is an unacceptable procedure, as practical testing requires the student to perform at the application and correlation levels. Failing to bring the student to those levels is incomplete instruction, and does not provide a complete training experience.

Transfer of Learning

During a learning experience, the student may be aided by things learned previously. On the other hand, it is sometimes apparent that previous learning interferes with the current learning task. Consider the learning of two skills. If the learning of skill A helps to learn skill B, positive transfer occurs. If learning skill A hinders the learning of skill B, negative transfer occurs. It should be noted that the learning of skill B may affect the retention or proficiency of skill A, either positively or negatively. While these processes may help substantiate the interference theory of forgetting, they are still concerned with the transfer of learning.

Many aspects of teaching profit by this type of transfer. It may explain why students of apparently equal ability have differing success in certain areas. Negative transfer may hinder the learning of some; positive transfer may help others. This points to a need to know a student's past experience and what has already been learned. In lesson and syllabus development, instructors should plan for transfer of learning by organizing course materials and individual lesson materials in a meaningful sequence. Each phase should help the student learn what is to follow.

Habit Formation

The formation of correct habits from the beginning of any learning process is essential to further learning and for correct performance after the completion of training. Remember that primacy is one of the fundamental principles of learning. Therefore, it is the instructor's responsibility to insist on correct techniques and procedures from the outset of training to provide proper habit patterns. It is much easier to foster proper habits from the beginning of training than to correct faulty ones later.

Due to the high level of knowledge and skill required in aviation for pilots, training has traditionally followed a building block concept. This means new learning and habits are based on a solid foundation of experience and/or old learning. As knowledge and skill increase, there is an expanding base upon which to build for the future.

Theories of Forgetting

A consideration of why people forget may point the way to help them remember. Several theories account for forgetting, including disuse and interference.

Disuse

The theory of disuse suggests that a person forgets those things which are not used. The high school or college graduate is saddened by the lack of factual data retained several years after graduation. Since the things which are remembered are those used on the job, a person concludes that forgetting is the result of disuse. But the explanation is not quite so simple. Experimental studies show that a hypnotized person can describe specific details of an event which normally is beyond recall. Apparently the memory is there, locked in the recesses of the mind. The difficulty is summoning it up to consciousness.

Interference

The basis of the interference theory is that people forget something because a certain experience has overshadowed it, or that the learning of similar things has intervened. This theory might explain how the range of experiences after graduation from school causes a person to forget or to lose knowledge. In other words, new events displace many things that had been learned. From experiments, at least two conclusions about interference may be drawn. First, similar material seems to interfere with memory more than dissimilar material; and second, material not well learned suffers most from interference.

The Teaching Process

Effective teaching is based on principles of learning which have been previously discussed in this chapter. The learning process is not easily separated into a definite number of steps. Sometimes, learning occurs almost instantaneously, and other times it is acquired only through long, patient study and diligent practice. The teaching process, on the other hand, can be divided into steps. Although there is disagreement about the number of steps, examination of the various lists of steps in the teaching process reveals that different authors are saying essentially the same thing: the teaching of new material can be reduced to preparation, presentation, application, and review and evaluation. *[Figure 10-6]*

When beginning the teaching process, it may be helpful if the instructor remembers that in order to provide a "holistic," or complete, approach to aviation training, it is necessary to teach more than just the "how-to" of flying. Too many times, the entire focus of flight training is spent on the mechanical aspects of performing a maneuver; it may be a better approach to also include and discuss the "why" of an action, to assist the student in gaining a better understanding of the flight process.

Preparation

For each lesson or instructional period, the instructor must prepare a lesson plan. Traditionally, this plan includes a

Figure 10-6. *The teaching process consists of four basic principles.*

statement of lesson objectives, the procedures and facilities to be used during the lesson, the specific goals to be attained, and the means to be used for review and evaluation. The lesson plan should also include home study or other special preparation to be done by the student. The instructor should make certain that all necessary supplies, materials, and equipment needed for the lesson are readily available and that the equipment is operating properly. Preparation of the lesson plan may be accomplished after reference to the syllabus or practical test standards (PTS), or it may be in a preprinted form as prepared by a publisher of training materials. These documents list general objectives that are to be accomplished. Objectives are needed to bring the unit of instruction into focus. The instructor can organize the overall instructional plan by writing down the objectives and making certain that they flow in a logical sequence from beginning to end. The objectives allow the instructor to structure the training and permit the student to see clearly what is required along the way.

Figure 10-7. *Performance-based objectives are made up of a description of the skill or behavior, conditions, and criteria.*

Performance-Based Objectives

One good way to write lesson plans is to begin by formulating performance-based objectives. The instructor uses the objectives as listed in the syllabus or the appropriate PTS as the beginning point for establishing performance-based objectives. These objectives are very helpful in delineating exactly what needs to be done and how it is done during each lesson. Once the performance-based objectives are written, most of the work of writing a final lesson plan is completed. One useful thought is the utilization of the "DAM principle"; objectives, as well as goals, should be difficult, attainable, and measurable.

Performance-based objectives are used to set measurable, reasonable standards that describe the desired performance of the student. This usually involves the term behavioral objective, although it may be referred to as a performance, instructional, or educational objective. All refer to the same thing, the behavior of the student.

Performance-based objectives consist of three parts: description of the skill or behavior, conditions, and criteria. Each part is required and must be stated in a way that leaves every reader with the same picture of the objective, how it is performed, and to what level of performance. *[Figure 10-7]*

Use of performance-based objectives also provides the student with a better understanding of the big picture, as well as knowledge of exactly what is expected. This overview can alleviate a significant source of frustration on the part of the student.

Presentation

Instructors have several methods of presentation from which to choose. The nature of the subject matter and the objective in teaching it normally determine the method of presentation.

The lecture method is suitable for presenting new material, for summarizing ideas, and for showing relationships between theory and practice. For example, it is suitable for the presentation of a ground school lesson on performance planning. This method is most effective when accompanied by instructional aids and training devices. In the case of a discussion on performance planning, a chalkboard, a marker board, or flip chart could be used effectively.

The demonstration-performance method is desirable for teaching a skill, such as instruction on most flight maneuvers. Showing a student pilot how to perform a steep descent, for example, would be appropriate for this method. The instructor would first demonstrate the maneuver, and then have the student attempt the same maneuver.

Another form of presentation is the guided discussion which is used in a classroom situation. It is a good method for encouraging active participation of the students. It is especially helpful in teaching subjects such as safety and emergency procedures where students can use initiative and imagination in addressing problem areas.

Review and Application

Application is where the student uses what the instructor has presented. After a classroom presentation, the student may be asked to explain the new material. The student also may be asked to perform a procedure or operation that has just been demonstrated. In most instructional situations, the instructor's explanation and demonstration activities are alternated with student performance efforts. The instructor makes a presentation and then asks the student to try the same procedure or operation.

Usually the instructor has to interrupt the student's efforts for corrections and further demonstrations. This is necessary, because it is very important that each student perform the maneuver or operation the right way the first few times. This is when habits are established. Faulty habits are difficult to correct and must be addressed as soon as possible. Flight instructors in particular must be aware of this problem since students do a lot of their practice without an instructor. Only after reasonable competence has been demonstrated should the student be allowed to practice certain maneuvers on solo flights. Then, the student can practice the maneuver again and again until correct performance becomes almost automatic. Periodic review and evaluation by the instructor is necessary to ensure that the student has not acquired any bad habits.

Teaching Methods

The information presented in previous sections has been largely theoretical, emphasizing concepts and principles pertinent to the learning process, human behavior, and effective communication in education and training programs. This knowledge, if properly used, enables instructors to be more confident, efficient, and successful. The discussion which follows departs from the theoretical with some specific recommendations for the actual conduct of the teaching process. Included are methods and procedures which have been tested and found to be effective.

Personal computers are a part of every segment of our society today. Since a number of computer-based programs are currently available from publishers of aviation training materials, a brief description of new technologies and how to use them effectively is provided near the end of this section.

Organizing Material

Regardless of the teaching method used, an instructor must properly organize the material. Lessons do not stand alone within a course of training. There must be a plan of action to lead instructors and their students through the course in a logical manner toward the desired goal. Usually the goal for students is a certificate or rating. A systematic plan of action requires the use of an appropriate training syllabus. Generally, the syllabus must contain a description of each lesson, including objectives and completion standards.

Although instructors may develop their own syllabus, in practice many instructors use a group-developed syllabus such as that developed and available through the Balloon Federation of America. Thus, the main concern of the instructor usually is the more manageable task of organizing a block of training with integrated lesson plans. The traditional organization of a lesson plan is: introduction, development, and conclusion.

Introduction

The introduction sets the stage for everything to come. Efforts in this area pay great dividends in terms of quality of instruction. In brief, the introduction is made up of three elements—attention, motivation, and an overview of what is to be covered. *[Figure 10-8]*

Introduction	
Element	**Purpose**
Attention	• Establish common ground between instructor and student • Capture and hold the attention of the class • Specify benefits the student can expect from the lesson
Motivation	• Establish receptive attitude toward lesson • Create smooth transition into lesson
Overview	• Indicate what is to be covered and relate this information to the overall course

Figure 10-8. *The introduction prepares the students to receive the information in the lesson.*

Attention

The purpose of the attention element is to focus each student's attention on the lesson. The instructor may begin by telling a story, making an unexpected or surprising statement, asking a question, or telling a joke. Any of these may be appropriate at one time or another. Regardless of which is used, it should relate to the subject and establish a background for developing the learning outcomes. Telling a story or a joke that is not related in some way to the subject can only distract from the lesson. The main concern is to gain the attention of everyone and concentrate on the subject.

Motivation

The purpose of the motivation element is to offer the student specific reasons why the lesson content is important to know, understand, apply, or perform. For example, the instructor may talk about an occurrence where the knowledge in the lesson was applied. Or the instructor may remind the student of an upcoming test on the material. This motivation should appeal to each student personally and engender a desire to learn the material.

Overview

Every lesson introduction should contain an overview that explains what is to be covered during the period. A clear, concise presentation of the objective and the key ideas gives the students a road map of the route to be followed. A good visual aid can help the instructor show the students the path that they are to travel. The introduction should be free of stories, humor, or incidents that do not help the students focus their attention on the lesson objective. Also, the instructor should avoid a long apologetic introduction, because it only serves to dampen the students' interest in the lesson.

Development

Development is the main part of the lesson. Here, the instructor develops the subject matter in a manner that helps the students achieve the desired learning outcomes. The instructor must logically organize the material to show the relationships of the main points. The instructor usually shows these primary relationships by developing the main points in one of the following ways: from past to present, simple to complex, known to unknown, and most frequently used to least frequently used.

Under each main point in a lesson, the subordinate points should lead naturally from one to the other. With this arrangement, each point leads logically into, and serves as a reminder of, the next. Meaningful transitions from one main point to another keep the students oriented, aware of where they have been, and where they are going. This permits effective sorting or categorizing chunks of information in the working or short-term memory. Organizing a lesson so the students grasp the logical relationships of ideas is not an easy task, but it is necessary if the students are to learn and remember what they have learned. Poorly organized information is of little or no value to the student because it cannot be readily understood or remembered.

Conclusion

An effective conclusion retraces the important elements of the lesson and relates them to the objective. This review and wrap-up of ideas reinforces student learning and improves the retention of what has been learned. New ideas should not be introduced in the conclusion because at this point they are likely to confuse the students.

By organizing the lesson material into a logical format, the instructor has maximized the opportunity for students to retain the desired information. However, each teaching situation is unique. The setting and purpose of the lesson determines which teaching method—lecture, guided discussion, demonstration-performance, cooperative or group learning, computer-based training, or a combination—is used.

Lecture Method

The lecture method is the most widely used form of presentation. Every instructor should know how to develop and present a lecture. *[Figure 10-9]* Lectures are used for introduction of new subjects, summarizing ideas, showing relationships between theory and practice, and reemphasizing main points. The lecture method is adaptable to many different settings, including either small or large groups. Lectures also may be used to introduce a unit of instruction or a complete training program. Finally, lectures may be combined with other teaching methods to give added meaning and direction.

Figure 10-9. *Instructors should try a dry run with another instructor to get a feel for the lecture presentation.*

The lecture method of teaching needs to be very flexible since it may be used in different ways. For example, there are several types of lectures such as the illustrated talk where the speaker relies heavily on visual aids to convey ideas to the listeners. With a briefing, the speaker presents a concise array of facts to the listeners who normally do not expect elaboration of supporting material. During a formal lecture, the speaker's purpose is to inform, to persuade, or to entertain with little or no verbal participation by the students. When using a teaching lecture, the instructor plans and delivers an oral presentation in a manner that allows some participation by the students and helps direct them toward the desired learning outcomes.

Demonstration-Performance Method

This method of teaching is based on the simple, yet sound principle that we learn by doing. Students learn physical or mental skills by actually performing those skills under supervision. An individual learns to write by writing, and to fly a balloon by actually performing flight maneuvers. Students also learn mental skills, such as speed reading, by this method. Skills requiring the use of tools, machines, and equipment are particularly well suited to this instructional method.

Every instructor should recognize the importance of student performance in the learning process. Early in a lesson that is to include demonstration and performance, the instructor should identify the most important learning outcomes. Next, explain and demonstrate the steps involved in performing the skill being taught. Then, allow students time to practice each step, so they can increase their ability to perform the skill.

The demonstration-performance method of teaching has five essential phases:

- Explanation Phase—explanations must be clear, pertinent to the objectives of the particular lesson to be presented, and based on the known experience and knowledge of the student. In addition to the necessary actions to be performed, the instructor should describe the end result of these efforts.
- Demonstration Phase—the instructor must show the student the actions necessary to perform a skill.
- Student Performance and Instructor Supervision Phases—these two phases, which involve separate actions, are performed concurrently, and are thus described under a single heading. The first action is the performance by the student of the physical or mental skill that had been explained. The second is the instructor's supervision, insuring that errors are immediately corrected to standards already prescribed.
- Evaluation Phase—in this final phase, the instructor judges student performance. The student performs whatever competence has been attained, and the instructor evaluates and discovers just how well the skill has been learned. Form this measurement of student achievement, the instructor determines the effectiveness of the instruction.

Computer-based Training

Many new and innovative training technologies are available today. One of the most significant is computer-based training (CBT)—the use of the personal computer as a training device. *[Figure 10-10]* CBT is sometimes called computer-based instruction (CBI). The terms CBT and CBI are synonymous and may be used interchangeably.

Figure 10-10. *The instructor must continually monitor student performance when using CBT, as with all instructional aids.*

Common examples of CBT with specific application to balloon flight training include the computer versions of the test prep study guides which are useful for preparation for the FAA knowledge tests. These programs typically allow the students to select a test, complete the questions, and find out how they did on the test. The student may then conduct a review of questions missed. An excellent resource for balloon training is the web site webexams.com; this provides the student with the ability to take practice exams, which assists in determining weaknesses in training.

While computers provide many training advantages, they also have limitations. Improper or excessive use of CBT should be avoided. Computer-based training should not be used by the instructor as stand-alone training any more than a textbook or video. Like video or a textbook, CBT is an aid to the instructor. The instructor must be actively involved with the students when using instructional aids. This involvement should include close supervision, questions, examinations, quizzes, or guided discussions on the subject matter.

Techniques of Flight Instruction

In this section, the demonstration-performance method is applied to the telling-and-doing technique of flight instruction, as well as the integrated technique of flight instruction.

The Telling-and-Doing Technique

This technique has been in use for a long time and is very effective in teaching physical skills. Flight instructors find it valuable in teaching procedures and maneuvers. The telling-and-doing technique is actually a variation of the demonstration-performance method. In the telling-and-doing technique, the first step is preparation. This is particularly important in flight instruction because of the introduction of new maneuvers or procedures. The flight instructor needs to be well prepared and highly organized if complex maneuvers and procedures are to be taught effectively. The student must be intellectually and psychologically ready for the learning activity. The preparation step is accomplished prior to the flight lesson with a discussion of lesson objectives and completion standards, as well as a thorough preflight briefing. *[Figure 10-11]*

Instructor Tells-Instructor Does

Presentation is the second step in the teaching process. It is a continuation of preparing the student, which began in the detailed preflight discussion, and now continues by a carefully planned demonstration and accompanying verbal explanation of the procedure or maneuver. It is important that the demonstration conforms to the explanation as closely as possible. In addition, it should be demonstrated in the same sequence in which it was explained to avoid confusion and provide reinforcement.

Student Tells-Instructor Does

This is a transition between the second and third steps in the teaching process. It is the most obvious departure from the demonstration-performance technique, and may provide the most significant advantages. In this step, the student actually plays the role of instructor, telling the instructor what to do and how to do it. Two benefits accrue from this step. First, being freed from the need to concentrate on performance of the maneuver and from concern about its outcome, the student should be able to organize his or her thoughts regarding the steps involved and the techniques to be used. In the process of explaining the maneuver as the instructor performs it, perceptions begin to develop into insights. Mental habits begin to form with repetition of the instructions previously received. Second, with the student doing the talking, the instructor is able to evaluate the student's understanding of the factors involved in performance of the maneuver.

Student Tells-Student Does

Application is the third step in the teaching process. This is where learning takes place and where performance habits are formed. If the student has been adequately prepared (first step) and the procedure or maneuver fully explained and demonstrated (second step), meaningful learning occurs. The instructor should be alert during the student's practice to detect any errors in technique and to prevent the formation of faulty habits.

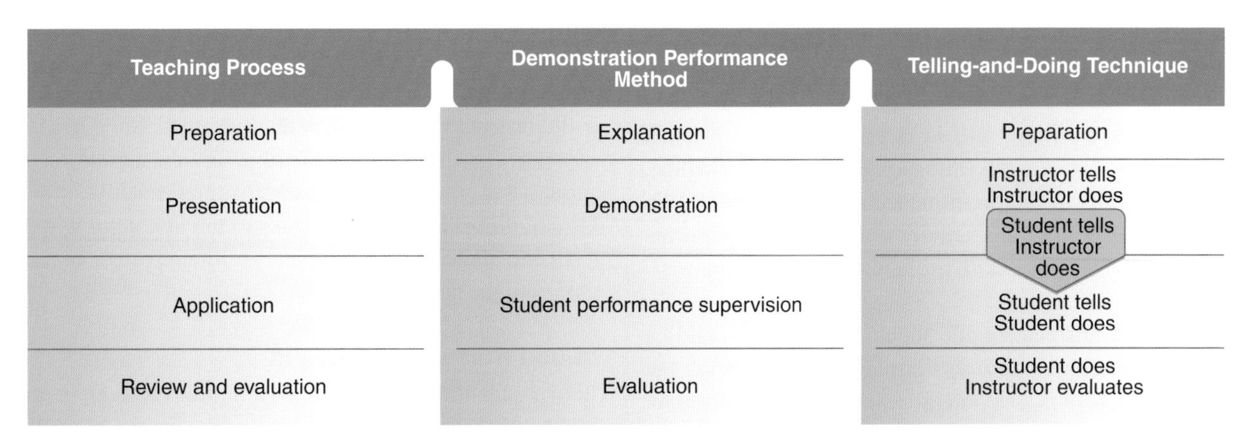

Teaching Process	Demonstration Performance Method	Telling-and-Doing Technique
Preparation	Explanation	Preparation
Presentation	Demonstration	Instructor tells Instructor does
		Student tells Instructor does
Application	Student performance supervision	Student tells Student does
Review and evaluation	Evaluation	Student does Instructor evaluates

Figure 10-11. *Comparison of steps in the teaching process, the demonstration-performance method, and the telling-and-doing technique. This comparison shows the similarities, as well as some differences. The main difference in the telling-and-doing technique is the important transition, student tells—instructor does, which occurs between the second and third step.*

At the same time, the student should be encouraged to think about what to do during the performance of a maneuver, until it becomes habitual. In this step, the thinking is done verbally. This focuses concentration on the task to be accomplished, so that total involvement in the maneuver is fostered.

Student Does-Instructor Evaluates

The fourth step of the teaching process is review and evaluation. In this step, the instructor reviews what has been covered during the instructional flight and determines to what extent the student has met the objectives outlined during the preflight discussion. Since the student no longer is required to talk through the maneuver during this step, the instructor should be satisfied that the student is well prepared and understands the task before starting. This last step is identical to the final step used in the demonstration-performance method. The instructor observes as the student performs, then makes appropriate comments.

At the conclusion of the evaluation phase, record the student's performance and verbally advise each student of the progress made toward the objectives. Regardless of how well a skill is taught, there may still be failures. Since success is a motivating factor, instructors should be positive in revealing results. When pointing out areas that need improvement, offer concrete suggestions that help. The instructor should make every effort to end the evaluation on a positive note.

Critique and Evaluation

Since every student is different and each learning situation is unique, the actual outcome may not be entirely as expected. The instructor must be able to appraise student performance and convey this information back to the student. No skill is more important to an instructor than the ability to analyze, appraise, and judge student performance. The student quite naturally looks to the instructor for guidance, analysis, appraisal, as well as suggestions for improvement and encouragement. This feedback from instructor to student is called a critique.

In most cases, a critique should be conducted in private. It should come immediately after a student's performance, while the details of the performance are easy to recall. An instructor may critique any activity which a student performs or practices to improve skill, proficiency, and learning.

Two common misconceptions about the critique should be corrected at the outset. First, a critique is not a step in the grading process. It is a step in the learning process. Second, a critique is not necessarily negative in content. It considers the good along with the bad, the individual parts, relationships of the individual parts, and the overall performance. A critique can, and usually should, be as varied in content as the performance being critiqued.

Purpose of a Critique

A critique should provide the student with something constructive upon which he or she can work or build. It should provide direction and guidance to raise their level of performance. Students must understand the purpose of the critique; otherwise, they are unlikely to accept the criticism offered and little improvement will result.

Methods of Critique

The critique of student performance is always the instructor's responsibility, and it can never be delegated in its entirety. The instructor can add interest and variety to the criticism through the use of imagination and by drawing on the talents, ideas, and opinions of others. There are several useful methods of conducting a critique, two of which have specific application to balloon flight instruction.

Student-Led Critique

The instructor asks a student to lead the critique. The instructor can specify the pattern of organization and the techniques or can leave it to the discretion of the student leader. Because of the inexperience of the participants in the lesson area, student-led critiques may not be efficient, but they can generate student interest and learning and, on the whole, be effective.

Self-Critique

A student is required to critique personal performance. Like all other methods, a self-critique must be controlled and supervised by the instructor. Whatever the methods employed, the instructor must not leave controversial issues unresolved, nor erroneous impressions uncorrected. The instructor must make allowances for the student's relative inexperience. Normally, the instructor should reserve time at the end of the student critique to cover those areas that might have been omitted, not emphasized sufficiently, or considered worth repeating. One variant of this method is for the instructor to ask the student to name three negative aspects of the flight training period, and discuss corrective action. Then, the instructor asks for three positive aspects of the training. This also indicates to the instructor whether or not the student is able to analyze their performance, in relation to the standards sought.

Ground Rules for Critiquing

There are a number of rules and techniques to keep in mind when conducting a critique. The following list can be applied, regardless of the type of critiquing activity.

- Avoid trying to cover too much. A few well-made points usually is more beneficial than a large number of points that are not developed adequately.
- Allow time for a summary of the critique to reemphasize the most important things a student should remember.
- Never allow yourself to be maneuvered into the unpleasant position of defending criticism. If the criticism is honest, objective, constructive, and comprehensive, no defense should be necessary.
- If part of the critique is written, make certain that it is consistent with the oral portion.

Although, at times, a critique may seem like an evaluation, it is not. Both student and instructor should consider it as an integral part of the lesson. It normally is a wrap-up of the lesson. A good critique closes the chapter on the lesson and sets the stage for the next lesson.

Characteristics of an Effective Critique

In order to provide direction and raise the students' level of performance, the critique must be factual and be aligned with the completion standards of the lesson. This, of course, is because the critique is a part of the learning process. Some of the requirements for an effective critique are shown in *Figure 10-12*.

- Objective—the effective critique is focused on student performance. It should be objective, and not reflect the personal opinions, likes, dislikes, and biases of the instructor. If a critique is to be objective, it must be honest; it must be based on the performance as it was, not as it could have been, or as the instructor and student wished that it had been.
- Flexible—the instructor needs to examine the entire performance of a student and the context in which it is accomplished. Sometimes a good student turns in a poor performance and a poor student turns in a good one. A friendly student may suddenly become hostile, or a hostile student may suddenly become friendly and cooperative. The instructor must fit the tone, technique, and content of the critique to the occasion, as well as the student. A critique should be designed and executed so that the instructor can allow for variables. An effective critique is one that is flexible enough to satisfy the requirements of the moment.
- Acceptable—before students willingly accept their instructor's criticism, they must first accept the

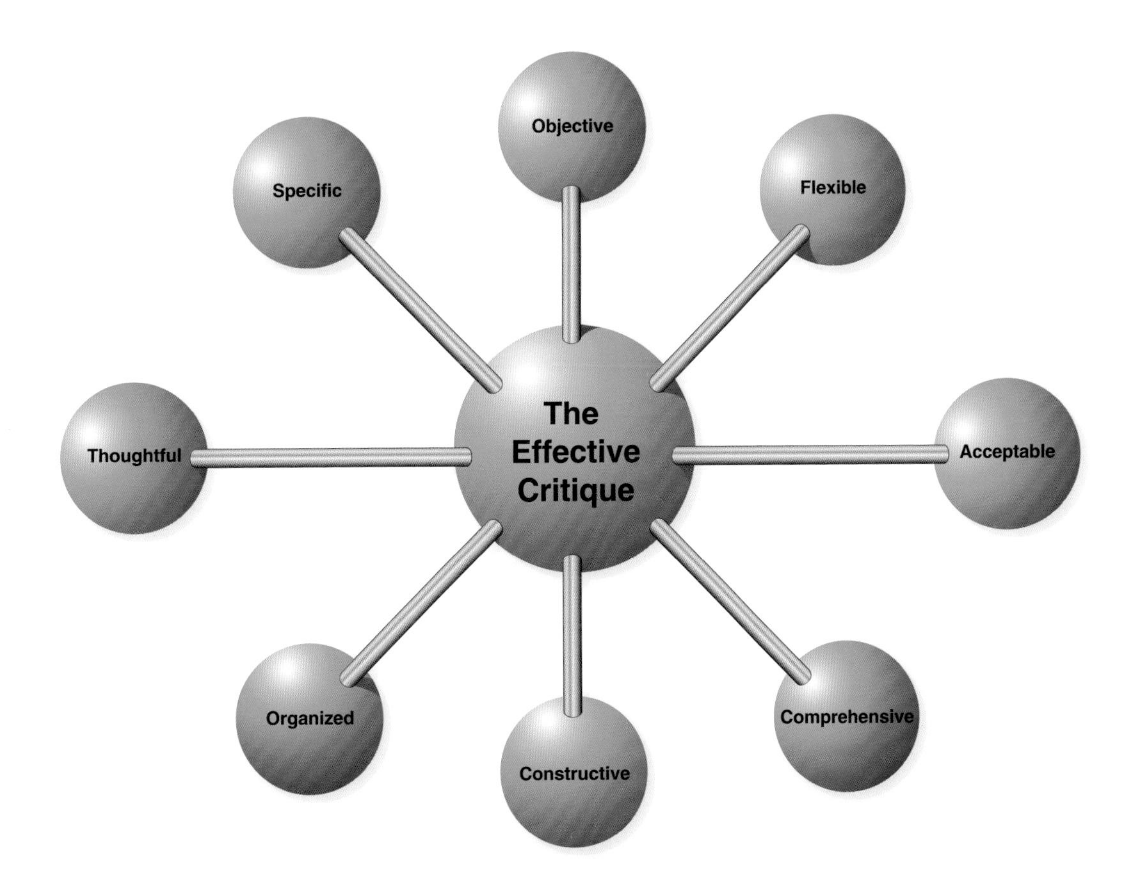

Figure 10-12. *Elements of an effective critique.*

instructor. Students must have confidence in the instructor's qualifications, teaching ability, sincerity, competence, and authority. If a critique is presented fairly, with authority, conviction, sincerity, and from a position of recognizable competence, the student probably accepts it as such. Instructors should not rely on their position to make a critique more acceptable to their students.

- Comprehensive—a comprehensive critique is not necessarily a long one, nor must it treat every aspect of the performance in detail. The instructor must decide whether the greater benefit comes from a discussion of a few major points or a number of minor points. The instructor might critique what most needs improvement, or only what the student can reasonably be expected to improve. An effective critique covers strengths as well as weaknesses.
- Constructive—a critique is pointless unless the student profits from it. The instructor should give positive guidance for correcting the fault and strengthening the weakness. Negative criticism that does not point toward improvement or a higher level of performance should be omitted from a critique altogether.
- Organized—unless a critique follows some pattern of organization, a series of otherwise valid comments may lose their impact. Almost any pattern is acceptable as long as it is logical and makes sense to the student as well as to the instructor. An effective organizational pattern might be the sequence of the performance itself. Sometimes a critique can profitably begin at the point where a demonstration failed and work backward through the steps that led to the failure.
- Thoughtful—an effective critique reflects the instructor's thoughtfulness toward the student's need for self-esteem, recognition, and approval from others. The instructor should never minimize the inherent dignity and importance of the individual. Ridicule, anger, or fun at the expense of the student have no place in a critique. While being straightforward and honest, the instructor should always respect the student's personal feelings.
- Specific—the instructor's comments and recommendations should be specific, rather than general. The student needs to focus on something concrete. If the instructor has a clear, well-founded, and supportable idea in mind, it should be expressed with firmness and authority in terms that cannot be misunderstood.

Evaluation

Whenever learning takes place, the result is a definable, observable, measurable change in behavior. The purpose of an evaluation is to determine how a student is progressing in the course. Evaluation is concerned with defining, observing, and measuring or judging this new behavior. Evaluation normally occurs before, during, and after instruction; it is an integral part of the learning process. During instruction, some sort of evaluation is essential to determine what the student is learning and how well they are learning it. The instructor's evaluation may be the result of observations of the students' overall performance, or it may be accomplished as either a spontaneous or planned evaluation, such as an oral quiz, written test, or skill performance test. *[Figure 10-13]*

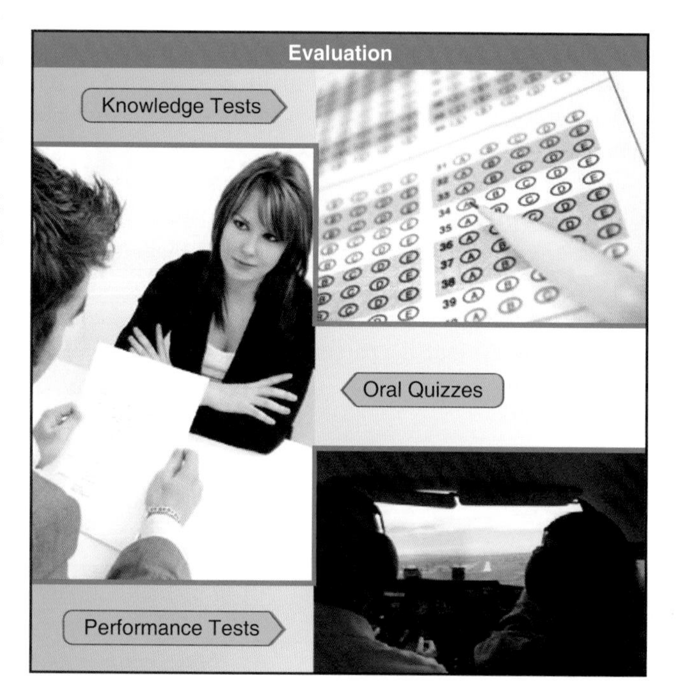

Figure 10-13. *There are three common types of evaluations that instructors may use.*

Oral Quizzes

The most used means of evaluation is the direct or indirect oral questioning of the student by the instructor. Questions may be loosely classified as fact questions and thought questions. The answer to a fact question is based on memory or recall. This type of question usually concerns who, what, when, and where. Thought questions usually involve why or how, and require the student to combine knowledge of facts with an ability to analyze situations, solve problems, and arrive at conclusions. Proper quizzing by the instructor can have a number of desirable results.

Characteristics of Effective Questions

An effective oral quiz requires some preparation. The instructor should devise and write pertinent questions in advance. One method is to place them in the lesson plan. Prepared questions merely serve as a framework, and as the lesson progresses, should be supplemented by such impromptu questions as the instructor considers appropriate. To be effective, questions must apply to the subject of instruction. Unless the question pertains strictly to the particular training being conducted, it serves only to confuse the students and divert their thoughts to an unrelated subject. An effective question should be brief and concise, but also clear and definite.

Effective questions must be adapted to the ability, experience, and stage of training of the students. Effective questions center on only one idea. A single question should be limited to who, what, when, where, how, or why, not a combination. Effective questions must present a challenge to the students. Questions of suitable difficulty serve to stimulate learning. Effective questions demand and deserve the use of proper English.

Answering Questions from Students

Responses to student questions must also conform with certain considerations if answering is to be an effective teaching method. The question must be clearly understood by the instructor before an answer is attempted. The instructor should display interest in the student's question and frame an answer that is as direct and accurate as possible. After the instructor completes a response, it should be determined whether or not the student's request for information has been completely answered, and if the student is satisfied with the answer.

Occasionally, a student asks a question that the instructor cannot answer. In such cases, the instructor should freely admit not knowing the answer, but should promise to get the answer or, if practicable, offer to help the student look it up in available references.

Written Tests

As evaluation devices, written tests are only as good as the knowledge and proficiency of the test writer. This section is intended to provide the flight instructor with only the basic concepts of written test design.

Characteristics of a Good Test

A test is a set of questions, problems, or exercises for determining whether a person has a particular knowledge or skill. A test can consist of just one test item, but it usually consists of a number of test items. A test item measures a single objective and calls for a single response. The test could be as simple as the correct answer to an essay question or as complex as completing a knowledge or practical test. Regardless of the underlying purpose, effective tests share certain characteristics. For a full list of these characteristics, refer to the Aviation Instructor's Handbook.

Test Development

When testing aviation students, the instructor is usually concerned more with criterion-referenced testing than norm-referenced testing. Criterion-referenced testing evaluates each student's performance against a carefully written, measurable, standard or criterion. Norm-referenced testing measures a student's performance against the performance of other students. There is little or no concern about the student's performance in relation to the performance of other students. The FAA knowledge and practical tests for pilots are all criterion referenced because in aviation training it is necessary to measure student performance against a high standard of proficiency consistent with safety.

Presolo Knowledge Tests

Title 14 of the Code of Federal Regulations (14 CFR) part 61 requires the satisfactory completion of a presolo knowledge test prior to solo flight. The presolo knowledge test is required to be administered, graded, and all incorrect answers reviewed by the instructor providing the training prior to endorsing the student pilot certificate and logbook for solo flight. The regulation states that the presolo knowledge test must include questions applicable to 14 CFR parts 61 and 91 and on the flight characteristics and operational limitations of the make and model aircraft to be flown.

The content and number of test questions are to be determined by the flight instructor. An adequate sampling of the general operating rules should be included. In addition, a sufficient number of specific questions should be asked to ensure the student has the knowledge to safely operate the aircraft in the local environment.

Specific procedures for developing test questions are covered in the Aviation Instructor's Handbook, but a review of some items as they apply to the presolo knowledge test are in order. Though selection-type (usually referred to as multiple-choice) test items are easier to grade, it is recommended that supply-type (or "fill in the blank") test items be used for the portions of the presolo knowledge test where specific knowledge is to be tested. One problem with supply-type test items is difficulty in assigning the appropriate grade.

Since solo flight requires a thorough working knowledge of the different conditions likely to be encountered on the solo flight, it is important that the test properly evaluate this area. In this way, the instructor can see any areas that are not adequately understood and can then cover them in the review of the test. Selection-type test items do not allow the instructor to evaluate the student's knowledge beyond the immediate scope of the test items. The supply-type test item measures much more adequately the knowledge of the student, and lends itself very well to presolo testing.

The instructor must keep a record of the test results for at least three (3) years, as required by the provisions of 14 CFR section 61.189. The record should at least include the date, name of the student, and the results of the test.

Performance Tests

The flight instructor does not administer the practical test for a pilot certificate. Flight instructors do get involved with the same skill or performance testing that is measured in these tests. Performance testing is desirable for evaluating training that involves an operation, a procedure, or a process. The job of the instructor is to prepare the student to take these tests. Therefore, each element of the practical test has been evaluated prior to an applicant taking the practical exam.

The purpose of the practical test standards (PTS) is to delineate the standards by which FAA inspectors and designated pilot examiners conduct tests for ratings and certificates. The standards are in accordance with the requirements of 14 CFR parts 61 and 91 and other FAA publications including the Aeronautical Information Manual and pertinent advisory circulars and handbooks. The objective of the PTS is to ensure the certification of pilots at a high level of performance and proficiency, consistent with safety.

Since every task in the PTS may be covered on the check ride, the instructor must evaluate all of the tasks before certifying the applicant to take the practical test. While this evaluation is not totally formal in nature, it should adhere to criterion-referenced testing. Although the instructor should always train the student to the very highest level possible, the evaluation of the student is only in relation to the standards listed in the PTS. The instructor, and the examiner, should also keep in mind that the standards are set at a level that is already very high. They are not minimum standards and they do not represent a floor of acceptability. In other words, the standards are the acceptable level that must be met and there are no requirements to exceed them.

Planning Instructional Activities

Any instructional activity must be well planned and organized if it is to proceed in an effective manner. Much of the basic planning necessary for the flight instructor is provided by the knowledge and proficiency requirements published in 14 CFR, approved school syllabi, and the various texts, manuals, and training courses available. This section reviews the planning required by the instructor as it relates to four key topics—course of training, blocks of learning, training syllabus, and lesson plans.

Course of Training

In education, a course of training may be defined as a complete series of studies leading to attainment of a specific goal. The goal might be a certificate of completion, graduation, or an academic degree. For example, a student pilot may enroll in a private pilot certificate course, and upon completion of all course requirements, be awarded a graduation certificate.

Other terms closely associated with a course of training include curriculum, syllabus, and training course outline. In many cases, these terms are used interchangeably, but there are important differences.

A curriculum may be defined as a set of courses in an area of specialization offered by an educational institution. A curriculum for a pilot school usually includes courses for the various pilot certificates and ratings. A syllabus is a brief or general summary or outline of a course of study. In aviation, the term "training syllabus" is commonly used. In this context, a training syllabus is a step-by-step, building block progression of learning with provisions for regular review and evaluations at prescribed stages of learning. The syllabus defines the unit of training, states by objective what the student is expected to accomplish during the unit of training, shows an organized plan for instruction, and dictates the evaluation process for either the unit or stages of learning. And, finally, a training course outline, within a curriculum, may be described as the specific content of a particular course. It normally includes statements of objectives, descriptions of teaching aids, definitions of evaluating criteria, and indications of desired outcome.

Objectives and Standards

The overall objective of an aviation training course is usually well established, and the general standards are included in various rules and related publications. For example, eligibility, knowledge, proficiency, and experience requirements for pilots and maintenance students are stipulated in the regulations, and the standards are published in the applicable practical test standards (PTS). It should be noted, though, that the PTS standards are limited to the most critical performance tasks. Certification tests do not represent an entire training syllabus.

Blocks of Learning

After the overall training objectives have been established, the next step is the identification of the blocks of learning *[Figure 10-14]* which constitute the necessary parts of the total objective. Just as in building a pyramid, some blocks are

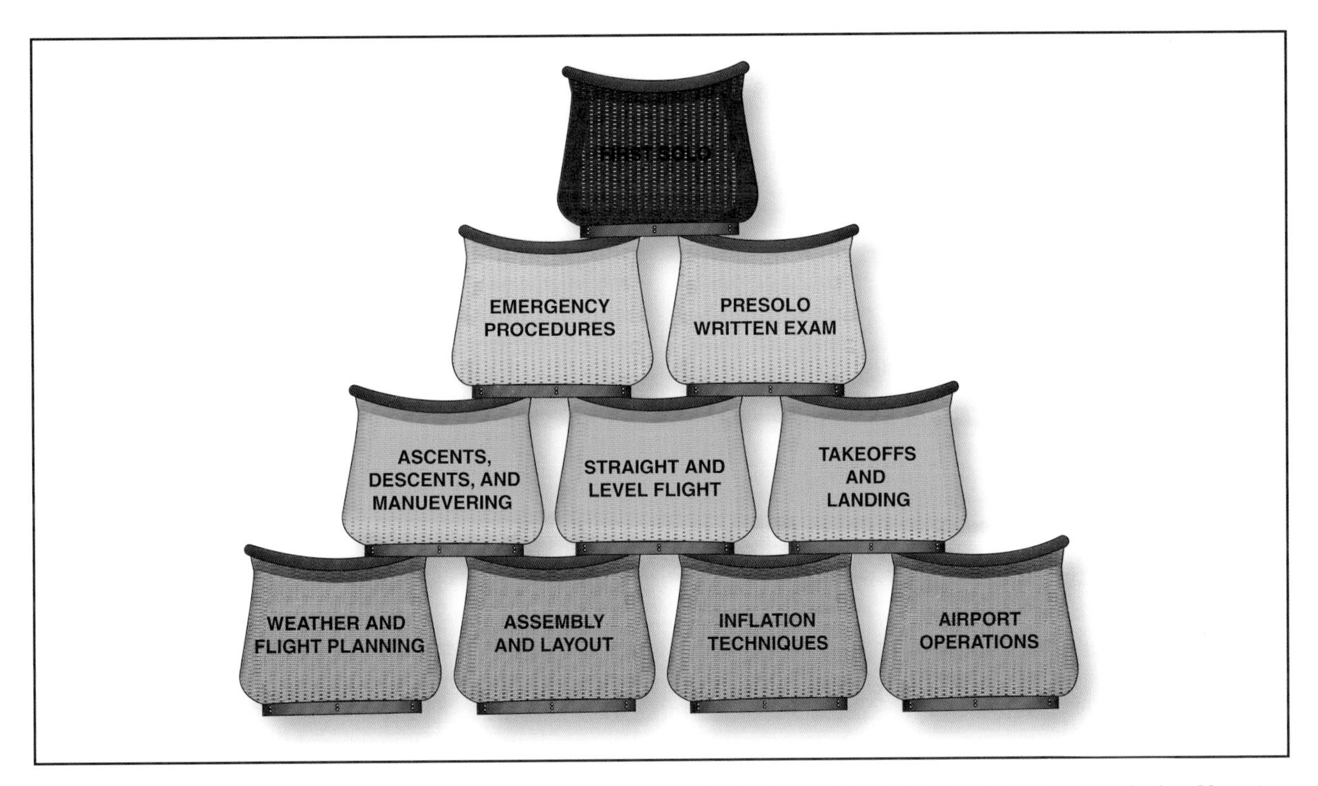

Figure 10-14. *The presolo stage, or phase, of private training is comprised of several basic building blocks. These blocks of learning, which should include coordinated ground and flight training, lead up to the first solo.*

submerged in the structure and never appear on the surface, but each is an integral and necessary part of the structure. During the process of identifying the blocks of learning to be assembled for the proposed training activity, the planner must also examine each carefully to see that it is truly an integral part of the structure.

While determining the overall training objectives is a necessary first step in the planning process, early identification of the foundation blocks of learning is also essential. Training for any such complicated and involved task as piloting or maintaining an aircraft requires the development and assembly of many segments or blocks of learning in their proper relationships. In this way, a student can master the segments or blocks individually and can progressively combine these with other related segments until their sum meets the overall training objectives.

Training Syllabus

There are a number of valid reasons why all flight instructors should use a training syllabus. As technology advances, training requirements become more demanding. At the same time, new and often more complicated rules continue to be proposed and implemented. In addition, the rules for instruction in other than an approved flight school are still quite specific about the type and duration of training. These factors, along with the continuing growth of aviation, add to the complexity of aviation training and certification. Instructors need a practical guide to help them make sure the training is accomplished in a logical sequence and that all of the requirements are completed and properly documented. A well organized, comprehensive syllabus can fulfill these needs.

Syllabus Format and Content

The format and organization of the syllabus may vary, but it always should be in the form of an abstract or digest of the course of training. It should contain blocks of learning to be completed in the most efficient order.

Since a syllabus is intended to be a summary of a course of training, it should be fairly brief, yet comprehensive enough to cover essential information. This information is usually presented in an outline format with lesson-by-lesson coverage. Some syllabi include tables to show recommended training time for each lesson, as well as the overall minimum time requirements.

Since effective training relies on organized blocks of learning, all syllabi should stress well-defined objectives and standards for each lesson. Appropriate objectives and standards should be established for the overall course, the separate ground and flight segments, and for each stage of training. Other details may be added to a syllabus in order to explain how to use it and describe the pertinent training and reference materials. Examples of the training and reference materials include textbooks, video, compact disks, exams, briefings, and instructional guides.

How To Use a Training Syllabus

Any practical training syllabus must be flexible, and should be used primarily as a guide. When necessary, the order of training can and should be altered to suit the progress of the student and the demands of special circumstances. For example, previous experience or different rates of learning often require some alteration or repetition to fit individual students. The syllabus also should be flexible enough so it can be adapted to weather variations and scheduling changes without disrupting the teaching process or completely suspending training.

Effective use of a syllabus requires that it be referred to throughout the entire course of training. Both the instructor and the student should have a copy of the approved syllabus. However, as previously mentioned, a syllabus should not be adhered to so stringently that it becomes inflexible or unchangeable. It must be flexible enough to adapt to special needs of individual students.

A syllabus lesson may include several other items that add to or clarify the objective, content, or standards. A lesson may specify the recommended class time, reference or study materials, recommended sequence of training, and study assignment for the next lesson. Both ground and flight lessons may have explanatory information notes added to specific lessons.

While a syllabus is designed to provide a road map showing how to accomplish the overall objective of a course of training, it may be useful for other purposes. As already mentioned, it can be used as a checklist to ensure that required training has successfully been completed. Thus, a syllabus can be an effective tool for recordkeeping. Enhanced syllabi, which also are designed for recordkeeping, can be very beneficial to the independent instructor.

A recordkeeping function may be facilitated by boxes or blank spaces adjacent to the knowledge areas, procedures, or maneuvers in a flight lesson, much as the syllabus designed by the Balloon Federation of America does. Most syllabi introduce each procedure or maneuver in one flight lesson and review them in subsequent lessons. Some syllabi also include provisions for grading student performance and recording both ground and flight training time. Accurate recordkeeping is necessary to keep both the student and the instructor informed on the status of training. These records also serve as a basis for endorsements and recommendations for knowledge and practical tests.

Another benefit of using a syllabus is that it helps in development of lesson plans. A well constructed syllabus already contains much of the essential information that is required in a lesson plan, including objectives, content, and completion standards.

Lesson Plans

A lesson plan is an organized outline for a single instructional period. *[Figure 10-15]* It is a necessary guide for the instructor in that it tells what to do, in what order to do it, and what procedure to use in teaching the material of a lesson. Lesson plans should be prepared for each training period and be developed to show specific knowledge and/or skills to be taught.

A mental outline of a lesson is not a lesson plan. A lesson plan should be put into writing. Another instructor should be able to take the lesson plan and know what to do in conducting the same period of instruction. When putting it in writing, the lesson plan can be analyzed from the standpoint of adequacy and completeness.

Lesson plans serve many purposes, such as:

- Assure a wise selection of material and the elimination of unimportant details.
- Make certain that due consideration is given to each part of the lesson.
- Aid the instructor in presenting the material in a suitable sequence for efficient learning.
- Provide an outline of the teaching procedure to be used.
- Serve as a means of relating the lesson to the objectives of the course of training.
- Give the inexperienced instructor confidence.
- Promote uniformity of instruction regardless of the instructor or the date on which the lesson is given.

Characteristics of a Well-Planned Lesson

The quality of planning affects the quality of results. Successful professionals understand the price of excellence is hard work and thorough preparation. The effective instructor realizes that the time and energy spent in planning and preparing each lesson is well worth the effort in the long run. *[Figure 10-16]*

A complete cycle of planning usually includes several steps. After the objective is determined, the instructor must research the subject as it is defined by the objective. Once the research is complete, the instructor must determine the

BFA Preflight Training Lesson Plan	
Lesson 1	**Flight Planning**
Objective:	To develop the student's skill in planning a flight
Elements:	• Observe wind and weather conditions from the weather channel, radio, and visually by use of pibal or other autolite • Call flight service station for weather briefing • Select launch site
Equipment:	• Weather information form • Sectional chart and road map • Plotter • Compass • Pibal
Instructor's Action:	• Discuss lesson • Obtain weather information • Observe surface winds with pibal • Determine launch site • Locate launch site on chart • Draw dead reckoning line in direction of flight and mark off distance or time on the line • Suggest landmarks to verify position in flight • Airspace considerations • Critique plan
Student's Action:	• Obtain weather briefing from FSS (1-800-WX-BRIEF) • Recommend launch site based on accurate weather information recorded on form • Draw flight plan approved by instructor
Completion Standards:	• Student should demonstrate ability to obtain complete weather briefing from the FAA Flight Service Station and correlate this information with observed weather conditions • Locate launch site on chart • Drew proposed course with time ticks on dead reckoning line • Understands need for flexibility to adjust flight plan • Selects landmarks in flight to verify position

Figure 10-15. *This ground lesson example shows a unit of ground instruction. In this example, neither the time nor the number of ground training periods to be devoted to the lesson is specified. The lesson should include three key parts—objective, content, and completion standards.*

method of instruction and identify a useful lesson planning format. Other steps, such as deciding how to organize the lesson and selecting suitable support material also must be accomplished. The final steps include assembling training aids and writing the lesson plan outline. One technique for writing the lesson plan outline is to prepare the beginning and ending first. Then, complete the outline and make revisions as necessary. The following are some of the important characteristics that should be reflected in all well-planned lessons.

- Unity—each lesson should be a unified segment of instruction. A lesson is concerned with certain limited objectives, which are stated in terms of desired student learning outcomes. All teaching procedures and materials should be selected to attain these objectives.
- Content—each lesson should contain new material. However, the new facts, principles, procedures, or skills should be related to the lesson previously presented. A short review of earlier lessons is usually necessary, particularly in flight training.
- Scope—each lesson should be reasonable in scope. A person can master only a few principles or skills at a time, the number depending on complexity. Presenting too much material in a lesson results in confusion; presenting too little material results in inefficiency.
- Practicality—each lesson should be planned in terms of the conditions under which the training is to be conducted. Lesson plans for training conducted in a balloon differ from those conducted in a classroom. Also, the kinds and quantities of instructional aids available have a great influence on lesson planning and instructional procedures.
- Flexibility—although the lesson plan provides an outline and sequence for the training to be conducted, a degree of flexibility should be incorporated. For example, the outline of content may include blank spaces for add-on material, if required.
- Relation to course of training—each lesson should be planned and taught so that its relation to the course objectives are clear to each student. For example, a lesson on short-field takeoffs and landings should be related to both the certification and safety objectives of the course of training.
- Instructional Steps—every lesson, when adequately developed, lends itself to the four steps of the teaching process: preparation, presentation, application, and review and evaluation.

How To Use a Lesson Plan Properly

Be Familiar With the Lesson Plan

The instructor should study each step of the plan and should be thoroughly familiar with as much information related to the subject as possible.

Use the Lesson Plan as a Guide

The lesson plan is an outline for conducting an instructional period. It assures that pertinent materials are at hand and

BFA Inflight Training Lesson Plan

Unit 1 **Introductory Dual Flight**

Objective: To introduce the student to balloon flight

Elements:
- Flight planning
- Balloon layout and inflation
- Fundamentals of level flight
- Level flight in relation to ground
- Ascents and descents
- Normal approach to land
- Postflight discussion

Equipment:
- Balloon Flight Manual
- Airworthy balloon

Instructor's Action:
- Before flight, discuss lesson objective by using preflight lesson plans for Flight Planning, Use of Blast Valve, Layout and Inflation, and Fundamentals of Level Flight
- Demonstrate all elements of planning, layout, inflation and flight allowing student to perform after each demonstration and coach student practice
- Postflight critique of student performance and make study assignment

Student's Action:
- Student should observe instructor's demonstrations and practice activity with coaching from instructor

Flight planning	D/E	1	2	3
a. Weather briefing—FSS	___	❑	❑	❑
b. Observe surface winds-pibal	___	❑	❑	❑
c. Select launch site—landowner permission	___	❑	❑	❑
d. Plot flight path	___	❑	❑	❑
e. Weight and pressure	___	❑	❑	❑
Blast valve operation				
Layout and inflation				
a. Comprehensive checklist	___	❑	❑	❑
b. Crew briefing	___	❑	❑	❑
c. Layout direction	___	❑	❑	❑
d. Rig basket and install burner	___	❑	❑	❑
e. Fuel system check	___	❑	❑	❑
f. Fuel quantity	___	❑	❑	❑
g. Inflation restraint	___	❑	❑	❑
h. Instruments—calibrate to elevation or pressure	___	❑	❑	❑
i. Rig envelope	___	❑	❑	❑
j. Cold inflation—fan operation	___	❑	❑	❑
k. Walk around inspection	___	❑	❑	❑
l. Recheck weather and weight	___	❑	❑	❑
m. Checklist—hot inflation	___	❑	❑	❑
n. Hot inflation	___	❑	❑	❑
Launch				
a. Equilibrium-launch checklist	___	❑	❑	❑
b. Weight-off	___	❑	❑	❑
c. Controlled ascent	___	❑	❑	❑
Fundamentals of flight				
a. Recognizes vertical direction	___	❑	❑	❑
b. Level flight—relation to ground	___	❑	❑	❑
c. Level flight—with instruments	___	❑	❑	❑
d. Ascents and descents with instruments	___	❑	❑	❑
e. Use of wings at various altitudes to steer	___	❑	❑	❑
f. Navigation	___	❑	❑	❑
g. Normal approach and landing using checklist	___	❑	❑	❑
Deflation				
a. Landowner relations	___	❑	❑	❑
b. Supervise crew	___	❑	❑	❑
Refueling	___	❑	❑	❑
Postflight discussion including use of the comprehensive	___	❑	❑	❑
Completion of pilot and aircraft logs	___	❑	❑	❑
____________	___	❑	❑	❑
____________	___	❑	❑	❑

Completion Standards: Student should have a general understanding of a balloon flight and be aware of standards required to be a competent pilot

Notes: ____________________

Key: Instructor's action: D/E - D = demonstrated
E = explained or discussed
Student's action: 1 = no assistance required
2 = some assistance required
3 = need help

Student ____________ **Date** ____________

Instructor ____________ **Date** ____________

Balloon make/model ____________ **N#** ____________

Flight time ____________

Figure 10-16. *An example of an inflight lesson plan for balloon training.*

that the presentation is accomplished with order and unity. Having a plan prevents the instructor from getting off the track, omitting essential points, and introducing irrelevant material. Students have a right to expect an instructor to give the same attention to teaching that they give to learning. The most certain means of achieving teaching success is to have a carefully thought-out lesson plan.

Adapt the Lesson Plan to the Class or Student

In teaching a ground school period, the instructor may find that the procedures outlined in the lesson plan are not leading to the desired results. In this situation, the instructor should change the approach. There is no certain way of predicting the reactions of different groups of students. An approach that has been successful with one group may not be equally successful with another.

A lesson plan for an instructional flight period should be appropriate to the background, flight experience, and ability of the particular student. A lesson plan may have to be modified considerably during flight, due to deficiencies in the student's knowledge or poor mastery of elements essential to the effective completion of the lesson. In some cases, the entire lesson plan may have to be abandoned in favor of review.

Revise the Lesson Plan Periodically

After a lesson plan has been prepared for a training period, a continuous revision may be necessary. This is true for a number of reasons, including availability or non-availability of instructional aids, changes in regulations, new manuals and textbooks, and changes in the state-of-the art among others.

Lesson Plan Formats

The format and style of a lesson plan depends on several factors. Certainly the subject matter has a lot to do with how a lesson is presented and what teaching method is used. Individual lesson plans may be quite simple for one-on-one training, or they may be elaborate and complicated for large, structured classroom lessons. Preferably, each lesson should have somewhat limited objectives that are achievable within a reasonable period of time. This principle should apply to both ground and flight training. As previously noted, aviation training is not simple.

In spite of need for varied subject coverage, diverse teaching methods, and relatively high level learning objectives, most aviation lesson plans have the common characteristics already discussed. They all should include objectives, content to support the objectives, and completion standards. Various authorities often divide the main headings into several subheadings, and terminology, even for the main headings, varies extensively. For example, completion standards may be called assessment, review and feedback, performance evaluation, or some other related term.

Chapter Summary

It is an unfortunate truth that many commercial balloon pilots are not as qualified and prepared for their instructional responsibility as they should be. All commercial pilots are required to study the necessary principles of instruction in order to pass their knowledge test, and must perform a lesson on a flight maneuver as a part of the practical test.

A commercial pilot who chooses to instruct needs to hold himself to the same standards as the rest of the aviation community. He should provide compete and thorough flight training, as well as insuring that proper ground training is conducted, so that the prospective pilot is well-grounded in all facets of aviation. To do less minimizes the instructor's role in the educational process, and perpetuates poor training in the ballooning community.

The information in this chapter provides a basis for an instructor's knowledge of the skills and techniques of the instructional process. It is the responsibility of each individual instructor to expand on that knowledge, and provide the best instruction possible.

Chapter 11

The Gas Balloon

Introduction

This chapter discusses topics relating specifically to manned gas ballooning. The understanding and flying of gas balloons is very similar to hot air ballooning, but there are also significant differences. This chapter generally discusses the differences and assumes that the reader is familiar with the topics in the other chapters of this handbook. Frequent comparisons to hot air ballooning are used to place discussions in a more familiar context for the hot air balloon pilot.

Much of this chapter relates to gas balloons with an envelope volume of 1,000 cubic meters (35,315 cubic feet). This is by far the most common size and is the maximum volume allowed for the major competitions. Several types of lifting gas are discussed, but the main emphasis is on helium and hydrogen. All discussions assume that the envelope is the "zero pressure" type. Zero pressure envelopes have an open appendix, or tube, at the bottom that maintains a zero pressure difference between the gas inside the envelope and the atmospheric air outside. Neither super-pressure balloons nor Rozière (a type of gas-hot air hybrid construction) balloons are discussed.

Gas ballooning was once the most common form of aviation in the United States. Today, due to the costs involved, gas flights are most frequently undertaken for training, competition, or record breaking purposes. Short, recreational gas flights, while not unheard of, are much less common than hot air.

The History of Gas Ballooning

The first gas balloon flight occurred just a few weeks after the first manned balloon flight of Jean-François Pilâtre de Rozier and the Marquis d'Arlandes. While the Montgolfiers were experimenting with hot air balloons, the Robert brothers and Jacques Charles were also experimenting with gas balloons.

The first gas balloon was small (13 feet in diameter) and was filled with what was called "inflammable air." *[Figure 11-1]* It generated 35 pounds of net lift and, when set free, remained aloft for 45 minutes and traveled 15 miles. Its final, sudden descent was attributed to a rupture in the balloon.

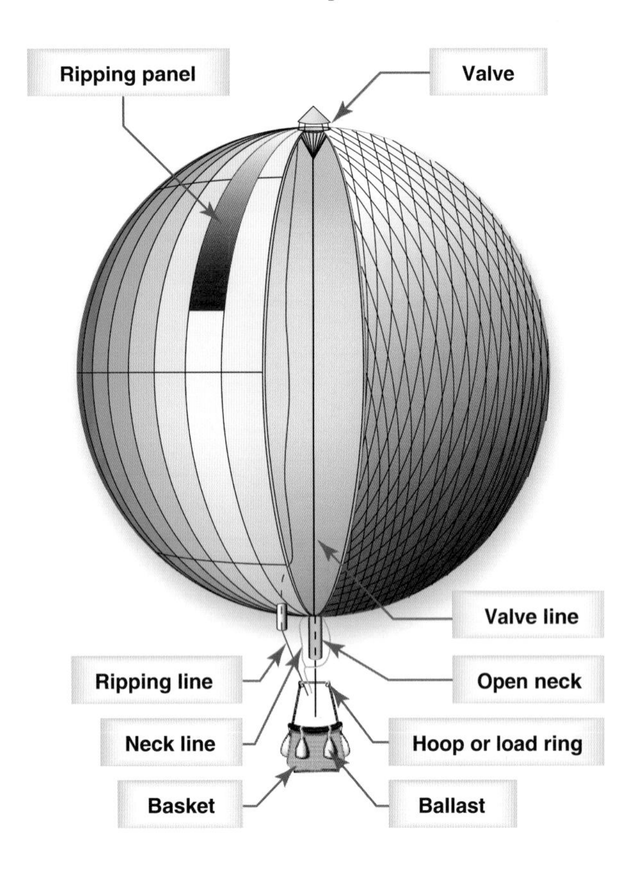

Figure 11-1. *Early style of gas balloon. Many of these terms still apply today.*

With the success of the demonstration flight, Charles next made a 25.5-foot diameter globe of rubberized silk. A net was fitted over the upper half of the balloon and tied to a loop around the middle of the bag. From this loop, a sort of car or boat was suspended. The bag was tied at the bottom to contain the gas. Dropping ballast caused the balloon to rise, and releasing gas through a valve at the top made it fall. On December 1, 1783, Professor Charles and Nicholas Robert made the first manned gas balloon flight. The first flight lasted for 1 hour and 45 minutes, rose to 1,800 feet, and covered more than 27 miles. On landing, Robert got out of the boat and Charles, now flying solo, ascended to almost 9,000 feet. The balloon had been flaccid on first landing, but filled out as it rose and gas was released to prevent it from bursting. Benjamin Franklin was present for this lift off and recognized the potential for balloons in military operations.

Other ascensions followed, mostly to gather scientific data. It was through these early experiments that the design/shape of gas balloons, lifting capacity, and weight of air were determined. One notable early flight was the crossing of the English Channel from Britain to France by Jean-Pierre Blanchard and Dr. John Jeffries on January 7, 1785. Their balloon was scarcely sufficient to carry them and they threw out their ballast, anchors, food, and clothing to complete the crossing.

In 1906, James Gordon Bennett needed news for his newspaper and having been successful in starting car and boat races, decided to initiate a gas balloon race. The first race was organized for Paris, France. Each country could send contestants to the race. This race, the Coupe Aéronautique de Gordon Bennett, is still held annually although there were years when no race occurred due to wars. Today, the race is the most prestigious gas race in the world. It is sanctioned by the Fédération Aéronautique Internationale (FAI). The country of the previous year's winner is entitled to host the race. In 2006, the 50th race was held flying out of Belgium (www.coupegordonbennett.org).

The Albuquerque International Balloon Fiesta (AIBF) started the America's Challenge Gas Balloon Race in 1995. It has been conducted yearly, with the exception of 1999 when AIBF instead hosted the Gordon Bennett Race. This has become a very prestigious and international race that offers pilots the opportunity for extremely long flights (www.balloonfiesta.com/Education/History/).

Balloon Systems

Gas balloons designs are generally classified as netted or quick fill.

Netted Balloon Systems

The netted design *[Figure 11-2]* dates from the 18th century. A spherical balloon envelope is encased in a diamond mesh net of light weight cord that runs around the balloon from the apex valve at the top of the envelope down to support points on the gondola.

Inflation is accomplished by first bringing the inflation hose to the center of a ground tarp. The envelope is next laid over the hose with the hose inserted into the appendix and with the apex of the balloon on top and centered. The net is deployed over and attached to the envelope and as the lifting gas is flowed, sandbags are hung on the diamonds of the net at ground level. As the balloon fills and forms a rising hemisphere, the crew moves the sandbags down the net to keep them at ground level. Enough bags are added to keep the balloon anchored to the ground.

After the envelope is completely filled, it is raised and the basket is brought underneath and attached to the net. The hemispherical shape is very stable in moderate to high ground winds as the flow is over the top of the shape. One drawback of netted balloons is that more crew is required to manage the sandbags during inflation.

Figure 11-2. *"Spirit of Springfield," a netted gas ballon.*

Quick Fill Balloon Systems

The quick fill design dates from the mid 20th century. A natural shape envelope is surrounded by vertical support tapes that run from the apex of the balloon down to the gondola. *[Figure 11-3]* Several horizontal tapes extend circumferentially around and are attached to the vertical tapes to stabilize the structure.

Inflation is accomplished by laying out the envelope on the ground along its full length with the apex at one end and the appendix (base) at the other. The gondola is attached to the load lines and is bagged down. As the lifting gas is flowed in through the appendix, several crew members hold the apex of the balloon down. When enough gas has entered the envelope to generate approximately 100 to 150 pounds of net lift, the apex is released, the envelope rises, and the inflation is completed.

Figure 11-3. *Quick fill design gas balloon.*

Lifting Gases

Modern gas balloons normally use helium or hydrogen as the lifting gas. Anhydrous ammonia and methane are two other less common options. Helium is a monotomic, inert gas and must be refined to reach a purity of greater than 99 percent from raw well gas at a purity of a few percent. Hydrogen typically exists in the molecular form H_2. It is flammable when combined in a mixture of 25 percent H_2 to 75 percent air. This means that this mixture supports an existing flame.

Helium is expensive but is more commonly used in the United States while hydrogen is more common in Europe. Helium provides slightly less lift than hydrogen but is the more stable gas. (See subsection titled Flying in Inversions on page 11-7 for a discussion of stability.) Balloon systems must not be prone to generating static electricity if they are to be used with hydrogen.

There are at least six factors to consider when choosing a lifting gas. These are:

1. Compatibility with the balloon system being used
2. Cost
3. Lifting capacity
4. Availability
5. Locale of flight
6. Inherent gas stability

Normally, the gas selection process starts with the type of balloon. Hydrogen can be used if the balloon system is hydrogen compatible, hydrogen is available, and local ordinances allow its use. If one of these conditions is not satisfied, as is often the case in the United States, then helium is the likely choice. If a very short training or pleasure flight of several hours is planned, the more economical alternative of ammonia or methane (natural gas) may be considered.

When inflating or landing with hydrogen or methane, care must be taken to ensure that no flame or material with the potential to generate sparks is present in the launch locale. Lighted cigarettes, cigars, nylon clothing, cell phones, and other electronic devices are examples of forbidden items. Only essential personnel should be allowed in the launch and landing areas. For long, competitive flights, the increased stability of helium is a factor in its favor.

Components of the Gas Balloon

Gas balloon systems can be broken into four parts:

1. Envelope to contain the lifting gas
2. Gondola for carrying pilots and equipment
3. Support system to connect the envelope to the gondola
4. Other equipment

Envelope

The maneuvering valve *[Figure 11-4]* is at the apex of the envelope. It allows for the controlled release of a small amount of gas to initiate a descent. It is usually spring actuated and controlled via a line that runs from the valve down through the envelope into the gondola. On some

Figure 11-4. *Maneuvering valve on quick fill gas ballon.*

balloons, a gas tight parachute top may be used instead of a valve. The envelope is also equipped with either a rip panel or deflation port for rapid, total deflation during high wind landings. More information on proper use of the valve and deflation ports is provided in the section The Practice of Gas Ballooning, page 11-8.

Gondola

The gas balloon's gondola *[Figure 11-5]* is typically somewhat larger than ones used for sport hot air ballooning, with four by five feet being a typical size. A foldable cot or sleeping pad normally runs along the long side of the gondola and a "kick-out" panel in the side wall at one end of the cot may be used to provide additional legroom for sleeping.

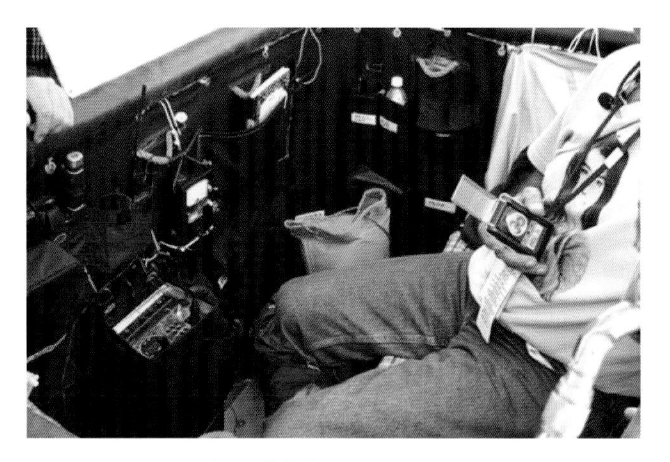

Figure 11-5. *Typical gondola layout.*

A trail rope is slung on the outside gondola along with much of the remaining support equipment. The trail rope is typically about 150 feet of natural fiber rope, one inch in diameter and weighs about 40 pounds. Its use is described in the landing paragraph.

Support Cabling

The connection between envelope and gondola may be made of rope, flat tape, or steel cable. The total strength of the cabling should have a breaking strength of at least five times the maximum gross load that is suspended on the cables. This must include sand ballast, as well as the weight of the gondola, the occupants, and all supplies and equipment. For hydrogen systems, the cabling must be electrically conductive. For ammonia systems, the cables should be long enough to separate the occupants of the gondola from gas fumes coming out of the appendix.

The bottom end of the cabling is gathered at a load ring which acts as the interface between the envelope and the gondola cables. The load ring also serves as the system's strong point for attachment of inflation harnesses or trail ropes.

Equipment

Equipment carried varies with the purpose of the flight. A minimum equipment list should include an altimeter/variometer, compass, global positioning system (GPS), aircraft position lights, several flashlights (for night flights), oxygen (for high altitude flights), aircraft radio, and aircraft sectional maps for navigation and communication. *[Figure 11-6]* Other items which may be included for safety and occupant comfort may include: a first aid kit, adequate food, water, warm clothing, and toilet facilities for both the planned flight and for the postflight period before the recovery crew arrives.

Figure 11-6. *Instruments and radios are sometimes carried in a pod such as this.*

Theory of Gas Ballooning

Complete books have been written on the theory of gas ballooning and a full discussion of the topic is not possible here. At the very least, three topics must be understood by the competent gas pilot: physics, weather aspects, and the significance of different lifting gases.

Physics of Gas Ballooning

Four main factors are instrumental in determining lift:

1. Type of lifting gas used (e.g., helium, hydrogen, or anhydrous ammonia)
2. Amount of lifting gas in the envelope (usually equal to the envelope's total capacity)
3. Outside air temperature
4. Ambient barometric pressure (which is directly related to altitude and local weather conditions).

Lift at Sea Level

The generation of buoyancy (commonly referred to as "lift") in a gas balloon is somewhat different from that of a hot air balloon. The basic principle is the same, but for different reasons.

One cubic meter of air weighs 2.702 pounds at sea level under International Standard Atmosphere (ISA) conditions (29.92 inches of mercury ("Hg) and 59 degrees Fahrenheit (°F) at sea level). At the same time, and also under ISA conditions, a cubic meter of helium weighs 0.3729 pounds. The difference between these two numbers, 2.329 pounds, is the gross lift of a gas balloon with a volume of one cubic meter. To determine the gross lift under ISA conditions, it then becomes a simple multiplication of 2.329 pounds times the volume of the envelope. For the standard 1,000 cubic meter gas balloon, the gross lift is 2,329 pounds.

The above calculation is valid for helium. If, however, hydrogen is used as the lifting gas, the factor is 0.189 pounds per cubic meter; anhydrous ammonia's weight is 1.583 pounds per cubic meter. As compared to helium, hydrogen has 8 percent more gross lift per cubic meter, while ammonia has approximately 50 percent less lift than helium under ISA conditions.

Lift at Altitude

As a balloon ascends, it is generally true that temperature, atmospheric pressure, and gross lift all decrease. Gross lift decreases as pressure decreases, but increases as temperature decreases. Thus, as a gas balloon rises in the atmosphere, the decreasing pressure and temperature oppose each other. The decreasing temperature increases lift while the decreasing pressure decreases lift. Atmospheric pressure changes are more significant than temperature changes. Thus, net lift decreases as altitude increases in a standard atmosphere.

When calculating the effect of changing pressure and temperature, it is necessary to multiply the sea level lift by the ratio of pressures and temperatures. For a nonstandard ambient pressure, multiply the lift at the ISA level either by 29.92 "Hg or 1,013.25 millibars (mb).

Temperatures at altitude must also be calculated and compensated for. The factor for temperature is a ratio of absolute temperatures expressed in either degrees Kelvin or Rankine. To get temperature in degrees Rankine, simply add 459 to the normal Fahrenheit temperature. For a new temperature, multiply the lift calculated at the ISA by the factor: (59 °F+ 459)/(new temperature + 459). When using temperature in degrees Centigrade (°C), add 273 to convert to absolute temperature (i.e., Kelvin). This is: (15 °C + 273)/(new temperature + 273). Various lift factors at differing altitudes, comparing helium and hydrogen, are illustrated in Appendix F.

Pressure Ceiling

The pressure ceiling is the altitude at which the lifting gas inside the envelope would expand to just completely fill the envelope, assuming the balloon rose to that altitude. Rising above the pressure ceiling causes lifting gas to be expelled from the appendix and establishes a new, higher pressure ceiling. Exceeding the current pressure ceiling causes loss of lifting gas, reduces gross lift, and typically causes the balloon to eventually begin to descend. Ballast must be expended to maintain the new higher altitude. However, maneuvers that result in altitude changes below the pressure ceiling, do not result in loss of lifting gas or gross lift. Very little ballast is required to ascend while below the pressure ceiling. For these reasons, the gas pilot should always be aware of what the approximate current pressure ceiling is and should consider the consequences of penetrating that ceiling.

The following approximations generally apply to a balloon below 18,000 mean seal level (MSL).

1. For a 1,000 cubic meter balloon at its pressure ceiling, an ambient pressure decrease of 1 "Hg causes a decrease in gross lift of about 80 pounds.
2. For a 1,000 cubic meter balloon at its pressure ceiling, an ambient and gas temperature decrease of 3.3 ºF causes a lift increase of about 16 pounds.
3. For a 1,000 cubic meter balloon at its pressure ceiling, a discharge of about 64 pounds of ballast results in approximately a 1,000 foot increase in altitude.

Additional Factors That Affect Lift

1. A balloon flying below its pressure ceiling (i.e., a flaccid balloon) responds differently from one flying at its pressure ceiling.
2. When the lifting gas inside the balloon is warmer (i.e., super heating) than the ambient air, additional lift is generated. The reverse happens when the lifting gas is colder than the ambient air.
3. Nonstandard atmospheric conditions, such as inversions, affect a balloon's stability.
4. The atmospheric humidity has a small effect on lift with more humidity resulting in slightly less lift.
5. The purity of the lifting gas directly affects lift. Most commercially produced gas is assumed to be greater than 99 percent pure, but purity can be reduced as a result of improper filling technique.

For further discussion of gas balloon calculations, the book *A Short Course on the Theory and Operation of the Free Balloon*, by C. H. Roth, Goodyear Tire and Rubber Company, is recommended reading. This manual provides a good overview of the physics and operation of gas ballooning as of 1917. It is long out of print, but photocopies are readily available.

Weather Considerations for Gas Ballooning

When studying weather for gas ballooning, one must look for trends both further into the future and higher above the ground. The best weather for any flight is determined by the flight's objectives. A flight to set a duration record (maximum time aloft) benefits from light winds and clear skies while a distance competition requires high winds aloft with lighter winds in the landing zone. A competitor in a long competition is likely to encounter several different weather patterns during flight simply due to the length of the flight. Examples of these include precipitation, snow, icing, thunderstorms, lightning, high winds, mountain winds, unstable air, or convective currents.

This discussion again focuses on the most common type of competitive flight, a Gordon Bennett type, with the objective to maximize great circle distance covered. Since winning distances can be well over a thousand miles at altitudes of up to 18,000 feet MSL with times aloft possibly exceeding seventy hours, a much larger area of the weather map must be studied than for a typical hot air flight.

Meteorological Differences From Hot Air Ballooning

In contrast to hot air flights, landing conditions are most likely to be different from those at launch and several weather patterns may be encountered during the flight. Freezing levels and the moisture content of the air should be checked to predict the possibility of icing. Any icing that occurs has multiple negative impacts on the flight. It adds weight to the balloon and can interfere with the functioning of the valve by preventing it from either opening properly or from sealing tightly after activation. The weight of the ice should initiate a natural descent to a lower altitude and warmer temperatures. If this does not happen naturally, the pilot may initiate a descent by valving. Be aware that the melting ice at lower altitudes lightens the system and additional valving is necessary to prevent a second ascend back above the freezing level. Falling ice shards have also been blamed for equipment damage on occasional flights. A second strategy against icing is to seek drier air above the saturated air but the higher altitude is unlikely to provide the warmer temperatures needed to melt the accumulated ice.

Encountering thunderstorm activity is much more common in gas versus hot air flights. This is especially dangerous when flying an explosive gas such as hydrogen. Isolated thunderstorms can develop due to solar heating of moist air in the afternoon over the great plains of the United States with very little warning. Prediction and avoidance are the best tools against thunderstorms. Real time contact between the pilots and a ground-based meteorologist with access to forecasting tools to predict the formation of thunderstorms can help to avoid these storms. During the different stages of a thunderstorm, rapid accelerations due to inflows or outflows, as well as rapid ascents or descents, are very likely. The presence of any of these effects, in conjunction with any thunder or lightning are indications that an immediate landing is prudent.

Meteorological Flight Planning

Flight planning for a gas balloon flight starts several days prior to the planned flight. The gas balloon pilot examines numerous meteorological tools, looking at frontal movement to get the big picture, wind speed and direction forecasts at several altitudes, and future times to predict a flight path. Then, they will study forecast precipitation probabilities and freezing levels over the flight period. Modern trajectory predictors, such as the HYSPLIT (www.arl.noaa.gov/ready.html) program, maintained by National Oceanic and Atmospheric Administration (NOAA) can be a great help. However, the HYSPLIT model gives no forecast for precipitation along the route, so the possibility of rain, snow, or icing must be assessed using other weather models. Often, ideal weather is found just after a frontal passage, after any moisture has cleared but while the air mass is still moving with the departing front.

For serious competitions, a professional meteorologist is an invaluable team member and is consulted before and during the flight. A good starting point for weather investigations is NOAA's aviation weather web site www.aviationweather.gov.

Flying in Inversions

Proper utilization of atmospheric temperature inversions during gas balloon flights can result in increased flight stability and ballast conservation. *Figure 11-7* shows a simplified atmospheric temperature lapse chart when no inversion is present. *Figure 11-8* is similar but with an inversion present. In both charts, altitude increases up the vertical scale and temperature increases going to the right on the horizontal scale. *Figure 11-7* shows temperature decreasing consistently with altitude, while *Figure 11-8* has an inversion zone (inside red circle) from 2,000 to 4,000 feet

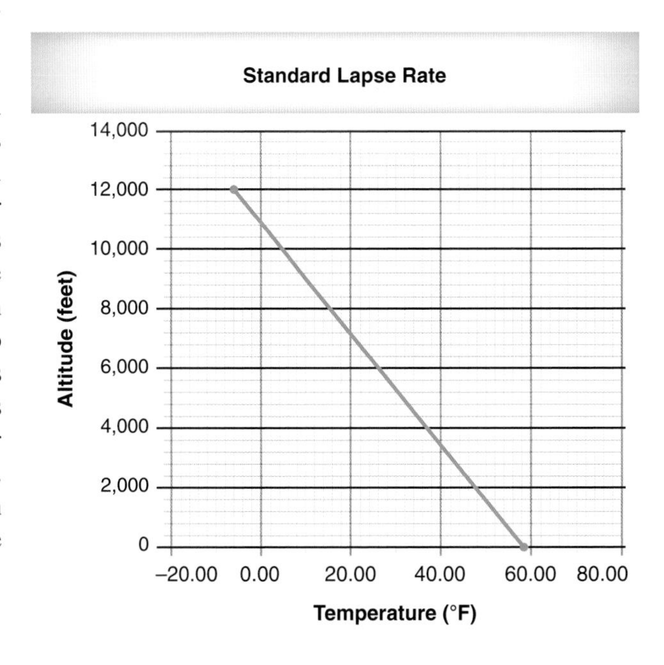

Figure 11-7. *Standard temperature lapse chart.*

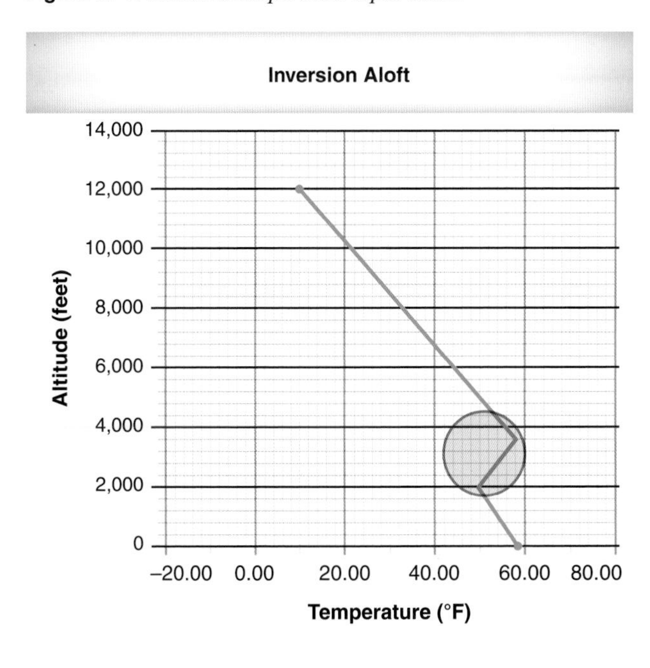

Figure 11-8. *Atmosphere exhibiting an inversion.*

in which the temperature increases with increasing altitude. A gas balloon flying in this inversion has the advantage of increased stability as compared to the altitudes above or below the inversion which exhibit normal lapse rates.

To explain this, it is important to understand the term "stability" with respect to a gas balloon. Stability can be imagined as an invisible hand that gently pulls the balloon down whenever it starts to rise or alternately pushes the balloon back up as it starts to fall. A balloon flying in stable weather tends to fly level with very little intervention from the pilot. However, stability is a weak condition and can be overcome by many factors, such as gain or loss of solar heating, orographic winds, and ballasting or valving.

To understand why an inversion creates stable flying conditions, think of a gas balloon flying at 3,000 feet in the middle of the inversion. If, for some reason, the balloon starts to ascend, two things happen. First, the balloon enters warmer ambient air. Second, the lifting gas inside the balloon expands and cools adiabatically as it reaches the slightly lower pressure atmosphere of the higher altitude. Both of these effects cause the balloon to lose lift and to descend. It is helpful to remember that, as a hot air balloon either cools or enters hotter air, the temperature differential between outside and inside air decreases and lift is lost. The same principle applies here, as applied to gas balloons. Subsequently, if the gas balloon descends from its starting point, it encounters cooler outside ambient air and its interior gas compresses and warms adiabatically, therefore gaining and ascending back to its original altitude. Remember, in an inversion, either motion (ascending or descending) tends to cause an opposing force to passively initiate a return to the original altitude.

With proper planning, the gas balloon pilot can take advantage of this scenario. Weather theory teaches that inversions often set up at night either right at the surface (sometimes referred to as "surface inversion") or at some altitude above the ground (generally referred to as an "inversion aloft") as shown in *Figure 11-8*. Visible signs of an inversion may be pollution trapped below the inversion causing reduced visibility and dirty looking air. Invisible signs of an inversion may be significantly more stable flying conditions in the inversion area. If a slow, steady initial ascent is initiated (best done by launching with a flaccid balloon), the balloon may find an inversion with no help from the pilot by leveling off as it enters the inversion zone. More likely the pilot has to hunt for the inversion; inversion levels can be determined from the use of the Skew-T charts as previously discussed in Chapter 4, Weather Theory and Reports.

As an example, during flight at night very near the ground, a pilot may feel he or she is continually ballasting to fly level. If a pilot suspects an inversion aloft may exist above, he or she can initiate a slow ascent. If the envelope is flaccid and the ambient air at this altitude is not inverted, then the balloon tends to continue rising until it either encounters an inversion or becomes full as it reaches its pressure ceiling. If the balloon levels out while it is still flaccid, this may indicate that it has entered an inversion zone. If the balloon continues to fly level passively, it is flying in an inversion.

The Practice of Gas Ballooning

Gas Balloon Regulations

For a pilot with a hot air rating, certification to fly gas balloons requires the removal of the "with airborne heater" limitation on his or her certificate. For a private pilot gas rating, aeronautical experience requirements in Title 14 of the Code of Federal Regulations (14 CFR) section 61.109(h)(1) require "...at least two flights of at least 2 hours each that consists of

(i) At least one training flight with an authorized instructor within 60 days prior to application for the rating on the areas of operation for a gas balloon;

(ii) At least one flight performing the duties of pilot in command in a gas balloon with an authorized instructor; and

(iii) At least one flight involving a controlled ascent to 3,000 feet above the launch site. "

The regulation for removal of the airborne heater restriction from an existing hot air pilot's certificate (14 CFR Section 61.115 Balloon ratings: Limitations) states:

(2) The limitation may be removed when the person obtains the required aeronautical experience in a gas balloon and receives a logbook endorsement from an authorized instructor who attests to the person's accomplishment of the required aeronautical experience and ability to satisfactorily operate a gas balloon.

NOTE: Only a logbook entry is required to complete the process (no check ride), since the gas authorization is a removal of a restriction from an existing rating rather than an issuance of a new rating.

The regulations for the aeronautical experience for a commercial rating are virtually identical according to 14 CFR section 61.129(h)(4)(i) except that the student must act as pilot in command on both flights and the controlled ascent must be to "...5,000 feet above the launch site."

Additional areas of specific interest to gas balloon pilots are the regulation for currency for night flight according to 14 CFR section 61.57(b)(1):

"...no person may act as pilot in command of an aircraft carrying passengers during the period beginning 1 hour after sunset and ending 1 hour before sunrise unless within the preceding 90 days that person has made at least three takeoffs and three landings to a full stop during the period beginning 1 hour after sunset and ending 1 hour before sunrise."

NOTE: This is similar to the currency for day flight and only applies to flights with passengers. It does not prohibit night flight when only pilots are on board.

Also of interest are the sections on aircraft lights (14 CFR section 91.209) for night flight; the use of supplemental oxygen (14 CFR section 91.211), and the use of transponders (14 CFR section 91.215).

Flight Planning

During the planning stage of the flight, the objectives of the flight should be established. Possibilities include: training, pleasure, competition, record setting, new equipment checkout, or possibly scientific investigation. Then, the expected flight parameters should be developed. These include the number of pilots and passengers, balloon parameters (type, size, and lifting gas to be used), launch time and location, expected duration, required weather, maximum altitude, and predicted landing zone. If a fixed date has been selected for the flight, an initial meteorological assessment should be made approximately three days out and subsequent weather developments are used to make a "go/no-go" decision. If the flight date is flexible over an extended time window, continual monitoring of the weather is required until proper conditions develop.

After this, an equipment list can be developed and a system weight and ballast calculation should be performed. Permission to use the desired launch site must be confirmed and the availability of adequate gas supply and launch and chase crew must be assured. A concise schedule should be sent to all crew members, defining when decisions are made and how these decisions are communicated to the crew.

Finally, all inflation and flight equipment must be assembled and checked out for proper functioning. The chase vehicle selected should be able cover the expected distance and bring all participants back home.

Layout and Inflation

When the day comes, all equipment is transported to the launch site. A site "walk-around" is performed to remove debris and trash and to check for obstructions to inflation or takeoff. Layout and assembly differ by balloon type, but should proceed according to the balloon's flight manual. For quick fill balloons, layout should be downwind, similar to hot air balloons. A crew briefing should be performed and a crew chief should be designated. Safety must be emphasized; new crew members must be given specific instructions and be assigned to a more experienced crew member for guidance. Gas balloon launches still tend to draw a crowd and some form of crowd control may be required. *[Figure 11-9]*

Figure 11-9. *America's Challenge launch at Fiesta Park, Albuquerque, New Mexico.*

A launch restraint should be secured and additional inflation ballast (weighing several hundred pounds more than the pilots and supplies that eventually are on board) should be added to the gondola.

The start of gas flow is a critical point in the inflation. A slow initial flow allows a last minute check of cable routings and crew positions. Surface weather conditions determine flow rate after the initial checkout. For quick fill systems, a partially filled envelope is much more subject to twisting in wind gusts. A flaccid envelope tends to present a concave (spoon-shaped) surface to the wind, producing a higher drag factor and placing more stress on the entire system. Once the envelope's shape has filled out to a convex (beach ball) shape, it is more able to stand by shedding wind gusts around it. Quick fill systems are much more subject to this effect than are netted balloons.

As described in the section on balloon systems, a netted balloon envelope is laid out flat on the ground and its shape during inflation resembles a sphere rising out of the ground. It always presents a convex (shedding) shape to the wind. A separate step is required after inflation to install the basket and attach it to the load lines coming down from the net.

A good rule of thumb is to take no more time than is necessary to complete the fill. This is especially true in windy conditions.

Launch

With the use of a checklist, confirm that all required equipment items and pilots are on board. A launch master is usually assigned to direct removal of excess inflation ballast until the system is neutrally buoyant. Desired ascent rate determines how much additional ballast is removed to attain the proper amount of positive buoyancy. The launch master should be an experienced gas balloonist and direct crew to allow the balloon to rise several feet off the ground several times to test the buoyancy before instructing the crew to bring it back to the ground one last time before final release. If the ascent rate is too slow, additional ballast are removed until the proper rate is achieved. After a final check for airspace clearance above, the "Hands off!" command is given and the balloon is allowed to fly free. *[Figure 11-10]*

Figure 11-10. *A gas balloon, shortly after the weigh off procedure, ascends into the sky.*

Inflight Procedures

All other things being equal, a gas balloon's natural tendency is to find its equilibrium altitude and to fly level at that altitude. By contrast, a hot air balloon's natural tendency to descend must be counteracted with periodic infusions of heat. In a gas balloon, pilot action is only required to initiate or arrest an ascent or descent or to counter atmospheric or other disturbances. A gas balloon may fly for an hour or longer with no intervention by the pilot. Pilot initiated altitude changes with gas balloons tend to occur at a slower rate over a longer time period as compared to hot air ballooning.

Ascents are initiated by jettisoning ballast (usually sand or water) while descents result from releasing lifting gas from a valve at the top of the envelope.

Significant midflight altitude changes are often undertaken as part of a long-term strategic plan rather than for short term tactical reasons. The consequences of any maneuver should be considered carefully before being undertaken. It is often stated that ballast is the fuel of a gas balloon and well before all ballast has been expended, the aircraft must be safely back on the ground. Use of large amounts of ballast to execute a major ascent invariably shortens the potential duration of a flight.

For example, in a distance competition, an ascent from the surface to 12,000 feet MSL may be executed to enter more favorable winds. This may take 1 hour to accomplish but that altitude may then be maintained for the next 8 hours if weather conditions are stable.

Two additional concepts that must be understood to pilot gas balloons are solar heating and lifting gas purity.

Solar Heating

Solar heating (also called super-heating) occurs when the heat of the sun is trapped inside the balloon's envelope and causes the temperature of the lifting gas to exceed the outside air temperature. As the heated lifting gas expands, one of two things will happen. If the envelope is flaccid, the less dense gas occupies a larger fraction of the envelope's volume and displaces more air and the system's gross lift temporarily increases. This causes the balloon to rise towards its pressure ceiling and it also temporarily causes the pressure ceiling to drop. When the envelope becomes full (i.e., reaches its pressure ceiling), gas is expelled and a new ceiling is established.

If the flight is continued through sunset, loss of solar heating results in a cooling of the lifting gas and a resultant loss of gross lift. Ballast must then be used to maintain buoyancy.

Purity of Lifting Gas

A final topic of interest is the effect of mixing air with the lifting gas. It may seem that since air is heavier than the lifting gas, the air would have a tendency to pool at the bottom of the envelope and be expelled through the appendix as the balloon rises into less dense air. However, this is not what happens. The forces of molecular attractions cause the air and lifting gas to become permanently mixed and a generally less pure mixture occurs. When this happens, some of the benefits of operating under the pressure ceiling are lost and every up/down maneuver causes the loss of lifting gas along with the expelled air.

For this reason, care should be taken to close the appendix when descending rapidly to avoid allowing air to force its way up into the envelope through the open appendix. Maintaining a high level of purity of the gas inside the envelope can extend flight duration. Care must be taken to open the appendix during rapid or prolonged ascents which bring the balloon close to its pressure ceiling. Ascent above pressure ceiling with a closed appendix increases internal pressure on the envelope's fabric and, in an extreme case, could cause the envelope to rupture.

Landing, Retrieval and Packup

The landing phase is the single most critical portion of the entire flight. The pilots may be fatigued from the long flight. They may be in unfamiliar geography and weather conditions. There may be time pressure to land in daylight if sunset is approaching. A landing after dark, under a full moon and in open territory may be performed rather routinely, but a daylight landing is still much preferred. A night time landing in dense woods under a new moon or with reduced visibility can be very stressful. Five minutes on oxygen prior to landing may help to relieve some fatigue and clear the senses.

The landing decision varies with each flight and should be discussed jointly among all the pilots. Landings should be initiated while there is still adequate ballast available to abort at least one approach, if necessary. The actual amount of ballast required varies with pilot experience, weather conditions, terrain and descent rate. For a 1,000 cubic meter balloon in relatively easy landing conditions and under a shallow descent, 50 pounds may suffice. Under more adverse conditions, 250 pounds or more may be advisable. Additional considerations for landing may include duration of flight, fatigue level of the pilots, accomplishment of the flight objectives, current and forecast weather, terrain, and time of day.

When the decision is made to land, all possible equipment should be securely stowed. Occupants should don helmets and any other protective gear. All antennas, solar panels, and other items hanging below the basket should be retrieved and stowed. Adequate ballast should be brought inside the basket where it is readily available to abort or round-out a landing.

The trail rope should be rigged and readied for deployment. The trail rope serves three purposes. First, it slows the descent rate. A fully deployed trail rope weighing about 40 pounds on a 1,000 cubic meter balloon normally arrests a descent of approximately 340 feet per minute (fpm) at 150 feet above ground level (AGL) to a descent rate of 0 fpm at ground contact. The descent from 150 feet will take about 45 seconds. These numbers are only approximations and local conditions at landing certainly cause these to vary somewhat. Secondly, the trail rope orients the balloon so the attachment point of the rope is on the upwind (or trailing) side of the basket. This may be important depending on the arrangement of the basket and/or deflation ports. Finally, as more of the rope contacts the ground, friction slows the horizontal speed of the balloon. The trail rope should be connected to the load ring with a quick release mechanism to allow release should the trail rope become permanently entangled on the ground.

If the gondola has rotated during flight, placing the trail rope on the gondola's downwind side, it does not deploy correctly unless the pilot guides it around to the upwind side of the gondola. Otherwise the rope deploys under the gondola and tends to pull the leading edge of the gondola down and may completely invert it. This is colloquially called "dog-housing" and puts the occupants in the uncomfortable position of being dragged along the ground trapped inside the basket. Unlike modern hot air balloons, most gas balloon systems do not have rigid uprights, so dog-housing is a real concern in a high wind landing; the best antidote is to keep as much weight as possible (including the occupants) on the upwind (trailing) side of the basket.

The trail rope also acts to stop any ascent since a rising balloon becomes heavier as more of the rope is lifted off the ground. This is why it is sometimes called "retrievable" or "reusable" ballast. For this reason, the trail rope should never be deployed until landing is completely certain. Aborting a landing with a deployed trail rope requires ballasting the weight of the trail rope (approximately forty pounds) in addition to the ballast normally required to achieve the desired ascent rate.

The anti-sail line should be pulled tight and secured to the load ring or other strong point. The purpose of this line is to hold the bottom of the envelope down taut to minimize drag in a high wind landing. If the envelope is allowed to ride free, it rides up and bows in the wind, forming a scoop that catches much more wind and increases the length of the drag along the ground.

At this point, the descent is initiated and one or more potential landing sites should be identified. If the descent continues as expected and an adequate landing site is attainable, the descent rate should be tailored toward that site. A last check should be made for powerlines and other obstructions on the path to the landing site. The area downwind of the site should also be checked in case landing runs long. Only then, at a height above ground equal to the length of the trail rope (usually about 150 feet), and only if there are no intervening obstacles between the balloon and the landing site, will the trail rope be deployed. *[Figure 11-11]*

Figure 11-11. *A successful gas balloon landing.*

As with a hot air balloon, if a fast layer has been encountered during the descent, it may be advisable to level off at an altitude below the fast layer to burn off some momentum. However, if this jeopardizes hitting the only likely landing site, a high wind landing is the better option.

The final phase of the landing is ground contact, and just as in piloting a hot air balloon, a decision must be made to perform either a rip-out or standup landing. Contrary to hot air ballooning, in gas, rip-out landings are the norm. Up to the final moment, ascent rate is controlled by actuating the valve (or parachute) to release small amounts of lifting gas. As in hot air ballooning, both pilots should be positioned for landing. Typically at about five feet AGL, the deflation port line (usually red) is pulled to open the port and release a large quantity of gas immediately. Some systems may have multiple deflation ports. Once the deflation port is activated, the flight is terminated with a rapid descent, so it is important to be close to the ground before activation. The height of activation can occasionally be as high as thirty to fifty feet in for emergency reasons, but a very hard landing is sure to ensue.

On the newer German (Wörner) balloons with sealing parachute tops, it is possible to reseat the parachute after a deep activation of several seconds, but gas equivalent to many bags of ballast has been lost. Shorter, shallower vents are used on these systems for maneuvering, but deflation is achieved with a deep constant pull on the parachute line.

Once the balloon has come to a stop and is no longer buoyant, the pilots exit and attempt to contact the chase. Any medical concerns should be dealt with immediately. Landowners should be located, if possible, and an inventory of the contents of the gondola should be done to check to see if anything bounced out during the landing. If darkness is near and no houses are in sight, flashlights, a compass, and GPS should be located immediately to ensure that bearings are not lost in the dark. On a long flight, it may be several hours until the chase crew arrives, but when they do, pack-up proceeds in reverse of assembly.

The America's Challenge Race, 2006

The 2006 America's Challenge gas balloon competition is considered by many in the gas balloon community to be one of the most exciting and controversial flights in recent years. It is presented here as an illustration of the skills and decision-making processes necessary for a successful gas balloon flight.

In July 2006, two gas balloon pilots from Georgia, Andy Cayton and Danni Suskin, were to participate in the Gordon Bennett challenge competition in Belgium. Upon preparation for launch, it was discovered that the balloon to be used had some mechanical problems that kept the Cayton-Suskin team from flying. Cayton came back to the United States with every intention of winning the America's Challenge in the hope of having the opportunity to pursue a victory in the following year's Gordon Bennett. As Suskin was unable to participate in the event, Cayton selected Kevin Knapp of North Carolina as his co-pilot for the America's Challenge.

Launch was originally scheduled for the evening of October 7, 2006. Don Day, a meteorologist who worked with Cayton on numerous world record hot air flights as well as other gas flights, was located at the launch field in Albuquerque, New Mexico. In conference with Cayton and Knapp, Day determined that the weather would be a significant factor in the launch and in any subsequent flight track across the United States. The initial track for the planned launch date would have put the team on a northeasterly track towards Canada; but, low freezing levels and thunderstorms east of the Sandia Mountains forced race officials to delay the planned launch. The next launch window, Tuesday, October 10, showed a potential track east, paralleling Interstate 40.

During the 3-day weather delay, Cayton had reason to reevaluate some strategic issues regarding the flight. One concern was the fact that both he and Knapp were fairly large men; he felt that this might place them at some disadvantage over teams with smaller pilots. Smaller pilots are able to carry more ballast, and thus can extend their flight time. Also, while disappointed with the no-fly situation at the Gordon Bennett, he realized that this might provide an advantage, as he would not be attempting a second duration flight while still fatigued from the Gordon Bennett. Cayton believed that these two issues balanced out, and continued with Knapp to prepare for the launch.

Tuesday, October 10, arrived clear and cold; the cold front and low-pressure system that had delayed the initial launch had passed through the Albuquerque area and was now ahead of them. A massive cold front would push through Canada into the central United States and move to the Gulf of Mexico during the second day of flight. The plan was to stay between the two systems to remain competitive and safe. With crew chief Ken Draughn and help from competitors Peter Cuneo and Bert Padelt, the inflation went smoothly. *[Figure 11-12]*

Figure 11-12. *Inflation of the America's Challenge balloons, Albuquerque, NM, October 2006.*

As launch position had been previously drawn by lot, the Cayton-Knapp team was the fifth balloon to launch. Early on the evening of the 10th, Cayton handed two bags of ballast to the balloonmeister, Stefan Handl, and they were in the air. The race was on! Knapp remembers many "good luck" calls from the crowd; the chase crew mounted their vehicle and departed the launch field not knowing they would have a role in one of the most controversial events in years

Cayton-Knapp tracked more easterly than the balloons that had previously launched and stayed well north of highway I-40 as they crossed the Sandia Mountains. Their altitude was well above 10,000 feet the first night and averaged 30 miles per hour (mph) with the temperature in the low 30's. Cayton and Knapp spent most of the night colder than expected and shivering to keep warm. They established radio contact with Lubbock, Texas Approach at 0630 Wednesday morning and shortly after experienced the magic of a sunrise from the air.

Most of Wednesday was spent flying over Texas, averaging 36 mph. Another team, that of Phil McNutt and Brian Critelli, flew 90 degrees directly below them passing to the north just before they reached the Dallas/Fort Worth metroplex. *[Figure 11-13]* Cayton-Knapp could see other balloons in the distance, but it was unusual to see another team's balloon so close in flight. Dallas Approach directed them to fly over the Class B airspace above 11,000 feet; at that altitude, they started tracking a more southerly direction.

Figure 11-13. *Dallas, Texas, from 11,000 feet, as seen from the Cayton-Knapp balloon in flight.*

Wednesday evening found the team over northern Louisiana. Most of the night was spent above 2,500 feet flying less than 12 mph. The strategy was to stay behind the weather system ahead and to position themselves for Thursday's flight. Cayton and Knapp fell behind several teams during the night, but that served to let them know they were where they needed to be.

Thursday morning's weather forecast was not favorable; winds on the surface, as well as at 3,000, 6,000, and 9,000 feet were all going out into the Gulf with no options but to land. Several teams saw this and landed. Meteorologist Day and Cayton had a long discussion via satellite phone, and confirmed the winds above 12,000 would hook out into the Gulf and bring them back to dry land near Panama City, Florida. All the available meteorological information told them it would work. If the team wished to stay competitive, they had to go high and out over the water.

They began their ascent around 10:00 AM and went out over the water just east of New Orleans. It was relatively slow going but Cayton and Knapp finally made it back to land as planned and began a slow descent around 16:30. Equipment was secured in anticipation of a possible landing near the Apalachicola National Forest in the Florida panhandle. After confirming their position in the race, and with no place to land, the decision was made to continue the flight through the night. The weather prognosis was good, the winds would be relatively calm during the night and remaining ballast was good, so they calculated the needed distance and time to win the race. *[Figure 11-14]*

Early on Thursday morning, it appeared that the team's patience had paid off. By 0430, they were moving east about 6 mph and were just over Cross City, Florida. At sunrise, they were 13 miles further east, and just north of Old Town and Fanning Springs. Surface winds were calm and there was a thin layer of ground fog. As the sun came up, the balloon experienced solar heating and ascended to just over 7,000 feet. The flight continued in an easterly direction at 15 mph and by 08:00 the team came to the realization that they were in a perfect position to take over the lead.

Figure 11-14. *Panama City Beach, Florida, as seen from the Cayton-Knapp balloon, 25 miles out over the Gulf of Mexico.*

As the Cayton-Knapp team crossed Interstate 75, just south of Gainesville, Florida, they received word that the German team of Eimers/Winker *[Figure 11-15,* AC-13*]* was on the ground, and near Cayton's home in Savannah, Georgia. Cayton and Knapp elected to continue flying until they had a cushion of 10 miles before beginning their descent.

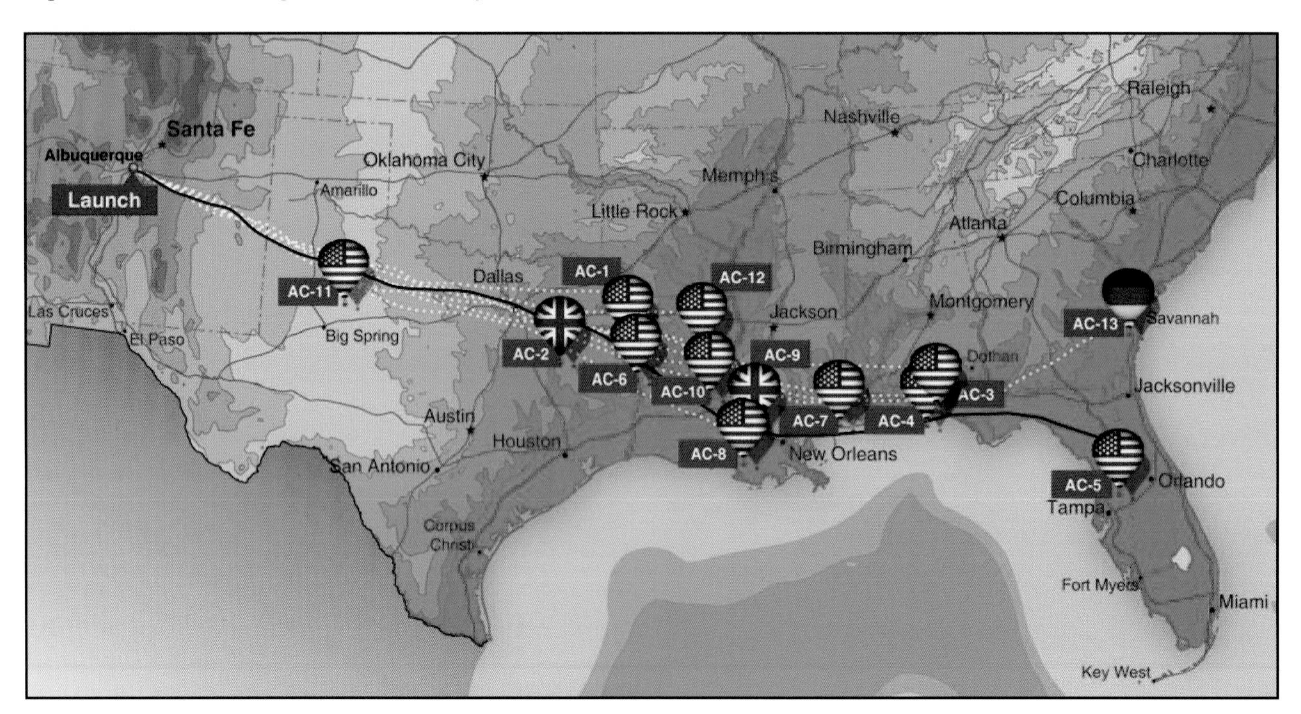

Figure 11-15. *Tracks of the 2006 America's Challenge competitors.*

With surface winds of approximately 3 mph and knowing the race was won, the team elected to land in Citra, Florida. Their chase crew was waiting for them in the yard of a cooperative landowner. After a 60 hour, 20 minute flight covering 1,478 miles, Andy Cayton and Kevin Knapp had beaten the odds to capture the 2006 America's Challenge Cup.

Chapter Summary

This chapter provides a short description of the unique aspects of gas ballooning in comparison to hot air ballooning. For more information, a local gas pilot and the Gas Division of the Balloon Federation of America (www.bfa.net) are good resources.

Appendix A

Vapor Pressures of LP Gases

Vapor Pressures of LP Gases		
Temperature (°F)	Approximate Pressure	Propane Butane
–40	2	
–30	7	
–20	12	
–10	18	
0	25	
10	34	
20	42	
30	53	
40	65	3
50	78	7
60	93	12
70	110	17
80	128	23
90	150	30
100	177	38
110	204	46

Appendix B

Pibal Plotting Grid

This grid may be used for the two practice pibal plotting exercises in Chapter 3. Additionally, it may be copied and used in the preflight planning process.

Appendix C

Balloon Flight Checklist

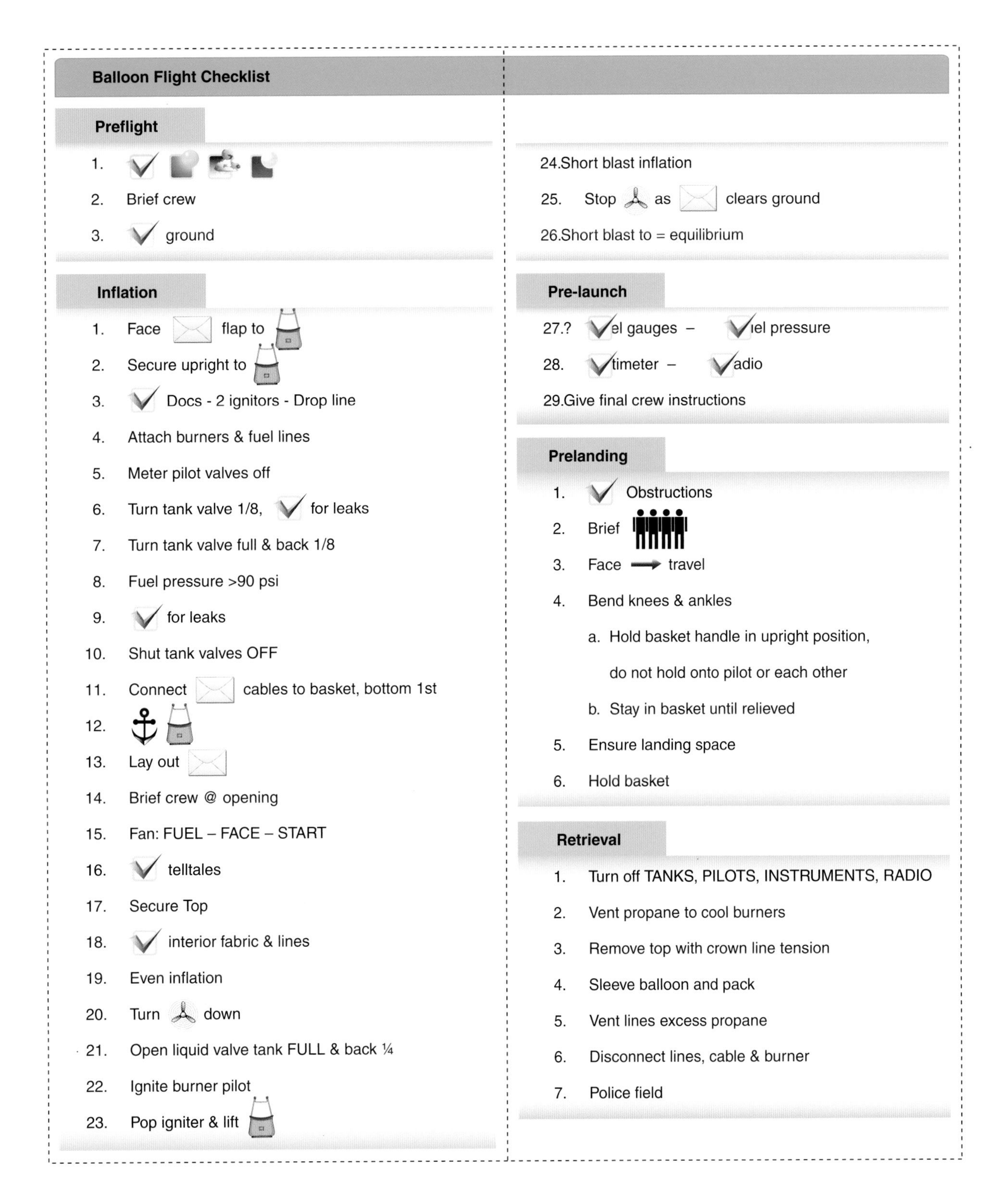

Balloon Flight Checklist

Preflight

1.
2. Brief crew
3. ground

Inflation

1. Face flap to
2. Secure upright to
3. Docs - 2 ignitors - Drop line
4. Attach burners & fuel lines
5. Meter pilot valves off
6. Turn tank valve 1/8, for leaks
7. Turn tank valve full & back 1/8
8. Fuel pressure >90 psi
9. for leaks
10. Shut tank valves OFF
11. Connect cables to basket, bottom 1st
12.
13. Lay out
14. Brief crew @ opening
15. Fan: FUEL – FACE – START
16. telltales
17. Secure Top
18. interior fabric & lines
19. Even inflation
20. Turn down
21. Open liquid valve tank FULL & back ¼
22. Ignite burner pilot
23. Pop igniter & lift

24.Short blast inflation

25. Stop as clears ground

26.Short blast to = equilibrium

Pre-launch

27.? el gauges – iel pressure

28. timeter – adio

29.Give final crew instructions

Prelanding

1. Obstructions
2. Brief
3. Face travel
4. Bend knees & ankles
   a. Hold basket handle in upright position, do not hold onto pilot or each other
   b. Stay in basket until relieved
5. Ensure landing space
6. Hold basket

Retrieval

1. Turn off TANKS, PILOTS, INSTRUMENTS, RADIO
2. Vent propane to cool burners
3. Remove top with crown line tension
4. Sleeve balloon and pack
5. Vent lines excess propane
6. Disconnect lines, cable & burner
7. Police field

Student Balloon Flight Checklist

Preflight

1. Weather briefings
2. Documents
3. Fuel check
4. Crew briefing
5. Passenger briefing
6. Layout & assembly
7. Inspection

Inflation

1. Crew briefing
2. Fan placement
3. Equilibrium
4. Board passengers
5. Basket check
6. Instrument check
7. Fuel

Launch

1. Final crew instructions
2. Obstacles—Clear
3. Traffic—Clear

Descent/Approach

1. Passenger briefing
2. Obstacles—Clear
3. High wind landing
4. Water landing

Landing

1. Monitor passengers
2. Secure fuel
3. Secure heater
4. Deflation

Recovery

1. Secure deflation system
2. Secure envelope
3. Secure heater
4. Secure basket
5. Pack-up
6. Fuel check

Postflight

1. Fueling
2. Report discrepancies
3. Storage

Appendix D

Pibal Velocity Versus Size

Pibal Velocity vs. Size		
Calculated Velocities for Round Balloons		
Diameter (in)	**Weight (gms)**	**Velocity (ft/min)**
9	2.50	241
10	2.50	284
11	2.50	319
12	2.50	350
9	3.00	221
10	3.00	270
11	3.00	308
12	3.00	342
9	3.50	201
10	3.50	255
11	3.50	297
12	3.50	333
14	3.50	391
12	4.00	324
13	4.00	357
14	4.00	386
15	4.00	412
16	4.00	436
12	4.40	317
13	4.40	351
14	4.40	381
15	4.40	408
16	4.40	433
12	5.15	303
14	5.15	372
15	5.15	400
16	5.15	426
17	5.15	451
18	5.15	473
12	7.77	252
14	7.77	339
15	7.77	373
16	7.77	404
17	7.77	431
18	7.77	457

----------------------------To Use This Chart----------------------------

Determine the weight of an empty balloon. Then, knowing the inflated diameter of the pibal, you can read the approximate vertical velocity of the pibal.

To get the weight of the empty balloon, weigh 10 empty pibals on a scale. Take the weight in ounces, and multiply by 28.35 to obtain the weight of the ten balloons in grams. Then, divide by 10 to obtain the weight of an individual balloon.

Appendix E

Log Book Endorsement Formats

Student Pilot Endorsements

Pre-solo Aeronautical Knowledge: 14 CFR section 61.87(b)

I certify that [First name, MI, Last name] has satisfactorily completed the pre-solo knowledge exam of 14 CFR section 61.87(b) for the [make and model aircraft].

Signature, Certificate #, Date

Pre-solo Flight Training: 14 CFR section 61.87(c) (The back of the Student Pilot Certificate must also be endorsed.)

I certify that [First name, MI, Last name] has received the required pre-solo training in a [make and model aircraft]. I have determined he/she has demonstrated the proficiency of 14 CFR section 61.87(d) and is proficient to make solo flights in a [make and model aircraft].

Signature, Certificate #, Date

Note: Other limitations may be included, such as maximum wind/weather conditions at launch, location, etc.

Solo Endorsement (each 90-day period): 14 CFR section 61.87(p)

I have given Mr./Mrs./Ms. [First name, MI, Last name] the instruction required by 14 CFR section 61.87(p). [He or She] has met the requirements of 14 CFR section 61.87(p) and is competent to make safe solo flights in a [make and model aircraft].

Signature, Certificate #, Date

Note: additional 90-day endorsements are recorded in the student's log book only, not on back of the Student Pilot Certificate.

Private Pilot Endorsements

Aeronautical Knowledge: 14 CFR sections 61.35(a)(1) 61.103(d), and 61.105

I certify that [First name, MI, Last name] has received the required training in accordance with 14 CFR section 61.105. I have determined [he or she] is prepared for the LTA-FB Private Pilot knowledge test.

Signature, Certificate #, Date

Flight Proficiency/Practical Test: 14 CFR sections 61.103(f), 61.107(b), and 61.109

I certify that [First name, MI, Last name] has received the required training in accordance with 14 CFR sections 61.107 and 61.109. I have determined [he or she] is prepared for the Private Pilot practical test.

Signature, Certificate, #, Date

Commercial Pilot Endorsements

Aeronautical Knowledge: 14 CFR sections 61.35(a)(1), 61.123(c), and 61.125.

I certify that [First name, MI, Last name] has received the required training in accordance with 14 CFR section 61.125. I have determined [he or she] is prepared for the LTA-FB Commercial Pilot knowledge test (CBH).

Signature, Certificate #, Date

Flight Proficiency/Practical Test: 14 CFR sections 61.123(e), 61.127, and 61.129

I certify that [First name, MI, Last name] has received the required training of 14 CFR sections 61.127 and 61.129. I have determine [he or she] is prepared for the Commercial Pilot practical test.

Signature, Certificate #, Date

Flight Review: 14 CFR section 61.56

I certify that [First name, MI, Last name] holder of pilot certificate #______________ has satisfactorily completed a flight review consisting of one hour of flight training and one hour of ground training as required by 14 CFR section 61.56 on [date].

Signature, Certificate #, Date

NOTE: This endorsement varies somewhat from the example provided in the Advisory Circular. While there is no regulatory requirement to indication the hours of training involved, many Flight Standards District Offices have interpreted the regulation in that manner. It may be best to use this example if an individual attends numerous balloon events around the country on a regular basis.

Endorsement for Retesting of a Written or Practical Test: 14 CFR section 61.49

I certify that [First name, MI, Last name] has received the additional [flight and/or ground] training as required by 14 CFR section 61.49. I have determined that [he or she] is prepared for the [name the knowledge/practical test to be retaken].

Signature, Certificate #, Date

Commercial Pilot Endorsements

Aeronautical Knowledge: 14 CFR sections 61.35(a)(1), 61.123(c), and 61.125.

I certify that [First name, MI, Last name] has received the required training in accordance with 14 CFR section 61.125.
I have determined [he or she] is prepared for the LTA-FB Commercial Pilot knowledge test (CBH).

Signature, Certificate #, Date

Flight Proficiency/Practical Test: 14 CFR sections 61.123(e), 61.127, and 61.129

I certify that [First name, MI, Last name] has received the required training of 14 CFR sections 61.127 and 61.129. I have determine [he or she] is prepared for the Commercial Pilot practical test.

Signature, Certificate #, Date

Flight Review: 14 CFR section 61.56

I certify that [First name, MI, Last name] holder of pilot certificate #______________ has satisfactorily completed a flight review consisting of one hour of flight training and one hour of ground training as required by 14 CFR section 61.56 on [date].

Signature, Certificate #, Date

NOTE: This endorsement varies somewhat from the example provided in the Advisory Circular. While there is no regulatory requirement to indication the hours of training involved, many Flight Standards District Offices have interpreted the regulation in that manner. It may be best to use this example if an individual attends numerous balloon events around the country on a regular basis.

Endorsement for Retesting of a Written or Practical Test: 14 CFR section 61.49

I certify that [First name, MI, Last name] has received the additional [flight and/or ground] training as required by 14 CFR section 61.49. I have determined that [he or she] is prepared for the [name the knowledge/practical test to be retaken].

Signature, Certificate #, Date

Log Book Endorsements

Ref. Appendix 1, Advisory Circular 61-65

14 CFR part 61 contains pilot training and testing procedures, instructor responsibilities and requirements, privileges, and limitations for each level of certificate. Pilot applicants, whether student, private or commercial pilot, being recommended for a written or practical test, solo flight, or completing a flight review must have a written endorsement in their log book by an authorized instructor. Each endorsement should include the instructor's signature, certificate number, and date.

Appendix F

Lift Table for Helium and Hydrogen at Standard Temperatures and Pressures

Lift Table for Helium and Hydrogen at Standard Temperatures and Pressures
(1,000 cubic meter envelope)

Helium density: 0.01056 pounds/cubic foot at STP
Hydrogen density: 0.00535 pounds/cubic foot at STP
Air density: 0.07651 pounds/cubic foot at STP
Envelope volume: 1,000 cubic meters (35,320 cu. ft.)

Altitude (feet MSL)	Pressure (in. Hg)	Air Temperature (°F)	Gross Lift (Helium)	Gross Lift (Hydrogen)
0	29.92	59.00	2,329.35	2,513.37
1,000	28.86	55.43	2,262.40	2,441.13
2,000	27.82	51.87	2,196.05	2,369.54
3,000	26.82	48.30	2,131.99	2,300.42
4,000	25.84	44.74	2,068.58	2,232.00
5,000	24.90	41.17	2,007.54	2,166.14
6,000	23.98	37.61	1,947.21	2,101.03
7,000	23.09	34.05	1,888.46	2,037.64
8,000	22.23	30.48	1,831.36	1,976.04
9,000	21.39	26.92	1,775.05	1,915.28
10,000	20.58	23.36	1,720.42	1,856.33
11,000	19.80	19.79	1,667.54	1,799.27
12,000	19.03	16.23	1,614.68	1,742.24
13,000	18.30	12.67	1,564.44	1,688.03
14,000	17.58	9.11	1,514.30	1,633.93
15,000	16.89	5.55	1,466.00	1,581.81
16,000	16.22	1.99	1,418.70	1,530.78
17,000	15.58	–1.58	1,373.34	1,481.83
18,000	14.95	–5.14	1,328.13	1,433.05
19,000	14.35	–8.70	1,284.89	1,386.39
20,000	13.76	–12.25	1,241.84	1,339.94
21,000	13.20	–15.81	1,200.85	1,295.72
22,000	12.65	–19.37	1,160.12	1,251.77
23,000	12.12	–22.93	1,120.57	1,209.10
24,000	11.61	–26.49	1,082.24	1,167.74
25,000	11.12	–30.05	1,045.15	1,127.72

Notes

1. Pressures and Temperatures taken from U. S. Standard Atmosphere Supplements, 1966, ESSA, NASA, USAF; Table 5.2, Geometric Altitudes, English Units
2. Tables assume no effects due to superheating or impurities in lifting gas supply.
3. Entire volume is assumed to be completely filled with lifting gas as would normally be the case on initial ascent.
4. For all descending altitudes, or for partially full envelopes, use the gross lift shown for the altitude at which the envelope would be completely full. This is called the pressure altitude and is equal to the maximum altitude achieved during the flight, if the envelope was full at that altitude. Gross lift does NOT change as balloon descends unless additional lifting gas is lost.

Glossary

Abort. To terminate an operation prematurely when it is seen that the desired result will not occur.

Advection. In weather, the term used for the horizontal transport of heat by the wind.

Absolute altitude. The actual distance between an aircraft and the terrain over which it is flying.

Advisory circular (AC). An FAA publication that informs the aviation public, in a systematic way, of nonregulatory material.

Accident. An occurrence associated with the operation of an aircraft which takes place between the time any person boards the aircraft with the intention of flight and all such persons have disembarked, and in which any person suffers death or serious injury, or in which the aircraft receives substantial damage. (NTSB 830.2)

Aeronautical Information Manual (AIM). A reference publication for pilots.

Airworthiness directive (AD). A regulatory notice sent out by the FAA to the registered owner of an aircraft informing him or her of a condition that prevents the aircraft from meeting its conditions for airworthiness. Compliance requirements will be stated in the AD.

Adiabatic process. In weather, the change of the temperature of air without transferring heat. In the adiabatic process, compression of the air mass results in the warming of the air; conversely, the expansion of the air mass results in cooling of the air.

Administrator. The FAA Administrator or any person to whom he or she has delegated authority in the matter concerned.

Aeronaut. A person who operates or travels in a balloon or airship.

Aeronautical Decision-Making (ADM). A systematic approach to the mental process, used by pilots to consistently determine the best course of action in response to a given set of circumstances.

Aeronautics. The branch of science that deals with flight and with the operations of all types of aircraft.

Aerostat. A device supported in the air by displacing more than its own weight of air.

Automated flight service station (AFSS). An air traffic facility that provides pilot briefings and numerous other services.

AGL. Above ground level.

Aircraft. A device that is used or intended to be used for flight in the air.

Airport. An area of land or water that is used for the landing and takeoff of an aircraft.

Altimeter. A pressure reading device that, when properly calibrated, indicates the height of the aircraft above mean sea level (MSL). An altimeter is a required instrument in a balloon, as directed by 14 CFR Part 31, Airworthiness Standards: Manned Free Balloons.

Altimeter setting. The station pressure (barometric pressure at the location the reading is taken) which has been corrected for the height of the station above sea level.

Ambient air. Air surrounding the outside of a balloon envelope.

Anabatic winds. In weather, a wind that blows up the slope of a hill or mountain due to increased heating along the valley walls.

Annual inspection. A maintenance term directed by 14 CFR part 91, section 91.409, which states that "no person may operate an aircraft unless, within the preceding 12 calendar months, it has had—(1) an annual inspection…and has been approved for return to service…"

Apex line. A line attached to the top of most balloons to assist in inflation or deflation. Also called crown line or top handling line.

Approved. Approved by the FAA Administrator or person authorized by the Administrator.

Archimedes' Principle. The Greek mathematician's principle of buoyancy, which states that an object (a balloon) immersed in a fluid (the air) loses as much of its own weight as the weight of the fluid it replaces.

ATC. Air Traffic Control.

Automatic Terminal Information Service (ATIS). The continuous broadcast (by radio or telephone) of recorded noncontrol, essential but routine, information in selected terminal areas.

Automatic Weather Observing System (AWOS). Continuous broadcast (by radio or telephone) of weather conditions at selected locations.

Ballast. Gas ballooning term; used to control buoyancy, and therefore altitude, during flight. Ballast, usually in the form of sand or water, is carried aloft by the gas balloon at launch. As the balloon pilot needs to adjust altitude, a small amount of ballast is jettisoned overboard, thereby reducing the gross weight of the balloon at that point in time. The balloon will then rise to a new pressure altitude, where it will remain until there is another dynamic change in the lift equation.

Balloon. A lighter-than-air aircraft that is not engine driven, and that sustains flight through the use of either gas buoyancy or an airborne heater.

Balloon Federation of America (BFA). A national association for balloon pilots and enthusiasts in the United States, and affiliated with the National Aeronautic Association. Information about the BFA can be found at www.bfa.net

Balloon flight manual. A manual containing operating instructions, limitations, weight, and performance information, which must be available in an aircraft during flight. Portions of the flight manual are FAA approved.

Basket. That portion of a hot air balloon that carries the pilot, passengers, cargo, fuel, and instruments.

Blast valve. The valve on a propane burner that controls the flow of propane burned to produce heat.

Bowline knot. Pronounced boh' lin. A common knot that is easy to tie and untie and will not slip.

Btu (British thermal unit). A measurement of heat. The amount of heat required to raise 1 pound of water from 60 to 61 °F.

Buoyancy. In ballooning, when the balloon is zero weight and is neither climbing nor falling.

Burn. A common term meaning to activate the main blast valve and produce a full flame for the purpose of heating the air in the envelope.

Burner. See Heater.

Capacity. See Volume.

Captive balloon. Commonly used to describe a balloon that is permanently anchored to the ground.

Category. According to Title 14 of the Code of Federal Regulations: (1) As used with respect to the certification, ratings, privileges, and limitations of airmen, means a broad classification of aircraft. Examples include: airplane; rotorcraft; glider; and lighter-than-air; and (2) As used with respect to the certification of aircraft, means a grouping of aircraft based upon intended use or operating limitations. Examples include: transport, normal, utility, acrobatic, limited, restricted, and provisional (14 CFR part 1).

Ceiling. The lowest broken or overcast layer of clouds or vertical visibility into an obscuration.

CFR. Code of Federal Regulations.

Charles' Law. If the pressure of a gas is held constant and its absolute temperature is increased, the volume of the gas will also increase. This principle is particularly relevant in gas ballooning.

Checklist. A tool that is used as a human factors aid in aviation safety. It is a systematic and sequential list of all operations that must be performed to properly accomplish a task.

Class. According to Title 14 of the Code of Federal Regulations: (1) As used with respect to the certification, ratings, privileges, and limitations of airmen, means a classification of aircraft within a category having similar operating characteristics. Examples include: single-engine; multiengine; land; water; gyroplane; helicopter; airship; and free balloon; and (2) As used with respect to the certification of aircraft, means a broad grouping of aircraft having similar characteristics of propulsion, flight or landing. Examples include: airplane, rotorcraft, glider, balloon, landplane, and seaplane." (14 CFR part 1)

Coating. A thin synthetic added to the surface of balloon fabric to lessen porosity and ultraviolet-light damage.

Cold front. In weather, the leading edge of a cold air mass displacing a warmer air mass.

Cold inflation. Forcing cold air into the envelope, giving it some shape to allow heating with the heater.

Commercial pilot. A person who, for compensation or hire, is certificated to fly an aircraft carrying passengers or cargo.

Controlled airspace. Airspace designated as Class A, B, C, D, or E within which air traffic control service is provided to some or all aircraft.

Convection. Generally, the transfer of heat energy in a fluid. As applied to weather, the type of heat transfer occurring in the atmosphere when the ground is heated by the sun.

Cooling vent. A vent, in the side or top of the balloon envelope, which opens to release hot air, and that closes after the release of air automatically.

Crew chief. A crewmember who is assigned the responsibility of organizing and directing other crewmembers.

Crown line. A line attached to the top of most balloons to assist in the inflation and deflation of the envelope. Sometimes referred to as apex line or top handling line.

Currency. Common usage for recent flight experience. In order to carry passengers, a pilot must have performed three takeoffs and three landings within the preceding 90 days. In order to carry passengers at night, a pilot must have performed three takeoffs and three landings to a full stop at night (the period beginning 1 hour after sunset and ending 1 hour before sunrise).

Dacron. The registered trade name for polyester fabric developed by DuPont.

Declination. A term useful in map reading. Declination is the difference between true north and magnetic north, and is usually defined on an aviation sectional map or a topographical map.

Density altitude. As defined in the Pilot's Handbook of Aeronautical Knowledge, FAA-H-8083-25, "pressure altitude corrected for nonstandard temperature." However, a more pertinent definition is that in the Airman's Information Manual, which explains density altitude as being nothing more than a way to comparatively measure aircraft performance. See paragraph 7-5-6 of the AIM for a complete discussion.

Deflation panel. A panel at the top of the balloon envelope that is deployed at landing to release all hot air (or other lifting gas) from the envelope. A parachute top is a form of deflation panel.

Designated pilot examiner (DPE). A person appointed by the Administrator who may accept application for certificates, administer practical exams, and issue pilot certificates. DPEs are considered to be technically qualified, and must have a good industry reputation for professionalism and integrity.

Drag line. A gas balloon term used to describe a large, heavy rope, deployed at landing, which orients the balloon (and rip panel) to the wind, and transfers weight from the balloon to the ground, creating a landing flare.

Drop line. A rope or webbing, which may be deployed by the pilot to ground crew to assist in landing or ground handling of a balloon.

Envelope. Fabric portion of a balloon that contains hot air or gas.

Equator. The widest diameter of the envelope.

Equilibrium. When lift equals gravity, as in level flight. Equilibrium at launch is typically that temperature at which after the balloon has been inflated and is standing up (erect), the ground crew is able to hold the balloon in place by resting their hands lightly on the basket.

Fabric test. Testing of the envelope fabric for tensile strength, tear strength, and/or porosity. Fabric tests are specified by each balloon manufacturer.

False lift. See Uncontrolled Lift.

Federal Aviation Administration (FAA). The federal agency responsible to promote aviation safety through regulation and education.

Federal Communications Commission (FCC). The federal agency which regulates radio communication and communication equipment in the United States.

Flameout. The inadvertent extinguishing of a burner flame.

Flare. The last flight maneuver by an aircraft in a successful landing, wherein the balloon's descent is reduced to a path nearly parallel to the landing surface.

Flight review. Required for all certificated pilots every 24 months in order to retain pilot in command privileges. A flight review consists of at least 1 hour of flight training and 1 hour of ground training.

Flight time. According to Title 14 of the Code of Federal Regulations, the time from the moment the aircraft first moves under its own power for the purpose of flight until the moment it comes to rest at the next point of landing.

Flight visibility. According to Title 14 of the Code of Federal Regulations, the average forward horizontal distance, of an aircraft in flight, at which prominent unlighted objects may be seen and identified by day and prominent lighted objects may be seen and identified by night.

fpm. Feet per minute.

Flight Standards District Office (FSDO). Field offices of the FAA, which deal with certification and operation of aircraft.

Gauge. A device for measuring. Required gauges on a hot air balloon are the envelope temperature gauge (pyrometer) and the fuel quantity gauge for each fuel tank. Most balloons also have fuel pressure gauges.

Gondola. Portion of a gas balloon that carries the pilot, passengers, cargo, ballast, and instruments.

Gore. A vertical section of fabric, often made of two vertical, or numerous horizontal panels, sewn together to make a balloon envelope.

Global Positioning System (GPS). The Global Positioning System is a series (or "constellation") of satellites circling the Earth, each broadcasting a unique signal. These signals, usually three or more, are detected by electronic units, and, thru triangulation and time delay sensings, provide an extremely accurate location readout of the GPS receiver. Many balloon pilots carry a small GPS unit on board the balloon to determine location, ground track and ground speed.

Ground crew. Persons who assist in the assembly, inflation, chase, and recovery of a balloon.

Ground visibility. According to Title 14 of the Code of Federal Regulations, prevailing horizontal visibility near the earth's surface as reported by the United States National Weather Service or an accredited observer.

Handling line. A line, usually ¼-to ½-inch diameter rope, attached to a balloon envelope or basket, used by the pilot or ground crew to assist in the ground handling, inflation, landing, and deflation of a balloon.

Heater. Propane-fueled device to heat air inside the envelope of a balloon, often referred to as a burner.

Helium. A light, inert gaseous chemical element mainly found as a natural gas in the southwestern United States. Used to inflate gas balloons and pilot balloons.

Helicopter Emergency Medical Service (HEMS) Weather Display. A product of the Aviation Digital Data Service, this computer program give weather information, both real-time and forecast, for a 5 kilometer square area nationwide. It can be viewed and downloaded at www.weather.aero/HEMS

Hydrogen. The lightest of all gaseous elements. Commonly used in Europe for inflating gas balloons. Flammable by itself and explosive when mixed with oxygen. As opposed to helium, hydrogen is easily manufactured.

Hypoxia. An aeromedical term; means "reduced oxygen" or "not enough oxygen."

International Civil Aviation Organization (ICAO). An agency of the United Nations, which has codified principles and techniques of international air navigation and fostered the planning and development of international air transport to ensure safe and orderly growth. ICAO has produced many international treaties to standardize aviation; the United States is a member of the ICAO, and as such, has agreed to many of the standardization efforts, most notably in the area of airspace nomenclature.

Igniter. A welding striker, piezo sparker, matches, or other means used to ignite the balloon pilot flame.

Incident. An occurrence other than an accident, associated with the operation of an aircraft, which affects or could affect the safety of operations.

Indicated altitude. The altitude shown on a properly calibrated altimeter.

Inoperative. Not functioning or not working.

Instructions for Continued Airworthiness. A manual published by an aircraft manufacturer specifying procedures for inspection, maintenance, repair, and mandatory replacement times for life-limited parts.

Integrated Airman Certification and Rating Application (IACRA). An online application system that allows for the issuance of student, private and commercial pilot certificates without generating paperwork; all certificate application and approved is performed through the use of electronic "signatures."

Instrument. According to Title 14 of the Code of Federal Regulations, A device using an internal mechanism to show visually or aurally the attitude, altitude, or operation of an aircraft or aircraft part. There are only two instruments required in a hot air balloon: vertical speed indicator (VSI) and altimeter.

Katabatic winds. A wind produced by the flow of cold, dense air down a slope in an area subject to radiational cooling. Mountain winds are the most common form of katabatic winds.

Kevlar® A registered trademark for a DuPont Corporation product, a synthetic fiber created in 1965 which has a strength factor 5 times that of steel. Frequently used in balloon systems for suspension cables and control lines of various types. Generally, Kevlar® is used for the core of a suspension line, and will have a cover or sheath over the core, as Kevlar® is ultraviolet light sensitive.

Life-Limited. An aircraft part whose service is limited to a specified number of operating hours or cycles. For example, some balloon manufacturers require that fuel hoses be replaced after a certain number of years.

Light Aircraft. Any aircraft with a maximum takeoff weight of less than 12,500 pounds. All presently FAA-certificated balloons are light aircraft.

Limitations. Restrictions placed on a balloon by its manufacturer. Examples are maximum envelope temperature and maximum gross weight.

Log. A record of activities: flight, instruction, inspection, and maintenance.

LTA. Lighter-Than-Air.

Maintenance. The upkeep of equipment, including preservation, repair, overhaul, and the replacement of parts.

Maintenance Manual. A set of detailed instructions issued by the manufacturer of an aircraft, engine, or component that describes the way maintenance should be performed.

Maintenance Release. A release, signed by an authorized inspector, repairman, mechanic, or pilot after work has been performed, stating that an aircraft or aircraft part has been approved for return to service. The person releasing the aircraft must have the authority appropriate to the work being signed off.

Master tank. The propane tank, usually tank number one, that offers all appropriate services, such as liquid, vapor, and backup system.

Maximum allowable gross lift. The maximum amount of weight that a balloon may lift under standard conditions. Usually a part of the balloon's design criteria, and may be found on the type Certificate Data Sheet for that particular balloon.

METAR. In weather, an acronym for Aviation Routine Weather Report, which is an observation of current surface reported in the standard international format. Routine METARs are transmitted hourly; there is a special report (as indicated by the acronym "SPECI") that may be issued at any time for rapidly changing weather conditions.

Metering valve. A valve on a balloon heater that can be set to allow propane to pass through at a specific rate.

Methanol. A type of alcohol, usually fermented from wood, required by most balloon manufacturers to be introduced into propane tanks annually to adsorb, and thus eliminate, small quantities of water from the fuel.

Mildew. A gray or white parasite fungus which, under warm, moist conditions, can live on organic dirt found on balloon envelopes. The fungus waste materials attack the coating on the fabric.

Mooring. Operation of an unmanned balloon secured to the ground by lines or controlled by anything touching the ground. See Title 14 of the Code of Federal Regulations, part 101.

Mouth. The bottom, open end of a hot air balloon envelope. Also called the "throat."

MSL. Mean sea level.

Neutral buoyancy. A condition wherein a balloon is weightless and is neither ascending nor descending.

Nitrogen charging. A technique of adding nitrogen gas to propane tanks to increase fuel pressure. Used in place of temperature to control propane pressure in hot air balloons during cold weather.

Nomogram. Technically, a calculating chart with scales that contain values of three or more mathematical variables, widely used in engineering. In ballooning, a balloon's performance chart is a nomogram.

Nonporous. The state of having no pores or openings which will not allow gas to pass through. New hot air balloon fabric is nearly nonporous.

Notice to Airmen (NOTAM). A notice containing information concerning facilities, services, or procedures, the timely knowledge of which is essential to personnel concerned with flight operations.

Nylon. The registered name for a polymeric fabric. Most balloon envelopes are made of nylon.

OODA Loop. An aeronautical decision-making model particularly suited to ballooning. Originally devised by Col. John Boyd for use by Air Force combat pilots, it has come to be widely utilized in the business community. See the discussion in chapter 1 of this handbook.

O-Ring. A doughnut-shaped packing, usually rubber, used between two moving parts to act as a seal. Balloon heater and tank valves usually have O-rings between the valve stem and valve bonnet.

Orographic. A term pertaining to mountains or anything caused by mountains, as in orographic wind (wind formed by mountains) and orographic cloud (a cloud whose existence is caused by disturbed flow of air over and around a mountain barrier).

Overtemp (or over temperature). The act of heating the air inside a hot air balloon envelope beyond the manufacturer's maximum temperature.

Oxygen starvation. The condition inside a balloon envelope where all available oxygen has been consumed by the heater flame and additional burning is impossible since propane must have oxygen to burn. In extreme cases, the blast flame and pilot light flame will extinguish after a long burn or series of burns and may not relight until the envelope has "breathed" additional air.

Parachute top. A deflation system wherein the deflation port is sealed with a disc of balloon fabric shaped like a parachute. Lines attached to the edge of the parachute disc gather into a single line that may be pulled down by the pilot in the basket.

Pibal. Pilot balloon; a small helium-filled balloon sent aloft to help determine wind direction, velocity, and stability.

Pilot in Command (PIC). The pilot responsible for the operation and safety of an aircraft during flight.

Piezo. (Pronounced pee-ate' zo). A piezoelectric spark generator that is built into many modern balloon heaters to ignite the pilot light.

Pilot light. A small, continuously burning flame used to ignite the main "blast" flame of a balloon heater.

Pilotage. Navigation by visual reference to landmarks.

Pinhole. Any small hole in a balloon envelope smaller than the maximum dimensions allowed for airworthiness.

Porosity. A condition of the envelope fabric that allows hot air to escape. Excessive porosity requires increased fuel use and results in higher envelope temperatures.

Positive control. According to Title 14 of the Code of Federal Regulations, control of all air traffic, within designated airspace, by air traffic control.

Preflight. All preparations, including gathering information, assembly, and inspection performed by the pilot before flight.

Pressure gradient. In weather, the difference between high and low pressure areas. Wind speed is directly proportional to the pressure gradient.

Pressure relief valve. A device in a propane tank designed to release excess pressure—which may be caused by overfilling, overheating, or excessive nitrogen pressurization—to prevent tank rupture.

Preventive maintenance. Simple or minor preservation operations and the replacement of small standard parts not involving complex assembly operations.

Prohibited area. According to Title 14 of the Code of Federal Regulations, designated airspace within which the flight of aircraft is prohibited.

Propane. A colorless and odorless gas. Ethyl mercaptan is added to propane to give it a detectable odor. Propane weighs 4.2 pounds per gallon at 60 °Fahrenheit.

Practical Test Standard (PTS). Book containing areas of knowledge and skill that a person must demonstrate competency in for the issuance of pilot certificates or ratings.

Pull test. A strength test in which a section of envelope fabric is pulled to a definite pound measurement to determine if it meets the certification requirements for airworthiness.

Pyrometer. An instrument used to measure air temperature inside the top of a balloon envelope.

Rapid descent. A relatively fast loss of altitude. A subjective term, but usually meant to describe a descent of more than 500 fpm.

Rating. According to Title 14 of the Code of Federal Regulations, a statement that, as part of a pilot certificate, sets forth special conditions, privileges, or limitations.

Red line. Refers to a line which activates the deflation panel of a balloon, or the maximum envelope temperature allowed, or the maximum on a gauge.

Repair station. A facility where specified aircraft and their parts may be inspected, repaired, altered, modified, or maintained. FAA approval is issued to a facility upon qualifications specified by the local FSDO.

Repairman certificate. An FAA certificate issued to a person who is employed by a repair station or air carrier as a specialist in some form of aircraft maintenance. A repairman certificate is also issued to an eligible person who is the primary builder of an experimental aircraft, to which the privileges of the certificate are applicable.

Required equipment. Equipment that must be aboard an aircraft, as required either by the FAA or balloon manufacturer, to maintain airworthiness.

Restricted area. Airspace of defined dimensions within which the flight of aircraft is restricted in accordance with certain conditions.

Return to service. A certificated mechanic or authorized inspector must approve an aircraft for return to service after it has been inspected, repaired, or altered. In addition, an aircraft that has been modified must be test flown by an appropriately certificated pilot before return to service.

Ridge. In weather, an elongated area of high pressure with no rotative motion.

Rip panel. A deflation panel, usually circular or triangular, at the top of a balloon envelope, which may be opened by pulling a line in the basket to allow hot air or gas to escape, and the envelope to deflate.

Rotator vent. See Turning Vent.

Rotor. May be found embedded in mountain waves. Formation usually occurs where wind speeds change in a wave, or where friction slows the wind near the ground. See *Figure 4-22* of this handbook for a graphical representation of rotors.

Sectional chart. Published on a routine basis, these charts are similar to automobile road maps, and provide useful information regarding airspace, reference points, tower frequencies, etc., to a balloon pilot. They are generally not very helpful for navigation, as the scale, 1:500,000, is too small to be of use to the balloon pilot.

SIGMET. Significant Meteorological Information.

Single Pilot Resource Management. A variant of the crew resource management model that is or more practical application to the balloon pilot. Defined as the "art and science of managing all resources available to the single pilot to ensure the successful outcome of the flight."

Skew-T plot. In weather, a graphic depiction of the data received from a radiosonde.

Small aircraft. Aircraft having a maximum certificated takeoff weight of 12,500 pounds or less. All currently type-certificated balloons are small aircraft.

Serial number (S/N). A number, usually one of a series, assigned for identification.

Step descent. A method of allowing a balloon to lower toward the ground by reducing the altitude, leveling off, and repeating the step, to lower the balloon in increments rather than one continuous motion.

Superheat. A gas balloon term, superheat occurs when the sun heats the gas inside the envelope to a temperature exceeding that of the ambient air, resulting in expansion of the gas.

Superpressure balloon. (1) A type of hot air balloon which has no openings to the atmosphere—the mouth is sealed with a special skirt—and is kept pumped full of air (at a higher pressure than the atmosphere) by an on-board fan. Used on moored balloons to allow operations in relatively strong wind. (2) In gas ballooning, a sealed envelope in which the internal envelope pressure exceeds that of a non-sealed envelope.

Suspension lines. Lines descending from the mouth of a balloon envelope from which the basket and heater are suspended.

Syllabus. An abstract or digest of training. It is intended to be a summary of a course of training, and should be brief, yet comprehensive enough to cover essential information.

Telling and Doing Technique. A four-step process teaching process particularly well suited to teaching physical skills.

Terminal Aerodrome Forecast (TAF). TAFs are valid for a 24-hour time period, and are updated four times daily. The TAF reporting system uses the same abbreviations as used in METAR reports.

Temperature gauge. The thermometer system, required in all type-certificated hot air balloons, that gives a constant reading of the inside air temperature at the top of the envelope. May be direct reading or remote, using a thermocouple or thermistor connected to a gauge in the basket or reading signals sent by a transmitter.

Temperature recorder. A small plastic laminate with temperature-sensitive paint dots that turn from white or silver to black, to record permanently the maximum temperature reached.

Tensile strength. The strength of a material that resists the stresses of trying to stretch or lengthen it.

Terminal velocity descent. A term used by balloonists for the speed obtained when the balloon is allowed to fall until it apparently stops accelerating, at which point the envelope acts as a parachute and its vertical speed is no longer affected by its lifting gas, but only by its shape (which is caused by design), load, and other factors.

Tethering. Operation of a manned balloon secured to the ground by a series of lines.

Thermal. A column of rising air associated with adjacent areas of differing temperature. Thermal activity caused by the sun's heating usually starts 2 to 3 hours after sunrise.

Time in service. According to Title 14 of the Code of Federal Regulations, with respect to maintenance time records, means the time from the moment an aircraft leaves the surface of the earth until it touches it at the next point of landing.

Touch-and-go landing. An operation by an aircraft that lands and takes off without stopping.

Topographic map. A map depicting area information on a smaller scale that an aviation sectional chart, and of much more value to the balloon pilot. Most topographic charts show areas of vegetation, roads, built-up areas, and the general topography (or terrain) of a given area.

Trough. In weather, an elongated area of low pressure with no rotative motion.

Turning vent. A vent on the side of a hot air balloon envelope which, when opened, allows escaping air to exit in a manner causing the balloon to rotate on its axis.

Type certification. Official recognition that the design and operating limitations of an aircraft, engine, or propeller meet the airworthiness standards prescribed by the Code of Federal Regulations for that particular category or type of aircraft, engine, or propeller.

Uncontrolled lift. Lift that occurs without specific action by the pilot. Often referred to as false lift.

Useful lift (load). The potential weight of the pilot, passengers, equipment and fuel. It is the basic empty weight of the aircraft subtracted from the maximum allowable gross weight.

VAD winds. In weather; velocity azimuth display winds are derived from the output of the 160 or more WRS-88 radar sites located throughout the United States. The WRS-88 is configured to produce radar returns off of dust and other particulate matter in the air, and in turn, those returns can be used to indicate wind direction and speed at different altitudes. Generally reported in 1000 foot increments.

Variometer. See Vertical Speed Indicator.

Vent. (1) The action of opening the vent to cool the air in the envelope. (2) An envelope opening that will automatically close.

Vertical Speed Indicator (VSI). An instrument that continuously records the rate at which an aircraft climbs or descends. Usually measured in FPM. A required instrument in a balloon.

Vent line. The line that activates the cooling vent.

Visual Flight Rules (VFR). Flight rules governing aircraft flight when the pilot has visual reference to the ground at all times.

Virga. Precipitation that falls from a cloud and evaporates before reaching the ground.

Volume. The total amount of air or gas (expressed in cubic feet) contained in a balloon envelope.

Warm front. In weather, the leading edge of a warm air mass displacing a colder air mass.

Warp. The threads in a piece of fabric that run the length of the fabric.

Weigh-off. Determine neutral buoyancy of a gas balloon or airship by taking weight off at launch.

Wind direction. The direction the wind is coming from.

Wind shear. A strong and sudden shift in wind speed or direction, which may be either vertical or horizontal. Wind shear should not be confused with normal wind change, which is gentler. Wind shear is often associated with the passage of a weather front, or a strong temperature inversion.

WINGS Program. A program that encourages general aviation pilots to continue training and provides an opportunity to practice selected maneuvers in a minimum of instruction time. Participation in the WINGS program relieves a pilot from compliance with flight review requirements, provided all WINGS requirements are met. Previously governed by Advisory Circular 61-91, the WINGS program is available as on online education program at www.faasafety.gov, effective mid-2007.

Index

D

E

F

M

N

O

P

Q

R

S

T

U

V

W